T0136867

Automotive Systems and Software Engineering

Yanja Dajsuren • Mark van den Brand
Editors

Automotive Systems and Software Engineering

State of the Art and Future Trends

 Springer

Editors
Yanja Dajsuren
Eindhoven University of Technology
Eindhoven, The Netherlands

Mark van den Brand
Eindhoven University of Technology
Eindhoven, The Netherlands

ISBN 978-3-030-12159-4   ISBN 978-3-030-12157-0   (eBook)
https://doi.org/10.1007/978-3-030-12157-0

This Springer imprint is published by the registered company Springer Nature Switzerland AG.
The registered company address is: Gewerbestrasse 11, 6330 Cham, Switzerland

Preface

Software is "conquering" the world. There is hardly any piece of equipment that does not have software in it. This is certainly true for the automotive domain. The amount of software has grown from a few lines of code in the 1970s to millions of lines of code in modern cars. This trend is estimated to continue in the next years given all the innovations in electric/hybrid cars, autonomous cars, and connected cars. Software is clearly an innovation engine in the automotive domain, it has led to safer and more efficient vehicles on one hand and more comfort on the other hand. There are also challenges related to the infiltration of software in vehicles, such as security, robustness, and trust. Unfortunately, software has also led to many recalls over the last years, and, recently, software was misused to meet emission regulations, the Dieselgate affair. The automotive industry is moving from the mechanical engineering domain to system/software engineering field, including the role of software as the glue to connect components and to provide functionality.

The Eindhoven University of Technology started in 2008 with a master program Automotive Technology, an interdisciplinary master program of various departments among other Mechanical Engineering, Electrical Engineering, and Mathematics and Computer Science. In the discussions on the curriculum, the importance of software was pointed out and a few software-related courses, real-time architecture, and software and system engineering were introduced. A few years later, a master's program on Automotive System Design leading to the degree of Professional Doctorate in Engineering (PDEng) was launched and in 2014 a bachelor's program on automotive started. This means that the Eindhoven University of Technology offers an entire curriculum on automotive and automotive software engineering as an important ingredient. Of course, apart from such educational programs, the university is involved in numerous research projects related to automotive and to automotive software engineering, for instance, the project Hybrid Innovations for Trucks (HIT) with DAF Trucks and other automotive suppliers and FP7/OPENCOSS (Open Platform for EvolutioNary Certification Of Safety-critical Systems), a large-scale integrated project with a consortium of seventeen companies from nine countries. These research projects have highlighted the importance of software in the automotive domain; furthermore, they were the key enablers for building up automotive software engineering expertise. The research projects have resulted in PhD theses in the area of quality of software architecture (see Chapter "Defining Architecture Framework for Automotive Systems"), modeling of functional safety standards (see Chapter "Safety-Driven Development and ISO 26262"), and an integrated design methodology of automotive software architectures and functional safety.

Based on these observations we thought that it was a good time to work on a book with an overview of the state of the art in automotive software development, from both an academic and an industrial point of view. The original idea for this book was to transform the PhD thesis of Yanja Dajsuren (editor of this book) into a text book. Later we decided to invite other authors to contribute to the book to broaden its scope. We composed a list of possible authors both from industry and academia following the book structure and we started inviting them. Although everybody reacted enthusiastically, obtaining contributions from the industrial authors was a challenge; projects have always had priority over papers. This is less of an issue in the academic world. Therefore, we had to drop a couple of chapters from the industrial authors to avoid further delay in publication.

The intended audience for the book are, on one hand, researchers from academia who are interested in learning the fundamental challenges related to software in automotive engineering, for instance, to security and safety. On the other hand, it is for practitioners who need an insight into the state-of-the-art developments in the area of research within academia. Although the book is not written as lectures notes, it can be used in advanced (post-)master's courses on software and system engineering. The book contains multiple interesting case studies that can be used for student projects.

The sixteen chapters cover all the important aspects of the field. Chapter "Automotive Software Engineering: Past, Present, and Future" discusses the evolution of automotive software engineering and future trends based on the past, present, and future of our research group. Chapter "Requirements Engineering for Automotive Embedded Systems" presents the notion of a requirement in general and describes the types of requirements used when designing automotive software systems. Chapters "Status Report on Automotive Software Development" and "State-of-the-Art Tools and Methods Used in the Automotive Industry" provide explore state-of-the-art methods in software development and testing from an industrial perspective and discuss the current challenges in the development process. The provided information can be used for optimal planning of development processes for future automotive systems and for further insights. Chapters "Software Reuse: From Cloned Variants to Managed Software Product Lines" and "Variability Identification and Representation for Automotive Simulink Models" present the novel tool-suites to enable software reuse in different granularities point of view. Chapter "Defining Architecture Framework for Automotive Systems" proposes an architecture framework for the automotive systems to facilitate the architecture-driven development process. Chapter "The RACE Project: An Informatics-Driven Greenfield Approach to Future E/E-Architectures for Cars" presents the results of the RACE project, which aims to redefine the architecture of future cars from an information processing point of view. Chapter "Development of ISO 11783: Virtual Terminal and Monitoring System for Agricultural Vehicles" summarizes the challenges of implementing modules such as sprayer and GPS using ISOBUS and proposes a format for implementing a virtual terminal for agricultural vehicles. Safety is one of the most important quality attribute of a vehicle that needs special attention in all the stages of the life cycle of a vehicle. In Chapter "Safety-Driven Development and ISO 26262" some of the most important aspects of functional safety from the perspective of ISO 26262 are discussed, namely, safety management, development process, architecture design, and safety assurance. Chapter "Introduction to Cooperative Intelligent Transportation Systems" introduces the overall system architecture and standards of European-wide Cooperative Intelligent Transportation/Transport Systems (C-ITS). This chapter is an introduction to the next three chapters that take three different perspectives on C-ITS, namely, intra-vehicle, inter-vehicle, and country-wide. The focus of this chapter lies on one hand on the architecture of C-ITS solutions and on the other hand on security and privacy of C-ITS. The final two chapters present a high level automotive trend watching on the analysis of electric and autonomous driving cars and market trends in ICT and Internet disrupting the transport sector.

Eindhoven, The Netherlands   Yanja Dajsuren
                                                                                        Mark van den Brand

Acknowledgments

Editing this book has been a great journey for us. We would like to express our gratitude to the following people, each of whom has contributed in a valuable way to the completion of this book.

Firstly, we would like to express our gratitude to the authors who took up the challenge besides their busy schedules. The authors shared challenges facing the automotive software engineering field and shared their vision for future research and development directions. We especially want to thank our contributors from the industry who worked extra hours to make this feasible and were very open and flexible toward feedback and comments from reviewers. Sometimes they would manage to spend time revising a chapter even with important deadlines for their projects. The authors from academy are also appreciated for their patience and collaboration for making this book editing process feasible. Each chapter is reviewed by an expert from industry and academy and underwent several revisions. All the authors were also involved in reviewing each others' works as well.

We also want to thank the publisher Arjen Sevenster (Atlantis Press), who first discussed the book opportunity in this field and motivated us for starting this project. Sharing the state of the art while discussing the challenges, research opportunities, and future trends will hopefully encourage more collaboration among researchers and practitioners in this field. The research and development opportunities are enormous. We hope the ever-changing automotive software engineering field will be joined by more and more researchers and engineers to advance it further.

Finally, we thank our editor at Springer, Ralf Gerstner, for making this book possible to be published in Springer and for his support and guidance during the editing process.

Contents

Part I
Introduction

Automotive Software Engineering: Past, Present, and Future

Yanja Dajsuren and Mark van den Brand

Abstract This book presents state-of-the-art technologies and future trends of automotive systems and software engineering. Fifteen chapters cover all important aspects of the field, such as automotive software architectures, software process and quality, safety and security, autonomous and cooperative driving vehicle technology, and intelligent transportation systems. Additionally, the development of and challenges provided by future vehicles such as solar and fully electric cars are discussed. This book provides challenges facing the automotive software engineering field and discusses future research directions.

1 Introduction

The amount of software found in vehicles has increased rapidly. The first lines of code in a vehicle were introduced in the 1970s; nowadays, having over 100 million lines of code is nothing extraordinary when talking about premium cars. An increasing amount of functionality is realized in software, and software is the main innovator in the automotive industry today.

Automotive systems can be categorized into vehicle-centric functional domains (including powertrain control, chassis control, and active/passive safety systems) and passenger-centric functional domains (covering multimedia/telematics, body/-comfort, and Human-Machine Interface). From these domains, powertrain, connectivity, active safety, and assisted driving are considered major areas of potential innovation. The amount of software will increase because of future innovations as well; think of adaptive cruise control, lane keeping, etc., which all leads to the ultimate goal of autonomous driving. The ever-increasing amount of software to enable innovation in vehicle-centric functional domains requires even more attention in order to assess and improve the quality of automotive software. This

Y. Dajsuren · M. van den Brand (✉)
Eindhoven University of Technology, Eindhoven, The Netherlands
e-mail: y.dajsuren@tue.nl; m.g.j.v.d.brand@tue.nl

© Springer Nature Switzerland AG 2019
Y. Dajsuren, M. van den Brand (eds.), *Automotive Systems and Software Engineering*, https://doi.org/10.1007/978-3-030-12157-0_1

is because software-driven innovations can come with software defects and failures and be vulnerable to hacker attacks. This can be observed by the enormous amounts of recalls lately; quite a few are software related.

Arguably, safety is one of the most important quality attributes of a vehicle that needs special attention during all the stages of the life cycle of a vehicle. Failures in software may be costly because of recalls, and may even be life threatening. The failure or malfunctioning of an automotive system may result in serious injuries or death of people. A number of functional safety standards have been developed for safety-critical systems; the ISO26262 standard is the functional safety standard for the automotive domain, geared toward passenger cars. A new version of the ISO26262 standard will cover trucks, buses, and motorcycles as well. The automotive industry is starting to apply these safety standards as guidelines for development projects. However, compliance with these standards is still very costly and time consuming due to huge amount of manual work. In chapter "Safety-Driven Development and ISO 26262," the most important aspects of functional safety from ISO26262 perspective are discussed.

2   Evolution of Automotive Software Engineering

The current method of building a vehicle is by integrating components developed and built by suppliers. The specifications of these components are defined by Original Equipment Manufacturers (OEMs) or car manufacturers, but the actual construction of the components is done by suppliers. Often the components come with their own Electronic Control Units (ECUs) and software stack. These individual software stacks add up to the huge amount of software in a modern vehicle.

This way of working has a few advantages; among others, it leads to a clear separation of concerns. Integration and communication are performed via a Controller Area Network (CAN) bus or FlexRay. This is the basis for a decentralized architecture. Components interacting with each other have to exchange messages via well-defined protocols. Adding or removing functionality boils down to connecting or disconnecting components from the CAN bus or FlexRay. The components can be designed in an isolated way as long as strict alignment with the interface protocols is respected. This way of developing leads to an explosion of ECUs and their corresponding software stacks.

The introduction of AUTOSAR (AUTomotive Open System ARchitecture) standard has led to a different way of working. AUTOSAR provides a generic layered architecture that shields the basic infrastructure of ECUs and provides a rather high-level interface to develop functionality. AUTOSAR takes care of mapping functionality to the available ECUs. AUTOSAR provides a basic software layer consisting of standardized software modules (mostly) without any specific functionality but offers services to implement the functional part of the application software. AUTOSAR provides a runtime environment (RTE), a middleware that shields off a network of ECUs and takes care of the information exchange between the application software components and between the basic software layer and the applications. The application layer consists of application software components that interact with the runtime environment. This leads in principle to less ECUs and a better balance of functionality over ECUs. The adoption of AUTOSAR by the automotive industry is a fact, and suppliers have to provide functional components based on AUTOSAR standard.

Although AUTOSAR is a major step in separating applications from the computing infrastructure, there is still room for improvement. There is an increasing need for computing power, especially in hybrid cars and cars with advanced driver support. The question is whether this computing power can be provided by a collection of ECUs or whether the introduction of general purpose hardware in the form of central processing units (CPUs) and graphics processing units (GPUs) is a better and sustainable development. A similar development happened to consumer electronics; dedicated hardware was replaced by general-purpose hardware, and functionality was realized in software instead of application-specific hardware. The introduction of general purpose hardware in the automotive domain will lead to a complete design philosophy. This will lead to a centralized automotive architecture with a few high-performance multicore processors and a vast collection of sensors and actuators connected to the central processing units. This design will impose severe constraints on the overall functional safety of the system.

In the last 6 years, we have been doing research in automotive software engineering. We have investigated how to evaluate the quality of automotive software architectures. Software in the automotive domain is mainly developed in Matlab/Simulink and more recently SysML; from the developed models, C code is generated. Of course, some functionality is developed directly in C. We have used general software quality metrics frameworks to establish the quality of Matlab/Simulink and SysML models [2] and automotive software architectures in general [1, 3]. In the area of functional safety, we have applied model-driven techniques to support functional safety development process and safety assurance. We have done research on how to apply ISO26262 for functional safety improvement [10]. Furthermore, we have developed a meta-model for the ISO26262 standard, generated based on this meta-model tooling for safety case construction and assessment [5–8].

In this book, we are not only shedding light on our research results, but we also want to share the current state of and the challenges and future trends in automotive software engineering based on the views of other researchers and practitioners. Chapter "Requirements Engineering for Automotive Embedded Software Systems" reviews the general notion of a requirement and describes the types of requirements used in designing automotive software systems. Chapter "Status Report on Automotive Software Development" illustrates the current challenges in automotive software development, gives an overview of the current development methods and tools, and shares the future directions on domain-specific languages and scenario-based virtual validation methods based on industrial projects.

3   C-ITS

Cooperative-Intelligent Transport[1] Systems (C-ITS) aim to facilitate cooperative, connected, and automated mobility. The C-ITS domain is composed of widely spread systems like traffic management systems, traffic light controllers, and vehicle onboard units. Such complex and heterogeneous systems have independent uses but demand a strategy to facilitate their convergence. In recent years, there has been great progress in the field of C-ITS in increasing energy efficiency and safety for specific transport modes. Several projects have been successfully carried out to define ITS architectures depending on the needs that the project wanted to cover or the technology that was being used. In parallel, car manufacturers and automotive suppliers are redefining efficient software development methods to support cooperative and autonomous driving cars such as Adaptive AUTOSAR. There is a need to bring together researchers and practitioners in the field of C-ITS and automotive software engineering in order to merge and harmonize the solutions proposed by both fields. This will enable harmonized solutions for a flexible integration of and interaction between C-ITS and automotive services.

The European Parliament in its directive 2010/40/EU defines Intelligent Transport Systems (ITS) as "systems in which information and communication technologies are applied in the field of road transport, including infrastructure, vehicles and users, and in traffic management and mobility management, as well as for interfaces with other modes of transport." ITS can be further described as systems that aim to make transportation safe and economical by combining data from vehicles and other sensors on the roadway with weather information.

It began during the 1990s [9] with projects in:

- the US named Intelligent Vehicle Highway System [4]
- various countries in Europe with the program Prometheus [12]
- Japan with a research committee Road/Automobile Communication System [11]

C-ITS enables communication between ITS systems by allowing road users and traffic managers to share information for the purpose of improving traffic safety and driver comfort and reducing traffic congestion. This interaction is where the term cooperative comes from. In this scenario, the vehicles can act as sensors as well. The C-ITS domain covers not only the field of software and systems engineering but also traffic engineering, civil engineering, and information technology, which require a unified architecture for the C-ITS domain. In Europe, many initiatives are taking place to advance C-ITS by aiming for a fully safe and efficient road transport without casualties and serious injuries on European roads.

Chapter "Introduction to Cooperative Intelligent Transportation Systems" and the chapters on C-ITS introduce the overall system architecture and standards of country-wide C-ITS systems and security from in-vehicle and V2X perspectives.

---

[1]Throughout this book we use Transportation and Transport interchangeably.

The focus lies on architecture and on security and privacy, protecting assets, safety, and functionality.

4 Towards Autonomous and Cooperative Driving

In a few years' time, autonomous and cooperative cars will become a reality. Developments such as Google-car, the autopilot functionality of Tesla, and Uber self-driving cars have accelerated the development of autonomous driving. The OEM-ers are introducing more and more advanced driving assistance systems (ADAS), which can be interpreted as paving the road to full autonomous driving. Further development of autonomous driving will involve further development of C-ITS systems. An autonomous car without a C-ITS support is a "blind car" because it has to "feel" its environment. Integration with to advanced C-ITS systems will support autonomous cars. This development can only be realized through standardization on both sides, meaning standardization in the car itself and standardization of C-ITS. Other challenges related to supporting autonomous cars via C-ITS are scalability, robustness, security, etc. These challenges have to be addressed before a large-scale introduction of autonomous driving can be done.

References

1. Dajsuren Y, van den Brand M, Serebrenik A, Huisman R (2012) Automotive adls: a study on enforcing consistency through multiple architectural levels. In: Proceedings of the 8th international ACM SIGSOFT conference on quality of software architectures, QoSA '12. ACM, New York, NY
2. Dajsuren Y, van den Brand MGJ, Serebrenik A, Roubtsov SA (2013) Simulink models are also software: modularity assessment. In: Proceedings of the 9th international ACM SIGSOFT conference on quality of software architectures, QoSA 2013, part of Comparch '13 federated events on component-based software engineering and software architecture, Vancouver, BC, 17–21 June 2013, pp 99–106
3. Dajsuren Y, Gerpheide CM, Serebrenik A, Wijs A, Vasilescu B, van den Brand, MGJ (2014) Formalizing correspondence rules for automotive architecture views. In: Proceedings of the 10th international ACM SIGSOFT conference on quality of software architectures, QoSA'14 (part of CompArch 2014), Marcq-en-Baroeul, Lille, 30 June–04 July 2014, pp 129–138
4. Heermann PD, Caskey DL (1995) Intelligent vehicle highway system: advanced public transportation systems. Math Comput Model 22(4–7):445–453
5. Luo Y, van den Brand M (2016) Metrics design for safety assessment. Inf Softw Technol 73:151–163
6. Luo Y, van den Brand M, Engelen L, Favaro JM, Klabbers M, Sartori G (2013) Extracting models from ISO 26262 for reusable safety assurance. In: Safe and secure software reuse - Proceedings of the 13th international conference on software reuse, ICSR 2013, Pisa, 18–20 June 2013, pp 192–207
7. Luo Y, van den Brand M, Engelen L, Klabbers M (2014) From conceptual models to safety assurance. In: Conceptual modeling - Proceedings of the 33rd international conference, ER 2014, Atlanta, GA, 27–29 October 2014, pp 195–208

8. Luo Y, van den Brand M, Engelen L, Klabbers M (2015) A modeling approach to support safety assurance in the automotive domain. In: Progress in systems engineering. Springer, Cham, pp 339–345
9. Osório AL, Afsarmanesh H, Camarinha-Matos LM (2010) Towards a reference architecture for a collaborative intelligent transport system infrastructure. In: Working conference on virtual enterprises. Springer, Berlin, pp 469–477
10. Saberi AK, Luo Y, Cichosz FP, van den Brand M, Jansen S (2015) An approach for functional safety improvement of an existing automotive system. In: 9th annual IEEE international systems conference (SysCon), pp 277–282
11. Takada K, Tanaka Y, Igarashi A, Fujita D (1989) Road/automobile communication system (RACS) and its economic effect. In: Vehicle navigation and information systems conference. Conference record. IEEE, Toronto, pp A15–A21
12. Williams, M (1988) Prometheus-the European research programme for optimising the road transport system in Europe. In: IEE colloquium on driver information. IET, London, p 1

Part II
Automotive Software Development

Requirements Engineering for Automotive Embedded Systems

Miroslaw Staron

Abstract Requirements engineering is both a phase of software development lifecycle and a subdomain of software engineering. In general, "requirements" is defined as the description of the functionality of software under design and its properties (functional and nonfunctional requirements). Requirements are often perceived as textual documentation. However, in automotive software engineering, requirements can have multiple forms—starting from the short textual descriptions of functionality to fully executable model-based specifications.

In this chapter, we overview the notion of a requirement in general, and describe the types of requirements used when designing automotive software systems. We use the V-model, prescribed by the ISO 26262 safety standard, which describes the way in which software is designed in the automotive domain. We consider the different types of requirements used in these phases.

1 Introduction

Contemporary cars, trucks, buses, and even bikes have software—some as much as 1 GB of onboard binary code excluding maps, music, and other downloadable data. As the history of software dates back to the 1970s with the first onboard Electronic Control Units (ECUs) in an engine, we could observe an enormous growth of software. Up until the end of the 1990s, the amount of onboard code was measured in megabytes, and only a few ECUs were present in the car. However, in the last decade, this amount has grown to over 130 ECUs per car and as much as the aforementioned 1 GB of code.

Moreover, software is included in more safety-critical areas, such as collision avoidance by breaking, automatic parking, or autonomous driving. Therefore, we need to enhance our expertise in working with software as one of the primary

M. Staron (✉)
Computer Science and Engineering, University of Gothenburg, Gothenburg, Sweden
e-mail: Miroslaw.Staron@cse.gu.se

© Springer Nature Switzerland AG 2019
Y. Dajsuren, M. van den Brand (eds.), *Automotive Systems and Software Engineering*, https://doi.org/10.1007/978-3-030-12157-0_2

development entities alongside mechanics and electronics. In this chapter, we focus on one of these areas—requirements engineering. We contribute by providing practical examples of how to efficiently utilize requirements engineering for automotive systems.

The area of requirements engineering is one of the disciplines in vehicle development on the one hand, and, on the other hand, it is a subdomain of software engineering and one of the initial phases of software development lifecycle. It deals with the methods, tools, and techniques for eliciting, specifying, documenting, prioritizing, and quality assuring the requirements. The requirements themselves are very important in enhancing the quality of software in various ways as quality is defined as *"The degree to which software fulfills the user requirements, implicit expectations and professional standards."* [16].

Requirements is often defined as *(1) a condition or capability needed by a user to solve a problem or achieve an objective; (2) a condition or capability that must be met or possessed by a system or system component to satisfy a contract, standard, specification, or other formally imposed documents; and (3) a documented representation of a condition or capability as in (1) or (2)* [16]. This definition stresses the link between the user of the system and the system itself, which is important for a number of reasons:

- Testability of the system—it should be clear how a requirement should be tested, for example, what is the usage scenario realized by the requirement.
- Traceability of the functionality to design—it should be possible to trace which parts of the software construction realize the requirement in order to provide safety argumentation and enable impact/change management.
- Traceability of the project progress—it should be possible to get an overview of which requirements have already been implemented and which are still to be implemented in the project.

It is a very technical definition for something that is intuitively well known—a requirement is a way of communicating what we, the users, want in our dream car. In this sense, it seems that the discipline of requirements engineering is simple. In practice, working with requirements is very complex as the ideas that we, users, have need to be translated to one of the millions of components of the car and its software. So, let us see how the automotive companies work with our requirements or dreams.

We discuss software requirements engineering because the automotive industry has recognized the need to shift its innovation from the mechanical parts of the car to its electronics and software. The majority of us, the customers, buy cars today because they are fast (sporty), safe, or comfortable. In many cases, these properties are realized by adjusting the software that steers parts of the modern cars. For example, we can have the same car with a software package that makes it extremely sporty—look at the Tesla's "Insane" acceleration package or the Volvo's Polestar performance package. These are just the two challenges which lead to two very important trends in automotive software requirements engineering:

1. Growing amount of software in contemporary cars—as the innovation is driven by software, the amount of software and its complexity grows exponentially. For example, the amount of software in the 1990s was a few megabytes of binary code (e.g., Volvo S80) and today reaches over one gigabyte excluding maps and other user data (e.g., Volvo XC90 of 2016).
2. Safety requirements posed by standards such as ISO 26262—as the software steers more parts of the car, there is a larger probability that it can interfere with our driving and cause accidents, and therefore, it has to be safety-assured just like the software in airplanes or trains. The contemporary standard for functional safety (ISO/IEC 26262, Road Vehicles—Functional Safety) prescribes methods and processes to specify, design, and verify/validate the software.

Automotive software requirements engineering needs rigid processes for handling the construction of the software in the car and therefore is much different from the other types of software requirements engineering, such as telecom or web design.

This chapter explores the theory of requirements engineering in automotive development by examining two types of requirements—textual specifications and models used as requirements. It also helps us to explore the evolution of requirements engineering in automotive software development to finally draw on the current trends and challenges for the future.

2 Requirements and Requirements Engineering

Requirements engineering in the automotive sector is increasingly about the software, since software is the source of innovations in the dramatically increasing tempo of changes. According to Houdek [15] and the report on the innovation in the car industry [8], the number of functions in an average car grows much faster than the number of devices, with the number of systematic innovations growing faster than the individual innovations. Systematic innovations are systems of software functions rather than individual functions. Therefore, the discipline of requirements engineering is more about engineering than it is about innovation.

The volume of an automotive requirement specification is in the range of 100000 pages for a new car model according to Houdek based on his study at Mercedes-Benz [15], with ca. 400 documents of 250 pages each at the lowest specification level (component specifications), which are sent over to a large number of suppliers (usually over 100 suppliers, one for each ECU in the car).

Weber and Weisbrod [42] expounded the complexity and size of requirement specifications in the automotive domain based on their experiences from DaimlerChrysler. Their large software development projects can have as many as 160 engineers working on a single requirement specification and producing over 3 GB of requirements data. Weber and Weisbrod describe the process of requirements engineering in the following way: "Textual requirements are only part of the game—automotive development is too complex for text alone to manage." This quote reflects the -state-of-the-art practice of requirements engineering—that the requirements form only one part of the construction database. However, let us look at how the requirements are specified in the automotive domain. Similar challenges of linking requirements to other parts of the construction database can be also found in our previous studies in [23].

3 Types of Requirements in Automotive Software Development

When designing software for a car, designers (who are often referred to as constructors) gradually break down the requirements from the car level to the component level. They also gradually refine them from textual requirements to models of behavior of the software. This gradual refinement is caused by the fact that the requirements have to be sent to Tier 1 suppliers for development and therefore should be as detailed as possible to enable their validation. Figure 1 presents the main phases of software development for automotive systems, roughly based on the software development process model prescribed by ISO/IEC 26262 Systems and Software Safety—Functional Safety standard [17].

In the figure, we also make a distinction between the responsibilities of original equipment manufacturers (OEMs) (vehicle manufactures) and their suppliers. This distinction is important as it is often the phase where the handshaking between the suppliers and the OEMs takes place, and therefore the requirements are used during the contract negotiations. In this context, a detailed, unambiguous, and correct requirement specification prevents potentially unnecessary costs related to

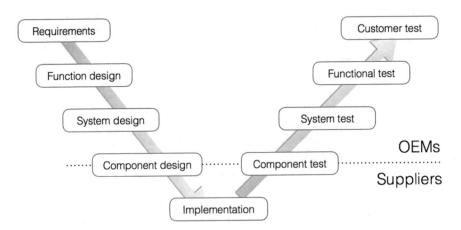

Fig. 1 V-shaped model of software development process in automotive software development

the changes in requirements caused by miscommunication between the OEMs and suppliers.

In the automotive domain, we have a number of tiers of suppliers:

- Tier 1—suppliers working directly with OEMs, usually delivering complete software and hardware subsystems and ECUs to the OEMs.
- Tier 2—suppliers working with Tier 1 suppliers, delivering parts of the sub-products that are then delivered by Tier 1 suppliers to the OEMs; Tier 2 suppliers usually do not work directly with OEMs, which makes it even more important for the requirements to be detailed so that they can be correctly broken down by Tier 1 suppliers for Tier 2.
- Tier 3—suppliers working with Tier 2 suppliers, similar to Tier 2 suppliers working with Tier 1 suppliers.

In this section, we describe these different types of requirements, which can be found in these phases.

3.1 Textual Requirements

AUTOSAR is a great source of inspiration for research in automotive software development, and therefore let us look at the requirements in this standard—it appears that they are mostly textual. An example of a requirement specified in this format, for a feature of keyless entry, is presented in Fig. 2.

The structure of the requirement is quite typical for requirements in general—it contains the description, rationale, and use cases. So far, we do not see anything specific. Nevertheless, if we look at the sheer size of such a specification—over

REQ-1: Keyless vehicle entry

Type	Valid
Description	It should be able to open the car with an RFID key or a mobile phone
Rationale	The majority of our competitors have a RFID sensors in the car that open and start the car based on the proximity of the designated driver who has the RFID sender (e.g. a card). To stay ahead of the competition, we need to provide the key as a mobile phone app for iOS and Android phones.
Use case	Keyless start-up
Dependencies	REQ-11: RFID implementation
Supporting material	---

Fig. 2 An example of a textual requirement, specified in a format used by AUTOSAR requirements

1000 pages—we can see that we might be at loggerheads, so let us discuss the kind of issues we can discover.

Why: The textual requirements are used when describing high-level properties of cars. These types of requirements are mostly used in two phases—the requirements phase when the specification of the car's functionality at a high level takes place and at the component design phase where large software requirement specification documents are sent to suppliers for development (although the textual requirements are often complemented by model-based requirements).

How: Specifying this kind of requirements rarely happens from scratch. Textual requirements are often specified based on models (e.g., UML domain models) and are intended to describe details of the innerworking of software systems. They are often linked to verification methods describing how the requirement should be verified—for example, describing the test procedure for validation that the requirement is implemented correctly. Quite often, it is the suppliers who do the verification as many requirements demand specific test equipment to test their implementation. If this is the case, the OEMs choose a subset of requirements and verify them to check the correctness of the verification procedure from their side.

What: The text for the requirement is specified in the format given in Fig. 2—tables with text. This format is effective for specific details, but ineffective when we want to communicate overviews and provide the context for the requirements. For this, we need other types of formats—use cases or models.

3.2 Use Cases

In software engineering, the golden standard to specify requirements is to adopt the use cases as defined by Jacobson together with this objectory methodology in the 1990s [18]. The use cases describe a course of interaction between an actor and the system under specification, for example, as shown in Fig. 3 where the actor interacts with the car in the use case "Keyless start/up." The corresponding diagram (called the use case diagram in UML) is used to present which interactions (use cases) exist and how many actors are included in these interactions.

In the automotive industry, this kind of requirements specification is the most common when describing functions of vehicles and their dependency. It is used to describe how the actors (drivers or other cars) interact with the designed vehicle (the system) in order to realize a specific use case. This kind of specification is often described using the sequence diagrams of UML, and we can see an example of such a specification in Fig. 4.

Fig. 3 An example use case specification with one use case

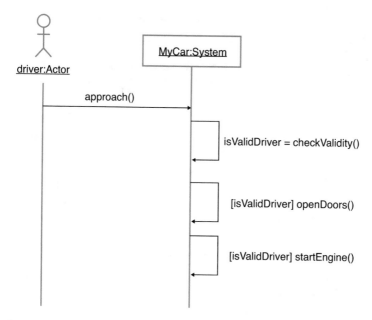

Fig. 4 An example specification of a use case utilizing the message sequence charts/sequence diagrams

Why: The use case specifications provide a high-level overview of the functionality of the designed system, such as a car, and therefore are very useful in the early phases of vehicle development. Usually, these early phases are functional design (use case diagrams) and the beginning of system design (use case specifications).

How: Using high-level descriptions of product properties, functional designers break down these properties into usage scenarios. These usage scenarios provide a possibility to identify which of the functions (use cases) are of value to the customers and which are too cumbersome.

What: These kinds of specifications consist of three parts—(1) the use case diagram, (2) the use case specification utilizing the sequence diagram, and (3) the textual specification of a use case detailing the steps of the interaction applying a somewhat structured natural language.

3.3 Model-Based Requirements

One method to provide more context to the requirements is to express them as models. This kind of representation can be done in two types of formalisms—UML-like models and Simulink models. In Fig. 5, we present an excerpt of a Simulink model for an ABS system from [32, 33, 37].

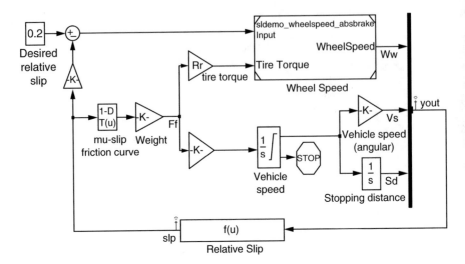

Fig. 5 An example Simulink model that can be used as a requirement to describe how to implement the ABS system

The model shows how to implement the ABS, but the most important property is that the model shows how the algorithm should behave and therefore how it should be verified.

Why: Using models as requirements has been recognized by practitioners, and in an automotive software project, up to 23% of the models are used as requirements according to our previous studies [26] and [25]. According to the same study, up to 13% of the effort is spent in the software project to design these kinds of requirements.

How: The simulation models used for requirements engineering are often used as part of the process of system design and function design where the software and system designers develop algorithms that describe how functions in modern cars are to be realized. These models can be automatically translated to C/C++ code using code generation, but it is rather uncommon. Hence, these models describe the entire functions that are often partitioned into different domains and spread over multiple components. Quite often, these kinds of requirements are translated into textual specifications as shown in the previous subsection.

What: The models are expressed using Simulink or a variation of statechart such as Statemate or Petri nets. These simulation models detail the functions described in the use cases by adding the system view of the interaction—the blocks and signals. The blocks and signals represent the realization of the functionality in a car and are focused on one function only. These models are often used as specifications, which are then detailed and often used to generate the source code automatically.

3.4   *Requirements as Models*

With the introduction of SysML, the models became more expressive than they were when modeled with UML. SysML introduced the notion of requirements diagram, as shown in Fig. 6.

Why: Considering the requirements as first-class entities in models provides the possibility to link them to construction elements of the design [39]. These links provide the possibility to trace requirements to implementation details and therefore speed up modifications.

How: The requirements and their rationale are modeled boxes and lines, just like any other modeling element in SysML. The requirements diagram is one of the most flexible diagrams in SysML, where we can place all kinds of structural elements.

What: The requirements capture the functions and properties of the products. They are linked to rationales and design intentions to increase awareness of the design and implementation constructs in the context.

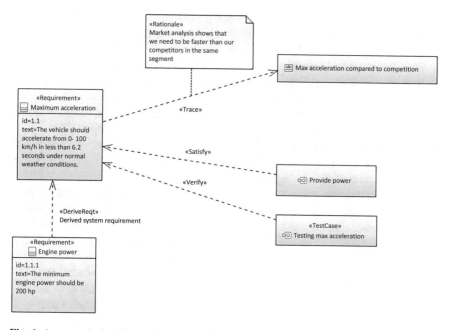

Fig. 6 An example SysML requirement model

4 Measuring Requirements and Requirement Specifications

Industry grade requirement specifications are significantly large—tens of thousands of requirements. Therefore, software engineers use quantitative assessments to understand the complexity and quality of software requirements.

Honig [14] provides a number of rudimentary measures for requirement specifications[1]:

- Requirement correctness—Is the individual requirement properly defining a genuine system function and need? In some cases, the measure may be determined by a formal system requirement verification process.
- Requirement unambiguity—Is the requirement clear and understandable to the expected users of the document? Are multiple, different interpretations of the requirement by different readers unlikely?
- Requirement completeness—Does this single atomic requirement include everything necessary to fully understand the desired function? Are all realizable types of input data, events, system environment covered? Are all terms used understandable or included in the glossary?
- Requirement verifiability—How adequately can this requirement be tested? Is it perfectly clear what test(s) are needed to confirm the requirement is met? Is it clear what should be considered a failure of a test of this requirement?
- Requirement modifiability—Is the individual requirement written so as to be easy to update, change, and eliminate in the future as system needs evolve?
- Requirement atomicity—Is the requirement all one, individual, atomic requirement, including limits, constraints, and all details of the functionality?
- Requirements completeness—Is the set of atomic requirements complete and providing a full definition of all necessary functionality for the entire system (or the current portion being reviewed)?
- Requirements consistency—Is the set of atomic requirements internally consistent, with no contradictions, no duplication between individual requirements?
- Requirements importance ranking—The set of atomic requirements are individually assigned to suitable importance categories (e.g., Essential, Desirable, Optional/Frill) and the assignment of values is appropriate.
- Requirements traceability—Are the individual atomic requirements uniquely identified with unchanging numbers? Are other existing documents or deliverables linked to individual requirements appropriately (e.g., use cases related to atomic requirements)?
- Requirements purity—Is the document free from system design and project schedule, staffing, etc.?
- Requirements count—Current number of individually identified and numbered atomic requirements.

---

[1]The definitions of the measures are quoted directly from the paper.

The abovementioned set of measures shows the major shortcoming of the requirement assessment practices—they are based on manual assessments. However, some studies show that requirements can be quantified automatically in a meaningful way. This quantification can be done based on the semantical analysis of the meaning of requirements (majority of the research), but it can also be approximated with the search-based techniques.

For example, Antinyan and Staron [2, 3] identified the following measures to be significant for assessing the complexity of requirements:

- Number of conjunctions
- Number of vague phrases
- Number of references
- Number of referenced documents
- Number of words

These measures can be combined into a requirement quality index. The index provides designers with the possibility to rank their requirements and improve their quality.

5   How All These Requirements Come Together

All these types of requirements need to come together somehow; hence, we have the process and the infrastructure for requirements engineering. Let us start with the infrastructure—usually named the design or construction database. In the light of the work of Weber and Weisbrod [42], it is called the common information model. Figure 7 presents the way in which this design database is used. The construction database contains all elements of the design of the electrical system of the vehicle—components, electronic control units, systems, controllers, etc. The structure of such a database is hierarchical and reflects the structure of the vehicle. Each of the elements in the database has a set of requirements linked to them. The requirements are also linked to one another to show how they are broken down. Such a database grows over time and is version controlled as different versions of the same elements can be used in different vehicles (e.g., different year models of the same car or different cars).

An example of such a system is described by Chen et al. [6] and has been developed by the company Systemite, which specializes in the databases for designs. Such a database structures all the elements of the construction of the integrated electronics of the vehicle and links all artifacts to the construction elements. An example of a construction element is the engine's electronic control unit, and all the functions that use this control unit are linked to it.

Such a database usually has a number of views that show the required set of details—functional view, architectural view, topological view, and software components' view. Each view provides the corresponding entry point and shows the relevant elements, but the database is always in a consistent state where all the links are valid.

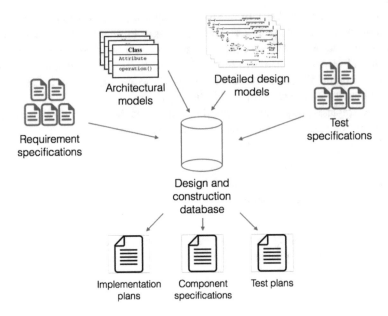

Fig. 7 Design database

The database is used to generate construction specifications for different actors. For each supplier who delivers an ECU, the database generates the set of all requirements that are linked to the ECU and all models that describe the behavior of the ECU. Sometimes, depending on the situation, the documentation contains even the simulation models for the functions that are to be included in the ECU.

6   Current Trends of Software Requirements Engineering in the Automotive Domain

Based on the observations of the evolution of the automotive embedded software, we could observe a number of trends in requirements engineering. In this section, we describe these trends.

Agility in Specification Development   Agile software development has been used in many domains outside the automotive domain, and now there is evidence that it is used increasingly in the automotive domain. In particular at the lower part of the V-model, the suppliers work more agile with their requirements engineering and software development [22]. We can also observe these trends scaling up to the complete vehicle development [11] and [20]. With this increased adoption of Agile principles, we can foresee increased ability to specify requirements along software development, especially as the trends in automotive electronics are that we use increasingly more commodity (or off-the-shelf) components. AUTOSAR also prescribes standardized approach to development, which eases the use of iterative development principles as the development of electronics/hardware is decoupled from the development of functions/software.

Increased Focus on Traceability Increased amount of software in cars and their increased presence in safety systems lead to stricter processes for keeping track of requirements for safety-critical systems. ISO 26262 (Road Vehicles—Functional Safety) is one example of this. In the automotive domain, this means that the increased complexity of software modules [34] leads to more fine-grained traceability management. One of the enablers of this increased traceability is the increased integration between tools and tool chaining [5] and [4].

Increased Focus on Non-functional Properties The increased use of software for active safety systems calls for the increased focus on non-functional properties of software. The increased traffic on communication buses within the car and the increased capacity of the communication buses call for more synchronization and verification. The safety analyses such as control path monitoring, safety bits, and data complexity control are just a few examples [38]. As the focus of requirements engineering research in the automotive domain was mainly (or implicitly) on the functional requirements, we foresee increased growth of research and emphasis on the non-functional requirements.

Increased Focus on Security Requirements A dedicated group of requirements is the security requirements. As our cars are increasingly connected, they are prone to hacker attacks [35] and [43]. The recent demonstration of the possibility to steer a Jeep Wrangler vehicle off-road showed that the threat is real and related to the safety of cars and transport systems. We therefore perceive that the ability to prevent attacks will be of focus for the automotive software development increasingly more in the coming decade.

7   Further Reading

This chapter provides an overview of the techniques used for requirements engineering in the automotive domain, and interested readers are encouraged to dive deeper into the topic. We provide a number of interesting entry points to more research in requirements engineering for automotive software systems.

Ott et al. [28] and [29] present a study on requirements engineering at Mercedes-Benz where they classified over 5800 requirement review protocols to their quality model. Their results showed that textual requirements (or natural language requirements as they are called in the publication) are prone to problems such as inconsistency, incompleteness, or ambiguity—with about 70% of the defects in requirements falling into these categories. In the light of this article, we can see the need for complementing the textual requirements with more context provided by use case models, user stories, and use cases.

Törner et al. [40] presented a similar study but of the requirements at Volvo Cars Group. In contrast to the study of Ott et al. [28], these authors studied the use case specifications and not the textual requirements. The results, however, are similar as the main types of defects are missing elements (correctness in Ott et al.'s model) and incorrect linguistics (ambiguity in Ott et al.'s model).

Eliasson et al. [12] described further experiences from Volvo Cars Group where they explored challenges with requirements engineering at large in a mechatronics development organization. Their findings showed that there is a lot of communication in parallel to the requirement specification. The stakeholders in the requirement specification frequently mentioned the need to have a good network in order to specify the requirements correctly. This indicates the challenges described previously in this chapter that the requirements need more context than it is usually provided in just the specification (especially the textual specification).

Mahally et al. [20] identified requirements to be the main barriers and enablers of moving toward Agile mechatronics organizations. Although today OEMs try to move toward fast development of mechatronics and reduce the cycle time by using Agile software development approaches, the challenges are that we do not know upfront whether a requirement needs the development of electronics or it is only a software requirement. According to Mahally et al., this kind of problem needs to be solved, and based on the prediction of Houdek [15], issues of this kind might be coming to an end as device development flattens out and most of the requirements will be software requirements. Similar challenges were presented by Pernståal et al. [31] who found that requirements engineering is one of the top improvement areas for the automotive OEMs. The ability to communicate via requirements was also an important part.

At Audi, Allmann et al. [1] presented the challenges in the requirements communication on the boundary between the OEMs and their suppliers. They have identified the needs for better communication and the challenges of communicating through textual representations. They recognized the needs for tighter partnerships as there is an inherent deficiency in communicating through requirements—transferring knowledge through an intermediate medium. Therefore, they recommend to integrate systems to minimize the knowledge loss via transfer of documents.

Siegl et al. [36] presented a method for formalizing requirement specifications using Time Usage Model and applied it successfully to a requirement specification from one of the German OEMs. The evaluation study showed an increased test coverage and increased quality of the requirement specification.

At BMW, Hardt et al. [13] demonstrated the use of formalized domain engineering models in order to reason about the dependencies between requirements in the presence of variants. Their approach provided a simplistic, yet powerful, formalism, and its strength was the industrial applicability.

A study of the functional architecture of a car project at BMW and the requirements linked to the functions by Vogelsanag and Fuhrmann [41] showed that 85% of the functions are dependent on one another and that these dependencies caused a significant amount of problems in software projects. This study shows the complexity of the functional decomposition of the vehicle's design and the complexity of its description.

At Bosch, Langenfeld et al. [19] the longitudinal study of a 5-year project showed that 61% of the defects in requirements come from the incompleteness or incorrectness of the requirement specifications.

One of interesting trends in requirements engineering is the automatization of tasks of requirement engineers. One of such tasks is the discovery of non-functional requirements. This task is based on reading the specifications of functional requirements and identifying phrases that should be transformed into non-functional requirements. A study on the automation of this task has been conducted by Cleland-Huang et al. [7]. The study showed that the automated classification of requirements could be as good as 90%, but at this stage it cannot replace the manual classifiers.

7.1 Requirements Specification Languages

A model for requirements traceability [10] DARWIN4Req has been proposed to address the challenges related to the ability to follow the requirements' lifecycle. The model allows to link requirements expressed in different formalities (e.g., UML, SySML) and link them to one another. However, to the best of our knowledge, the model and the tool have not been adopted on a wider scale yet.

EAST-ADL [9] is an architecture specification language, which contains the elements to capture requirements and link them to the architectural design. The approach is similar to SySML but with the difference that there is no dedicated requirement specification diagram. EAST-ADL has been demonstrated to work in industry; however, it is not a standard for automotive OEMs yet. Mahmud [21] presented a language ReSA that complements the EAST-ADL modeling language with the possibility to analyze and validate requirements (e.g., basic consistency checks).

For the non-functional requirements in the domain of safety, Peraldi-Frati and Albinet [30] have proposed another extension of the EAST-ADL language that allows for increased traceability of requirements and their linking to the non-functional properties of the designed embedded software (e.g., safety).

Mellegård and Staron [24] and [27] conducted an empirical study on the impact of using hierarchical graphical requirement specification on the quality of change impact assessment. For the purpose, they designed a requirements' specification language based on the existing formalism—Requirements Abstraction Model. The results showed that the graphical overview of the dependencies between requirements introduces a significant improvement.

8   Conclusions

Correct, unambiguous, and consistent requirement specifications are the foundations of high-quality software, in general, and in automotive embedded systems, in particular. In this chapter, we introduced the most common types of requirements used in this domain and provided their main strengths.

Based on the current state of evolution of the automotive software, we could observe three trends in the requirements engineering for the automotive embedded systems—(1) agility in requirement specification, (2) increased focus on non-functional requirements, and (3) increased focus on security as a domain for requirements. Toward the end of this chapter, we also provided an overview of the requirements practices in some of the vehicle manufacturers (Mercedes-Benz, Audi, BMW, and Volvo) based on the published experiences from these companies. We have also pointed out a number of directions for further reading for the interested.

In our future work, we plan to make a review of the requirements engineering practices in the main automotive OEMs and identify their commonalities and differences.

References

1. Allmann C, Winkler L, Kölzow T, et al (2006) The requirements engineering gap in the oem-supplier relationship. J Univers Knowl Manag 1(2):103–111
2. Antinyan V, Staron M (2017) Proactive reviews of textual requirements. In: IEEE 24th international conference on Software Analysis, Evolution and Reengineering (SANER), 2017. IEEE, Piscataway, pp 541–545
3. Antinyan V, Staron M (2017) Rendex: a method for automated reviews of textual requirements. J Syst Softw 131:63–77
4. Armengaud E, Biehl M, Bourrouilh Q, Breunig M, Farfeleder S, Hein C, Oertel M, Wallner A, Zoier M (2012) Integrated tool chain for improving traceability during the development of automotive systems. In: Proceedings of the 2012 embedded real time software and systems conference
5. Biehl M, DeJiu C, Törngren M (2010) Integrating safety analysis into the model-based development toolchain of automotive embedded systems. In: ACM sigplan notices, vol 45. ACM, New York, pp 125–132
6. Chen D, Törngren M, Shi J, Gerard S, Lönn H, Servat D, Strömberg M, Årzen KE (2006) Model integration in the development of embedded control systems-a characterization of current research efforts. In: 2006 IEEE international conference on control applications, computer aided control system design. IEEE, Piscataway, pp 1187–1193
7. Cleland-Huang J, Settimi R, Zou X, Solc P (2007) Automated classification of non-functional requirements. Requir Eng 12(2):103–120
8. Dannenberg J, Burgard J (2015) Car innovation: a comprehensive study on innovation in the automotive industry. Oliver Wyman Automotive, New York
9. Debruyne V, Simonot-Lion F, Trinquet Y (2005) East-adlan architecture description language. In: Architecture description languages. Springer, New York, pp 181–195

10. Dubois H, Peraldi-Frati MA, Lakhal F (2010) A model for requirements traceability in a heterogeneous model-based design process: application to automotive embedded systems. In: 2010 15th IEEE International Conference on Engineering of Complex Computer Systems (ICECCS). IEEE, Piscataway, pp 233–242
11. Eliasson U, Heldal R, Lantz J, Berger C (2014) Agile model-driven engineering in mechatronic systems-an industrial case study. In: Model-driven engineering languages and systems. Springer, Cham, pp 433–449
12. Eliasson U, Heldal R, Knauss E, Pelliccione P (2015) The need of complementing plan-driven requirements engineering with emerging communication: experiences from volvo car group. In: 2015 IEEE international Requirements Engineering conference (RE). IEEE, Piscataway, pp 372–381
13. Hardt M, Mackenthun R, Bielefeld J (2002) Integrating ECUs in vehicles-requirements engineering in series development. In: 2002 IEEE international Requirements Engineering conference (RE). IEEE, Piscataway, pp 227–236
14. Honig WL (2016) Requirements metrics - definitions of a working list of possible metrics for requirements quality. Retrieved from Loyola eCommons, Computer Science: Faculty Publications and Other Works
15. Houdek F (2013) Managing large scale specification projects. In: 19th international working conference on Requirements Engineering Foundations for Software Quality, REFSQ 2013, Essen, Germany, 8–11 April 2013
16. IEEE (1990) IEEE standard glossary of software engineering terminology (IEEE std 610.12-1990). IEEE Computer Society, Los Alamitos
17. ISO I (2011) 26262–road vehicles-functional safety. International Standard ISO/FDIS 26262
18. Jacobson I, Booch G, Rumbaugh J (1997) The objectory software development process. Addison Wesley, Boston. ISBN: 0-201-57169-2
19. Langenfeld V, Post A, Podelski A (2016) Requirements defects over a project lifetime: an empirical analysis of defect data from a 5-year automotive project at Bosch. In: Requirements engineering: foundation for software quality. Springer, Cham, pp 145–160
20. Mahally MM, Staron M, Bosch J (2015) Barriers and enablers for shortening software development lead-time in mechatronics organizations: a case study. In: Proceedings of the 2015 10th joint meeting on foundations of software engineering. ACM, New York, pp 1006–1009
21. Mahmud N, Seceleanu C, Ljungkrantz O (2015) Resa: an ontology-based requirement specification language tailored to automotive systems. In: 10th IEEE international Symposium on Industrial Embedded Systems (SIES), 2015. IEEE, Piscataway, pp 1–10
22. Manhart P, Schneider K (2004) Breaking the ice for agile development of embedded software: an industry experience report. In: Proceedings of the 26th international conference on software engineering. IEEE Computer Society, Washington, pp 378–386
23. Mellegård N, Staron M (2008) Methodology for requirements engineering in model-based projects for reactive automotive software. In: 18th ECOOP doctoral symposium and PhD student workshop, p 23
24. Mellegård N, Staron M (2009) A domain specific modelling language for specifying and visualizing requirements. In: The first international workshop on domain engineering, DE@ CAiSE, Amsterdam
25. Mellegård N, Staron M (2010) Characterizing model usage in embedded software engineering: a case study. In: Proceedings of the fourth European conference on software architecture: companion volume. ACM, New York, pp 245–252
26. Mellegård N, Staron M (2010) Distribution of effort among software development artefacts: an initial case study. In: Enterprise, business-process and information systems modeling. Springer, Berlin, pp 234–246
27. Mellegård N, Staron M (2010) Improving efficiency of change impact assessment using graphical requirement specifications: an experiment. In: Product-focused software process improvement. Springer, Berlin, pp 336–350

28. Ott D (2012) Defects in natural language requirement specifications at mercedes-benz: an investigation using a combination of legacy data and expert opinion. In: 2012 20th IEEE international Requirements Engineering conference (RE). IEEE, Piscataway, pp 291–296
29. Ott D (2013) Automatic requirement categorization of large natural language specifications at mercedes-benz for review improvements. In: Requirements engineering: foundation for software quality. Springer, Berlin, pp 50–64
30. Peraldi-Frati MA, Albinet A (2010) Requirement traceability in safety critical systems. In: Proceedings of the 1st workshop on critical automotive applications: robustness & safety. ACM, New York, pp 11–14
31. Pernstå011 J, Gorschek T, Feldt R, Florén D (2013) Software process improvement in inter-departmental development of software-intensive automotive systems–a case study. In: Product-focused software process improvement. Springer, Berlin, pp 93–107
32. Rana R, Staron M, Berger C, Hansson J, Nilsson M, Törner F (2013) Improving fault injection in automotive model based development using fault bypass modeling. In: GI-Jahrestagung. Chalmers University of Technology, Gothenburg, pp 2577–2591
33. Rana R, Staron M, Mellegård N, Berger C, Hansson J, Nilsson M, Törner F (2013) Evaluation of standard reliability growth models in the context of automotive software systems. In: Product-focused software process improvement. Springer, Berlin, pp 324–329
34. Rana R, Staron M, Berger C, Hansson J, Nilsson M, Törner F (2013) Increasing efficiency of ISO 26262 verification and validation by combining fault injection and mutation testing with model based development. In: ICSOFT 2013, pp 251–257
35. Sagstetter F, Lukasiewycz M, Steinhorst S, Wolf M, Bouard A, Harris WR, Jha S, Peyrin T, Poschmann A, Chakraborty S (2013) Security challenges in automotive hardware/software architecture design. In: Proceedings of the conference on design, automation and test in Europe, EDA consortium, pp 458–463
36. Siegl S, Russer M, Hielscher KS (2015) Partitioning the requirements of embedded systems by input/output dependency analysis for compositional creation of parallel test models. In: 9th annual IEEE international Systems Conference (SysCon), 2015. IEEE, Piscataway, pp 96–102
37. SimulinkDemo (2012) Modeling an anti-lock braking system. The MathWorks, Inc, Natick. Copyright 2005-2010
38. Sinha P (2011) Architectural design and reliability analysis of a fail-operational brake-by-wire system from iso 26262 perspectives. Reliab Eng Syst Saf **96**(10), 1349–1359
39. Staron M (2017) Automotive software architectures: an introduction. Springer, Cham
40. Törner F, Ivarsson M, Pettersson F, Öhman P (2006) Defects in automotive use cases. In: Proceedings of the 2006 ACM/IEEE international symposium on empirical software engineering. ACM, New York, pp 115–123
41. Vogelsanag A, Fuhrmann S (2013) Why feature dependencies challenge the requirements engineering of automotive systems: an empirical study. In: 2013 21st IEEE international Requirements Engineering conference (RE). IEEE, Piscataway, pp 267–272
42. Weber M, Weisbrod J (2002) Requirements engineering in automotive development-experiences and challenges. In: 2002 IEEE international Requirements Engineering conference (RE). IEEE, Piscataway, pp 331–340
43. Wright A (2011) Hacking cars. Commun ACM 54(11):18–19

Status Report on Automotive Software Development

Florian Bock, Christoph Sippl, Sebastian Siegl, and Reinhard German

Abstract Due to rapid changes in the development of modern automotive systems, the involved development methods, processes, and toolchains are constantly changed, modified, and improved to be able to handle the increasing complexity of the development procedure. In this chapter, the main current challenges in the development itself as well as in the modification of the implied processes are summarized, and both a textual and a graphical overview of the main currently involved tools are given. The provided information can be used for optimal planning of development processes for future automotive systems.

1 Introduction

The automotive domain and, therefore, the development of modern vehicles are becoming increasingly complex with each development iteration, each new model series, and each evolutionary technological step. This is due to different reasons, which are worth being looked at in more detail. From our point of view, the main reasons are:

- Hardware complexity: During the last decades, the capabilities of hardware in modern vehicles have vastly increased. For example, environment-detecting sensors, such as radio detection and ranging (*RADAR* [40]) and light detection and ranging (*LIDAR* [32]), gained much more attention and a higher distribution rate within the last few years. On the one hand, this offers many new possibilities to recognize the environment but on the other hand requires a much more

F. Bock (✉) · R. German
Department of Computer Science 7, Friedrich-Alexander-University Erlangen-Nuremberg, Erlangen, Germany
e-mail: florian.inifau.bock@fau.de; german@informatik.uni-erlangen.de

C. Sippl · S. Siegl
AUDI AG, Ingolstadt, Germany
e-mail: christoph.sippl@audi.de; sebastian.siegl@audi.de

© Springer Nature Switzerland AG 2019
Y. Dajsuren, M. van den Brand (eds.), *Automotive Systems and Software Engineering*, https://doi.org/10.1007/978-3-030-12157-0_3

detailed interpretation of the gathered data and a more efficient analysis of the dramatically increased amount of collected data. Generally speaking, the ongoing developments offer possibilities for new functionalities and improvements, but also lead to challenges and open issues when applying such enhancements.

- The focus of the development efforts switches from initially mainly mechanical and electronic inventions and technologies to more computer science- and software-related topics. Information technology undergoes the biggest gain in importance (estimated 400% per iteration according to [4]). This trend introduces many already addressed issues from the software engineering domain to the automotive development processes.
- Customer expectations are vastly increasing due to tight competition between the manufacturers, cross-domain interrelations such as the comparison with development cycles in the mobile device domain, and recognizable marketing efforts of the manufacturers.
- Global mega trends like digitalization, sustainability, and urbanization affect the automotive domain and its products. With the advance of technologies, computing power, and intense research efforts, the electrification of connected vehicles or the development of automated vehicles becomes possible. Especially, vehicle networks and automated driving functions have complex requirements, which have to be considered. Therefore, new development processes and methods are required to overcome the growing complexity of future vehicles.

Besides these reasons, many other topics have influence on the current situation in automotive software engineering, but the resulting motivation of the domain is quite clear: the need for more efficient and effective development methods to face the upcoming problems and challenges.

From a historical point of view, the development of automobiles started with exclusively mechanical topics. The introduction of the first complex control units during the 1980s concurrently led to the emergence of electronic challenges. The deployed software amount was quite low and restricted to the embedded software from separate units. The interconnection of different units was rudimentary. Therefore, the development methods used were based in the mechanical engineering domain and focused on hardware aspects. Manually created specification documents defined the systems to develop. The engineer in charge then constructed the control unit by hand and wrote the required code in a machine-oriented low-level programming language.

The introduction of more complex control units was accompanied by the launch of new and more sophisticated programming languages, for example, *C++* or *Java*. As a result, the software and hardware complexity and, therefore, the number of involved engineers increased. The communication between the different control units and the communication between the engaged engineers became increasingly complex, especially due to manual system design and implementation. The amount of code could no longer be maintained manually without unrealistic personnel costs. Therefore, the established development methods were improved and extended with in-house and external approaches. For example, the specification documents were no longer created as simple text documents, but stored in a database such as *IBM Rational DOORS*[1] and further enriched with additional artifacts like diagrams or code fragments. This allows adding traceability information, which is required to satisfy established quality and standardization requirements.

This extension of the established methods has come to a crucial point: the current automotive systems are becoming so complex that a mere extension and improvement of the established methods is not sufficient anymore. Instead, completely new approaches are required and are taken into account, for example, from other domains or out of research. A popular example of this process is the introduction of the *Unified Modeling Language (UML)*[2] and the *Systems Modeling Language (SysML)*[3] as foreign modeling languages in the automotive development context.

The following sections give an overview of the current situation by listing recent challenges in the development process in detail, summing up information about currently used development tools and toolchains, classifying those in a previously published taxonomy, and giving an outlook over upcoming innovations and new features.

2 Recent Challenges in Automotive Software Engineering

Due to the previously described changing situation in the automotive domain, many challenges are still present or have appeared in the past years. The sophisticated development cycles and the expectations of the customers are hard to satisfy if no solutions for these challenges are found. In the following, different types of existing challenges, grouped by their context, are given. Wherever possible, a rating from the current point of view is added.

2.1 Virtual Development and Validation

Virtual methods for development and validation became more popular and feasible over the last few years. This is due to a number of reasons: decreasing development cycles, test and validation at early stages of development, complexity of functions and systems, cost reduction, and additional issues. Nonetheless, virtual development and validation methods and related tools are still in their early stages. To get qualified feedback and identify open challenges in developing future vehicles, Bock et al. [8] conducted a survey by addressing experts of the automotive domain. One

---

[1] https://www.ibm.com/de-de/marketplace/requirements-management [Accessed: 21-Sept-2018].
[2] http://www.uml.org [Accessed: 21-Sept-2018].
[3] https://sysml.org [Accessed: 21-Sept-2018].

survey finding in the field of fully automated driving is that 77.4% of experts assess tools and methods for developing self-driving vehicles as not sufficient. Software challenges are named as main topic, whereas environmental interpretation and decision-making are rated as major tasks. Especially for such topics, adequate virtual development and validation methods are needed.

A sufficient virtual development and validation process is dependent on the tools deployed. Many different approaches and tools try to address virtual development and validation methods, but no common framework, which covers all target needs, is known. Hence, linking existing tools to a holistic, distributed development and testing framework may enable interdepartmental coordination throughout all development phases, as departments are often working with own solutions. Interdivisional work often fails due to missing common toolchains and standardized processes. Moreover, continuous development over the whole *V-model* [9] can only be achieved with great effort and costs.

As solution for automotive applications, the so-called Functional Engineering Platform (FEP) can be named. *FEP* is a continuous approach to functional engineering and testing [36]. This includes the possibility for development throughout all phases and departments by providing common models. Besides that, *FEP* provides a generic communication interface called *FEP Library* to link various development and simulation tools. Hence, *FEP* is not a runtime environment which executes simulations; it enables communication through participating tools. To enable co-simulations, *FEP* includes a synchronization mechanism. Furthermore, *FEP* allows to integrate hardware components like real control units, as in later development phases hardware-in-the-loop simulations must be conducted. A similar approach was already used in the aerospace industry with *Avionics Development System 2G (ADS2)*[4] for the development of airplanes.

Besides linking various tools to a holistic, distributed development framework, further challenges exist during the development phases. With rising complexity of vehicle functions, it is becoming more and more important to represent the complexity of the real world in the virtual environment. Conducting tests with *advanced driver-assistance systems (ADAS)* or automated driving vehicles at a certain point of the development process in the virtual environment is safer than in the real world, especially when it comes to use cases, where vehicles have to interact and respond to further road participants (i.e., third-party vehicles or pedestrians). Thus, participating tools need sufficient, realistic, and complex models for simulation. This includes physical models like sensor models, meta-models for communication (i.e., propagation phenomena of radio waves), environment models (i.e., road network, road surface, weather models, etc.), and behavior models (i.e., models for traffic, pedestrians, cyclists, etc.).

---

[4]https://techsat.com/technology-and-products/ads2/ [Accessed: 21-Sept-2018].

2.2 New Development Techniques

Besides new tools and methods for virtual development, new development techniques are coming in. Above all, applied *artificial intelligence (AI)* (i.e., pattern recognition and machine learning, etc.) is one of the most important and comparatively new techniques. *AI* is used in various fields of automotive development processes and in different vehicle functions. As *AI* improves the development and opens up new avenues for complex vehicle functions, it is difficult to validate decision algorithms at the same time. Beyond that, if *AI* techniques are not only used for the development of vehicles but are implemented in vehicles to perform certain functions like automated driving, the error-free operation can—if at all—be proven with great difficulty. The question therefore arises, how functions can be tested when it comes to vehicle functions that are critical according to the *Automotive Safety Integrity Level (ASIL)* [24], as the ISO 26262 norm [24] dictates test methods for ASIL functions. This includes that *ASIL*-critical functions have to be testable with *finite state machines (FSMs)* [26].

2.3 Feasible Development Methods

Nowadays, used development methods are not suitable for future vehicle systems, as function- and model-driven development methods can only be applied partly for developing *ADAS* and automated driving functions. With growing complexity of driving functions, requirements can only be described in an abstract way. Existing data for deriving requirements, like accident research databases, do not include all relevant situations for automated driving. Missing ground truth data for all relevant situations, which automated vehicles must be able to handle, complicate the formulation and derivation of requirements and, thus, also of test cases. Therefore, new development methods need to be developed to address the complexity of future driving functions and to overcome the challenge of deriving requirements and test cases. Scenario-based development and test methods are gaining importance and might be one way to cover these challenges.

2.4 Validation and Release Process

ADAS and automated driving functions perform on the basis of a high amount of information. These information include high-definition maps, precise position localization, and knowledge about the environment recorded by onboard sensors or shared through *vehicle-to-everything (V2X)* communication. Testing these highly connected systems is very challenging and time-consuming, as the complexity constantly increases. Theoretically, the error-free execution of such vehicle systems must be proven for every situation, which might occur [30]. Testing and validating *ADAS* automated driving systems solely through real test drives is economically not feasible [39]. Therefore, new concepts for deriving and performing test cases must be found. Scenario-based and virtual validation methods may address the challenges in testing future vehicle systems.

Besides that, a quality criterion for automated driving functions must be found [41]. Then, the question "How safe is safe enough" [27] must be answered to establish a generally accepted release process and related tools. These topics are addressed, inter alia, by the research projects PEGASUS[5] or Enable-S3.[6]

2.5 Cyber Security

Highly connected vehicles require a high amount of data to perform automated driving or provide further customer services. Additionally, driving data are used to improve vehicle functions and help to develop new systems. Therefore, pieces of recorded driving data may be sent to a back-end system. This implies challenges in the field of data security and privacy. When sending driving data, it is necessary to ensure that the data cannot be accessed by unauthorized persons. Data which are sent over wireless communication technologies have to be encrypted by end-to-end encryption, and vehicles must have sufficient localization privacy techniques, so that no third party can locate and retrace the highly connected vehicles [18].

Also, interfaces for reading out the vehicle's memory must have an adequate access protection. Besides the safe storage and encryption of the data, unauthorized remote control has to be prevented. As present vehicles are already highly connected and various control units interact, the system is vulnerable through different weak points. If an attacker gets access to the vehicle bus system, control over safety-critical functions such as steering or braking could be achieved. Hence, it is a challenging issue that vehicles are well protected by suitable security mechanisms [37]. As attackers adjust and improve their approaches constantly, it may be necessary that the installed security mechanisms in the vehicle are updated frequently. Functional safety and security are strongly related; however, the ISO 26262 norm does not address security issues.

---

[5]http://www.pegasus-projekt.info [Accessed: 21-Sept-2018].

[6]https://www.enable-s3.eu/domains/automotive/ [Accessed: 21-Sept-2018].

3 Related Work

For the three topics discussed in this chapter—development methods, classification taxonomy, and open challenges in the domain—different publications are already available. A non-exhaustive representative excerpt is summarized and rated below.

Ali et al. [1] propose a taxonomy for computer-based critiquing systems which uses both textual categorization scheme and graphical visualization, which allows a fast comparison of different tools. The involved categories are strictly tailored to the given domain.

Azuma et al. [2] evaluate the usage of the Bloom's taxonomy for software engineering approaches, which uses six levels of intellectual behaviors. The authors extend this base pattern by adding a second dimension: the categorization of knowledge areas. As a result, a classification of different tools according to the user's knowledge is possible.

Babar and Gorton [3] compare different software architecture analysis methods by using 17 evaluation questions grouped in four categories as comparison taxonomy.

Blum [6] uses a two-dimensional classification scheme with the distinction between product-oriented and problem-oriented and between conceptual and formal methods.

Broy et al. [10] describe an approach for a seamless model-based development process in the automotive domain. For this, the general process with its challenges and open topics is discussed. Furthermore, typical tools from the domain are mentioned and put in the right context.

Delgado et al. [15] propose a taxonomy for runtime software fault-monitoring approaches, which uses a classification scheme with three layers. The four top categories are derived from specific domain-related criteria such as *event handler*. Different tools are summarized and rated with regard to each relevant subcategory. As a result, the tools are comparable with limitation to specific domain.

Di Natale et al. [16] give a general overview of used development methods in the automotive domain and existing open challenges.

Guo and Jones [21] compare two different example toolchains containing various tools from the automotive domain. The included tools are described and their relevance and usage with regard to development processes are explained. Therefore, the authors give an overview of typical development approaches in the automotive domain.

Kornecki and Zalewski [28] propose a taxonomy for the evaluation of software development methods for safety-critical real-time systems. Three different types of tool evaluation are applied: *meta-*, *macro-*, and *micro-evaluation*. The taxonomy is based on the different contexts, in which the tools are utilized, and on the usage of several comparison criteria. In the course of this, typical development processes with their corresponding tools are described.

Tyndale [38] reviews several tools from the knowledge management domain and classified them by using a suggested tool categorization scheme. This scheme is based on several other publications with their individual taxonomies. Further a distinction between *New Tool* and *Old Tool* is applied to introduce more structure to the categorization scheme.

With regard to the three mentioned topics, it can be said that:

- A taxonomy is proposed in [1–3, 6, 15, 28, 38]. Although all approaches are applicable in principle, the lack of focus on the automotive domain hampers a direct adoption. Additionally, only [1] offers a graphical visualization as requested in Sect. 5.
- Common development methods are described in [10, 15, 16, 21, 28]. As discussed in Sect. 4, the used tools differ depending on the development context, which is why the results from these publications cannot be generalized. Instead, an individual inspection is necessary for our context.
- Open challenges and research topics are evaluated in [10, 16]. Although partly matching to our description in Sect. 2, challenges change through time, so they have to be reassessed over and over.

Because the published taxonomies cannot be adapted for our context, the development tool descriptions are insufficient and the open challenge list is outdated; each of the topics is reevaluated in this work.

4   Common Tools and Toolchains

In the automotive domain, several tools and toolchains have been established during the last decades. The used tools differ from company to company and from department to department. A general inter-company overview is difficult due to the strict confidentiality policy of most of the involved legal entities. In many cases, neither the used tools, nor the applied modifications or specific configurations are made public. As a trade-off, in the following, the mainly used tools and toolchains from our point of view are summarized and subsequently classified in a taxonomy. Some of these methods are applicable or have their roots in other domains, but the focus of each description is the automotive domain. Because the new field of automated driving functions introduced a complete set of new methods and approaches to the domain, the special area of traffic and environment simulation is additionally considered. The information of the subsections already have been published partly in [7].

4.1   *Function Development and Simulation*

The first group of methods and tools of the automotive domain, which should be summarized and reviewed, is the group that covers all aspects from function development and simulation. This list is specifically tailored from the authors' point

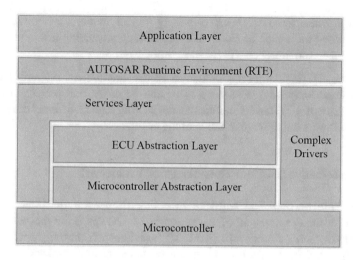

Fig. 1 *AUTOSAR* overview (based on [20])

of view. The individual list in a concrete project may differ. Nevertheless, the given representatives are established and commonly used.

4.1.1 Automotive Open System Architecture

General Information The *Automotive Open System Architecture (AUTOSAR)*[7] [20] is a software architecture standard widely used in the automotive domain and developed by the *AUTOSAR development partnership*. Its main focus is the design, implementation, and realization of automotive systems. For this, a layered software architecture is used (cf. Fig. 1). All software artifacts, which are necessary for the target system, are located at the *application layer*. Each artifact consists of so-called software components (SWCs), which include the algorithms (which are further enclosed in *runnables*) as well as the wrapper code for the function itself. As workaround for projects with a great amount of plain code, this code can also be encapsulated in such *runnables* and be used in an *AUTOSAR* environment. All information of the *AUTOSAR* artifacts use a well-defined XML scheme as data format, which simplifies the exchange and version tracking of the created models. Tests or test strategies are not specified. Recently, the original *AUTOSAR* standard was enhanced to *adaptive AUTOSAR* [19], which is especially useful for the development of automated driving functions.

---

[7]http://www.autosar.org [Accessed: 21-Sept-2018].

Domain Application *AUTOSAR* is widely spread in the automotive domain; methods based on it such as *EAST-ADL* are already advancing fast, and therefore the acceptance and presence will increase further.

Rating As *AUTOSAR* does not include an implementation, a third-party tool (e.g., *IBM Rational Rhapsody*) is required to be able to create real projects. In return, the popularity of *AUTOSAR* simplifies model sharing among different development teams and lowers the average initial learning effort.

4.1.2 Automotive Data and Time-Triggered Framework

General Information The software modeling framework that was originally designed for the development of driver-assistance systems is the *Automotive Data and Time-Triggered Framework (ADTF)*.[8] Initially, such driving functions were created by hand in the form of plain source code. Various modeling tools were introduced to manage the complexity of the development, but the technical background with the information about sensor data, interfaces, and physical conditions sustained. Therefore, a form of presentation based on this was consequential, and so the models created in *ADTF* use so-called filters with the corresponding inputs and outputs (e.g., signals) extracted from various data sources, such as a *controller area network (CAN)* [23] and camera data, as main elements. The *filters* are displayed as graphical elements, and the interaction between the different objects are modeled as lanes. All data sources can be used simultaneously and synchronized. This allows real-time data playback and offers visualization features, which are the basis for a simulation of the created models and a subsequent evaluation, especially with respect to timing constraints. All these aspects ensure the congruency of the simulation and the behavior of the real system.

Domain Application *ADTF* was initially developed in the automotive domain in Germany in 2011. The focus on driver-assistance systems with their related sensor data and the corresponding system design is mainly appropriate for this domain, although similar possibly related domains such as the avionic sector can also profit from the included concepts by adding some required extensions or modifications. *ADTF* includes architecture steps as straightforward process definition, an implementation in the form of a tool and the possibility to generate code for the target system. Testing is limited to short manual tests. For the model creation, a graphical *domain-specific language (DSL)* is used.

Rating Although fairly new in the domain, *ADTF* has already been established as development tool across several automotive companies and the related suppliers [8]. In fact, a quite high familiarity of the respondents with *ADTF* and a utilization rate of 50% can be stated [8]. This, in conjunction with the specific design for the

---

[8]https://automotive.elektrobit.com/products/eb-assist/adtf/ [Accessed: 21-Sept-2018].

implementation of driver-assistance systems, states *ADTF* as an appropriate solution for automotive development, although the absence of a detailed process definition and the insufficient testing possibilities are drawbacks.

4.1.3 Electronics Architecture and Software Technology-Architecture Description Language

General Information The *Electronics Architecture and Software Technology-Architecture Description Language (EAST-ADL)* [5] is developed and supported by the *EAST-ADL Association*. As foundation, *AUTOSAR* is used, although additional concepts such as nonfunctional requirements, vehicle features, and functional/hardware architecture details are covered. For model creation, a four-level abstraction system is used (cf. Fig. 2). The initially created rough vehicle model describes the system in an abstract way but is further enriched during the development. Finally, a high detailed *AUTOSAR* model is accomplished, which can then be used to create the target code.

Domain Application *EAST-ADL* has been specifically designed for the automotive domain, which is stated by the vehicle-based development layers. Although some of the used development processes are based either partly or largely on *EAST-ADL*, an explicit usage of the language is rather uncommon, particularly because of its missing implementation.

Rating As no realization of the language in the form of a tool is included, the process can hardly be used directly. Nonetheless, due to the fact that *EAST-ADL*

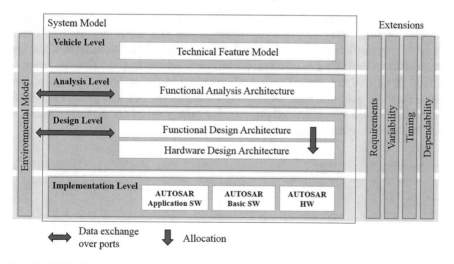

Fig. 2 *EAST-ADL* overview (based on [5])

was initially designed for the automotive domain, no explicit modifications have to be applied to use it in the domain.

4.1.4 MATLAB/Simulink and TargetLink

General Information *MathWorks MATLAB* is a numerical computing framework designed to handle complex mathematical problems and to calculate and display the corresponding results. As extension to this base framework, *Simulink*[9] is a graphical data flow modeling language that offers the possibility to create system models and the corresponding software models. For this, so-called blocks (functional entities) are used with links in the form of associations. All model items (e.g., bus-, mux-/demux-, or gain-blocks) are taken out of a predefined block library, whose elements are leaned on concepts of electrical circuits and electronic control unit design, so to speak on hardware aspects. Testing, verification, and validation methods are included. The models created in *Simulink* are used to generate source code for the target system by using *dSPACE TargetLink*.[10] This tool is focused on code generation and involves various verification and validation techniques to guarantee the reliability of the generated code [13].

Domain Application In the current automotive development projects, the *MATLAB/Simulink/TargetLink* toolchain is widely used for system design, implementation, and code generation. This is due to historical reasons. The toolchain has already been used for several decades; therefore, the majority of the engineers are familiar with it [8]. Nevertheless, upcoming projects such as driver-assistance systems for self-driving vehicles have reached a complexity that is hardly manageable with this toolchain, and therefore, alternative options are investigated.

Rating Although the toolchain is one of the major software engineering frameworks currently used in the automotive domain and offers a great variety of modeling possibilities, no lines of action are included. Additionally, it lacks the possibility to design the system architecture or to create, add, and manage requirements at an abstract level. Because of its origin in the embedded programming sector and its usage of hardware-related modeling entities, the toolchain is well suitable for hardware-depended projects, but poorly suited for high-level projects or, for example, entertainment system projects.

---

[9]http://www.mathworks.com/products/simulink/ [Accessed: 21-Sept-2018].

[10]http://www.dspace.com/en/pub/home/products/sw/pcgs/targetli.cfm [Accessed: 21-Sept-2018].

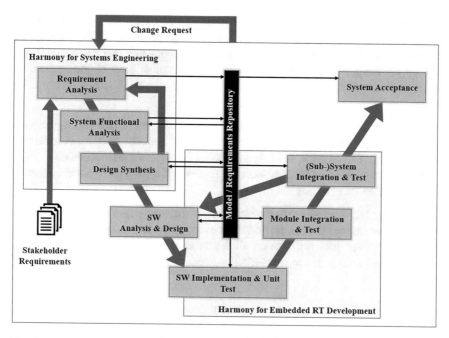

Fig. 3 IBM rational harmony process (based on [22])

4.1.5 Rational Rhapsody/Harmony

General Information *IBM Rational Rhapsody* is a *UML* modeling tool that supports all types of *UML* diagrams with the corresponding elements and provides simulation features and code generation techniques. It can be extended with additional profiles such as a *Systems Modeling Language (SysML)*[11] profile and has predefined interfaces to other tools (e.g., *IBM Rational DOORS*). Because it does not include lines of action, *IBM Rational Harmony* [22] is natively included. It is an iterative software modeling process based on the *V-model* [9], a generic profile that extends *Rhapsody* with additional wizards, modeling possibilities, and a consistent modeling process. *Harmony* is divided into two separate subsequent sections (cf. Fig. 3). The first section (*Harmony for Systems Engineering*) uses *SysML* for the creation, definition, and design of the system architecture. For this, use cases and the related requirements are either manually created or imported from external data sources. Afterward, activity and state diagrams are modeled by hand, and sequence diagrams can be generated automatically. With the help of these artifacts, the system behavior can be simulated and therefore tested and validated. After further refinement and enrichment procedures, the final model functions as handover artifact to the second part of the *Harmony* process: *Harmony for Embedded RT*

---

[11] http://www.sysml.org [Accessed: 21-Sept-2018].

Development. Here, the information from the *SysML* model are transferred into a *UML* model containing the architecture of the target software. After this step, the final target code is generated, which can be both simulated and run on the target device parallely to ensure the reliability and the consistency of the system. To increase the usability, semiautomatic wizards assist with all the different modeling steps.

Domain Application Although not specifically designed for the use in the automotive domain, the tool and process have become quite common in the domain over the last years. The creation of the target code solely out of graphical models differs distinctly from the common hand-coding process, which requires a certain extent of familiarization for each engineer.

Rating The required initial learning expense for new engineers and the higher amount of modeling effort at the first steps of a new project model are compensated by the consistent process through all levels of development and by the easier validation and testing due to simulation and code generation features.

4.1.6   Safety-Critical Application Design Environment

General Information *Esterel Safety-Critical Application Design Environment (SCADE)*[12] is a software development framework initially designed in the avionics industry. It consists of four separate tools. One of these tools is the *SCADE Suite*, which is focused on model-based software development. For the creation of the graphical models, a formal, synchronous, and data flow-oriented domain-specific language [11] is included and used. All development phases are covered, from the initial specification until the conclusive acceptance test. For this, validation and verification methods are provided to guarantee consistency of the created models. Code generation techniques are used to automatically generate the target code out of the designed models. As guidance through the development process, so-called lines of action are included.

Domain Application Although initially developed for the avionics industry, the *SCADE Suite* is designed modularly enough to be adapted for the automotive domain. Because both domains share many aspects and requirements such as reliability, constraints, and sensor-data-based functions, *SCADE* has been introduced to the automotive industry several years ago and became increasingly prominent recently.

Rating The stricter requirements in the avionics industry ensure a high reliability of the created models and the involved development process. Therefore, the requirements of the automotive domain should be easily met. Nevertheless, many modifications are still pending or currently ongoing, so that the *SCADE Suite* can be

---

[12]http://www.esterel-technologies.com/products/scade-suite/ [Accessed: 21-Sept-2018].

used throughout the complete development process. Additionally, the interworking with and the interface to other established tools are not yet sufficient, and a complete switch of the automotive development process to *SCADE* as a single development tool is not likely in the near future.

4.1.7 Simulation and Test of Anything

General Information *Simulation and Test of Anything (SimTAny)* [34, 35] (formerly known as *VeriTAS* [17]) is a framework that provides the *test-driven agile simulation (TAS)* process and an implementation in the form of a toolchain based on *Eclipse*[13] and the related UML modeling environment Papyrus.[14] According to the process, the system and the usage model are semiautomatically derived from the requirements in the form of individual *UML* models (see Fig. 4). A corresponding simulation model is automatically generated from the system model and test cases are automatically derived from the usage model. Both the simulation and the test cases can be executed and therefore allow a mutual validation. A separate implementation of the system or hardware is not necessary. A surrounding process and lines of action are part of *SimTAny*. All these aspects allow to identify and find errors or inconsistencies in the involved models as early as possible in the development process.

Domain Application *SimTAny* has its origin in the academics and research domain, so no specific focus on the automotive domain was implied initially. Nevertheless, the process is easily adaptable and was introduced several years ago into the first automotive projects, which resulted in an increasing prominence due to the simulation capabilities and the use of the already well-established framework *Eclipse*.

Rating The creation of new requirements by hand is possible, but it is not the specific focus of *SimTAny*, which is why the import of existing requirements is advised. The main emphasis is the system simulation without the use of dedicated production code. Therefore, this method is not suitable for code generation and has some flaws regarding the abstract system specification, which is compensated by the valuable simulation possibilities.

---

[13] https://eclipse.org/ [Accessed: 21-Sept-2018].
[14] https://eclipse.org/papyrus/ [Accessed: 21-Sept-2018].

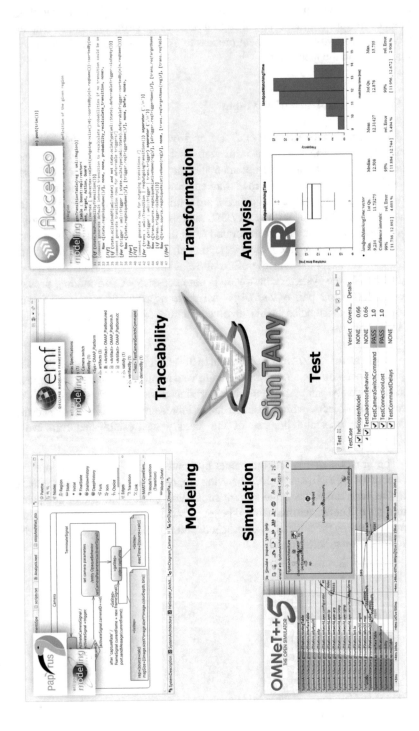

Fig. 4 *SimTAny* overview [35]

4.2 Traffic Simulation

Increasing traffic, especially in urban environment, new concepts for future mobility in public and private transport, the increasing number of available driver-assistance systems, as well as research and development activities for automated driving vehicles require a vast amount of simulation activities with different scopes. There already exists a large amount of studies and publications dealing with traffic modeling, traffic flow optimization, transportation planning, modeling and simulation of shared spaces, etc. Various simulation and modeling frameworks are available and used in the automotive domain, in research, and in academics. They are utilized to develop and evaluate new concepts for mobility, conducting studies on different use cases or validating new business models or products through all development stages. For example, [31] depicts a literature review on traffic flow simulators, [29] reviews existing traffic simulation frameworks, [33] and [12] compare different simulation frameworks, and [14] discusses advantages and disadvantages of driving simulation. Additional topics like driving simulation, traffic modeling and simulation, transportation planning, and virtual development and testing of transportation systems are handled and investigated separately.

In the following, we give an overview of the traffic simulation frameworks *Aimsun Next*, *SUMO*, *Vissim*, and *Viswalk*. Additionally, we give a brief presentation of *VTD* and *CarMaker*, which represent tools for driving simulation for virtual vehicle function development and testing. In our point of view, these are the most familiar and widely used tools, but the list is not intended to be exhaustive.

4.2.1 Aimsun Next

General Information The commercial traffic modeling and simulation software *Aimsun Next*[15] is designed for processing mesoscopic, microscopic, and hybrid traffic simulations. The main use cases of *Aimsun Next* are the evaluation of traffic flow and traffic guidance, such as the analysis of a given infrastructure design, an environmental impact analysis, an evaluation of travel demand management strategies, a signal control optimization, and a safety analysis. Additionally, *Aimsun Next* allows the evaluation of intelligent transportation systems, diverse toll and road pricing scopes, and further tasks. As it is a hybrid simulation framework, large areas can be simulated, while it is possible to consider smaller areas in more detail at the same time. In microscopic areas, *Legion for Aimsun*[16] enables pedestrian simulation within the framework. *Aimsun Next* includes modeling, simulation, and analysis in a single environment and provides an intuitive user interface. Besides simultaneous 2D/3D views and graphical outputs for the visualization, this

---

[15]https://www.aimsun.com/aimsun-next/ [Accessed: 21-Sept-2018].

[16]http://www.legion.com/legion-for-aimsun-%E2%80%93-vehicle-pedestrian-simulation [Accessed: 21-Sept-2018].

framework provides *Python* scripting for automation and evaluation and comes with a *Python* or *C++*-based *application programming interface (API)* for including intelligent transportation systems. The *Aimsun Next microSDK* allows users to overwrite behavior models and implement own solutions and to create entirely new applications or plug-ins with separate menus and dialogues, based on the industrial standard *C++*.

Domain Application *Aimsun Next* is mainly used for case studies dealing with traffic optimization and infrastructure design. In the automotive domain, it can be used for business modeling such as shared mobility with autonomous vehicles.

Rating Widely used by public and academic institutions, simulation frameworks like *Aimsun Next* are becoming increasingly important for the automotive domain for modeling future mobility. *Aimsun Next* provides an *API* that supports common programming languages and a *microSDK*; thus, it can be adapted to fit own requirements.

4.2.2 Simulation of Urban MObility

General Information *Simulation of Urban MObility (SUMO)*[17] is an open source project, mainly developed by the *Institute of Transportation Systems* at the *German Aerospace Center* with various extensions provided by external parties. It is a framework and platform for microscopic and inter- and multimodal traffic flow simulation. It helps perform space-continuous microscopic traffic scenario simulations. Every single vehicle is simulated individually and has its own route. As a free tool, *SUMO* can be extended with own algorithms. Since 2001, *SUMO* has evolved into a full-featured suite of traffic modeling utilities, which includes several different aspects. For simulating *V2X* communication, the discrete event simulation package *OMNeT++*[18] for simulating computer networks and distributed systems was extended by *Veins*,[19] an open source vehicular network simulation framework. *Veins* includes a suite of realistic vehicular network simulation models. These models are executed by *OMNeT++* while interacting with *SUMO*. For the bidirectionally coupled simulation of traffic flow and network traffic, the simulators are connected using a *transmission control protocol (TCP)* socket and a standardized protocol named *traffic control interface (TraCI)*.

Domain Application *SUMO* addresses microscopic simulation of vehicles, pedestrians, traffic lights, and public transport. It is efficiently usable for traffic flow optimization. Linked with *OMNeT++* and *Veins*, it is applicable for realistic vehicular network simulations and is widely spread in the scientific community.

---

[17]http://sumo.dlr.de/ [Accessed: 21-Sept-2018].
[18]https://omnetpp.org/ [Accessed: 21-Sept-2018].
[19]http://veins.car2x.org/ [Accessed: 21-Sept-2018].

Rating As an open source project implemented in *C++* and supporting various formats for road description, there is a strong community enhancing *SUMO*. The framework comes with a large suite of applications, documentations, manuals, tutorials, and examples. A larger set of publications have already been published in this context (e.g., at the *SUMO User Conference*).

4.2.3   Vissim and Viswalk

General Information *PTV Group* provides a suite for traffic modeling and simulation that includes both the tools: *Vissim* and *Viswalk*. *Vissim*[20] is the global leader on the market of commercial microscopic simulation frameworks and comes with a powerful graphical user interface for scenarios and map editors. Vissim includes motorized private transport, freight transport, rail transport, pedestrians, and cyclists. For microscopic pedestrian flow simulation, *Viswalk*[21] covers scenarios for routing and traffic jam analysis, waiting time analysis, capacity and concept planning of buildings, and concepts for evacuation scenarios of buildings and major events. With the help of *Vissim*, *Viswalk* simulates pedestrians in multimodal environments. *Vissim* comes with various interfaces to integrate existing networks and models of other *PTV* modules. It is also possible to import further strategic traffic models, to integrate external signal controlling, or to transfer emission data for getting detailed analysis using simulation. *Vissim* includes a large graphical user interface including 2D and 3D visualization.

Domain Application Common uses of *Vissim* are the optimization of routes, the simulation of virtual scenarios, and the modeling of city networks. Hence, it is used for traffic, transport, and mobility optimization. *Vissim* also supports integrating *Viswalk*, so microscopic traffic simulations in urban environment can be realized. Application for the automotive domain is business modeling, but also testing vehicle systems in virtual environment is conceivable, whereby simulation resolution is not capable for simulating vehicle dynamics. Thus, a suitable driving simulation has to be available.

Rating Due to the large software suite that *PTV* provides, as well as the fact that data exchange is possible between participating simulators, *Vissim* and *Viswalk* are powerful tools for various applications such as traffic analysis, business modeling for future mobility, and cross-functional applications like autonomous driving in shared space for planning parking garages. However, it is not possible to simulate vehicle dynamics, so adjustments and linking to other tools have to be realized to test automated driving functions.

---

[20] http://vision-traffic.ptvgroup.com/de/produkte/ptv-vissim/ [Accessed: 21-Sept-2018].

[21] http://vision-traffic.ptvgroup.com/de/produkte/ptv-viswalk/ [Accessed: 21-Sept-2018].

4.2.4 Virtual Test Drive

General Information   *VIRES Virtual Test Drive (VTD)*[22] is a submicroscopic environment-sensitive behavior driving simulation. *VTD* covers open-loop as well as closed-loop operations including a real-time mode, a simulation configuration control with scenario management, a control of simulation stages, and a master and a slave mode for linking multiple simulators in one simulation. The open interfaces *runtime data bus (RDB)* and *simulation control protocol (SCP)* enable bidirectional high-frequency data communication and event-driven communication. *VTD* allows realistic traffic density and features in urban and motorway environments and the integration of multiple externally controlled entities. It implies an image generator for visualization, the possibility for simulating simplified sensor models or physically driven sensors, and is capable of simulating vehicle dynamics (e.g., braking system, steering, suspension, tires, etc.). *VTD* comes with an editor for creating road networks based on the *OpenDRIVE Specification*[23] and a scenario editor for setting up different traffic scenarios.

Domain Application   *VTD* is a submicroscopic traffic simulator able to simulate complex scenarios in submicroscopic resolution. *RDB* and *SCP* can be connected to vehicle buses and stimulate *electronic control units (ECUs)*. Moreover, it can be used for developing and testing prototypical vehicle functions, especially during early development phases.

Rating   *VTD* is a toolkit to simulate vehicle behavior in high spatial and temporal resolution in virtual environments. It enables the user to develop and test vehicle functions in SiL, DiL, ViL, and HiL[24] methods. For environment descriptions and scenario modeling, it comes with editors for road network modeling and scenario modeling. Here, well-known formats like the *OpenDRIVE Specification* is used. *VTD* can be operated as co-simulation supporting third-party and custom packages like *ADTF* (cf. Sect. 4.1.2). *VTD* is also very manageable for scenario-based testing. Nevertheless, *VTD* is highly dependent on the operating system and a well-equipped hardware configuration.

4.2.5 CarMaker

General Information *IPG CarMaker*[25] is a commercial software solution for virtual test driving, supporting MiL,[26] SiL, HiL, and ViL methods. It is used for

---

[22]https://www.vires.com/ [Accessed: 21-Sept-2018].

[23]http://www.opendrive.org/ [Accessed: 21-Sept-2018].

[24]Software-, driver-, vehicle-, hardware-in-the-loop.

[25]https://ipg-automotive.com/products-services/simulation-software/carmaker/   [Accessed: 21-Sept-2018].

[26]Model-in-the-Loop.

modeling and testing real-world test scenarios and the entire surrounding environment, including vehicle dynamics and other road participants. *CarMaker* enables the user to test virtual prototypes during all stages of development. It includes open- as well as closed-loop testing during virtual tests and software/hardware platforms. Additionally, it offers a visualization feature for the simulations. Test automation and evaluation of simulation results are also integrated in the simulation framework. *CarMaker* includes an adaptive driver model and allows modeling driver activities and system interventions using maneuvers. It integrates a wide range of model and tool interfaces and is designed as an open integration and test platform.

Domain Application *CarMaker* is used for the development and test of virtual vehicle prototypes through all development phases and used by all common *original equipment manufacturers (OEMs)* and a large number of suppliers. Furthermore, research institutes and universities utilize *CarMaker*.

Rating *CarMaker* allows scenario-based, reproducible tests in virtual environments and increases the efficiency during the development and all involved tests. As it can be linked to well-known models and tools, it is a powerful tool for testing vehicle systems. Nonetheless, the used hardware has to be individually tailored.

4.2.6 Pedestrian and Cyclist Simulation

Besides the high number of simulation frameworks for different tasks, the simulation of vulnerable road users is getting more and more important to validate vehicle driving functions. Therefore, frameworks are needed which are capable of simulating driving functions in virtual environments including shared spaces and also provide validated behavior models for various types of vulnerable road users. To this date and to our knowledge, there is no simulation framework available, which is able to simulate both at the same time. Existing frameworks like *Viswalk*, *Legion for Aimsun*, and *MomenTUM* [25] are able to simulate pedestrians, but the scope of these frameworks is not on driving function testing. Distributed simulation frameworks, which are linking well-established driving simulators to frameworks for simulating vulnerable road users and development frameworks like *ADTF*, might be an option to pave the way to holistic urban environment simulation frameworks to validate driving functions.

4.3 System Specification and Documentation

Besides the functional development methods and the traffic simulation approaches, there are some tools that are specifically designed for the abstract specification of the automotive system or for the creation of documentation artifacts during the automotive development process. The two main tools from our point of view are described in this section.

4.3.1 Office

General Information *Microsoft Office*[27] is an office suite that consists of several different types of tools: a word processor (*Microsoft Word*), a spreadsheet tool (*Microsoft Excel*), a presentation software (*Microsoft PowerPoint*), and other tools. They are designed to be used by technical experts as well as by normal users. The suite is widely used in different domains, companies, and departments, and the majority of the users are familiar with the created documents and the presentation of the included information. This allows the suite to be used as central sharing possibility for information between persons or departments with no similar knowledge base. Although many alternative office solutions are available—open source frameworks as well as proprietary ones—the mainly used framework is still *Microsoft Office*. The main reasons for this are the availability of a professional support and the widespread distribution.

The documents are technically stored as compressed files with the *Extensible Markup Language (XML)* format and therefore are suitable to be used within a version control system. *Word* is typically used for the storage of, for example, textual information, requirements, project reports, or meeting transcripts. The layout is not very strict and can be extended or varied in many ways. *Excel* uses spreadsheets for the information storage and is, due to its well-defined structure, especially suitable for listings, mathematical calculations (via the use of embedded formulas), and technical detail information such as log files or datasets. *PowerPoint* is mainly used as a presentation tool; the layout is restricted and optimized for the use of projectors. Main usage scenarios are presentations for meetings, short project information for the management, or any other presentation of use case.

Parts of the documents created in one of the mentioned systems is migratable/transferable between the tools, so, for example, diagrams out of an *Excel* file can be embedded in a *PowerPoint* presentation. Further assistant systems like a thesaurus system, document validation methods, the extension via macros or diverse printing, and export functions empower the user in many ways.

Domain Application In the automotive domain, the *Office* suite is used as a tool to express the requirements and specifications of the developed systems and collect and sort the data and log files from the live system and as presentation/documentation software.

Rating Advantageous are the simple access to all tools, the lucidity of the documents, and the high diffusion rate. Nevertheless, the embedment of technical details, code, and simulation possibilities are restricted or nonexistent. This results in the need for further implementation methods.

---

[27]https://www.office.com/ [Accessed: 21-Sept-2018].

4.3.2 Rational DOORS

General Information Besides the plain textual specification based on files such as with *Microsoft Office*, especially in case of large projects with thousands of requirements and constraints, the complexity is hardly manageable and requires a solution that implies any type of a database. For this, *IBM Rational DOORS*[28] was designed and introduced in the automotive domain about two decades ago. It is server-based, so the user does not necessarily have direct access to the data container. It allows the creation, management, and export of textual requirements to several external formats (e.g., *Microsoft Excel* or *Word*). Graphical artifacts such as diagrams and files can be attached, which allow certain types of graphical documentation and architecture information beyond plain textual descriptions. Regarding traceability, *DOORS* uses unique identifiers for each requirement and allows the creation of links between requirements and to external artifacts. This offers the possibility to track changes and to find orphaned requirements.

Domain Application *DOORS* has been used in the automotive domain as specification tool for several decades now and is still the major requirement management and creation tool. Many interfaces to other tools commonly used in the domain exist, so the interconnection is quite good and the information stored in *DOORS* can easily be transferred to other engineers and departments and then be used in their creation or generation process.

Rating *DOORS* has a long usage history in the domain, so most of the engineers are quite familiar with it and therefore it can be easily used in new projects. The good interconnection with other tools also supports this usage. On the contrary, the limitation to textual requirements and attached files does not allow a direct creation or use of models, so the system and software architecture has to be modeled elsewhere, which is adverse.

5 Classification in the Automotive Development Process

The previously mentioned and described tools differ in their characteristics, as well as in their valid usage scenarios. The development of modern vehicles and driving functions is split into several process steps, from specification and planning, over concept and architecture, through implementation until test and approval. Each step includes several substeps with individual assigned persons, tools, requirements, and expected results. The main difficulty when planning a new automotive function or system is to select the appropriate development tools with regard to project constraints, budget, required technical characteristics, and project goals. The partial

---

[28]https://www.ibm.com/de-de/marketplace/requirements-management [Accessed: 21-Sept-2018].

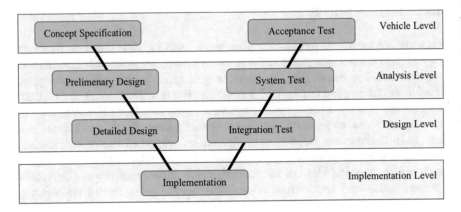

Fig. 5 General overlay [7]

great differences between the tools impede a direct comparison, and a textual overview is quite difficult to read and understand.

As a result, Bock et al. designed and proposed a graphical taxonomy for the comparison of automotive development methods [7]. It combines two separate level definitions out of the automotive domain to be able to provide a certain structure: the individual process steps defined in the V-model [9] and the levels used in the EAST-ADL [5]. The resulting classification scheme is depicted in Fig. 5.

The coverage of the different levels for each individual tool under investigation is illustrated as a diagonal bar corresponding to the *V-model* shape. To include more details about the tools and therefore to simplify a proper choice of the best tool for a certain project setup, additional formatting options are intended and used. Unlike the original formatting information described in the original publication which was focused on the type of underlying programming language, a modified version is reasonable in this case. For the intended overview, we chose three characteristics as most useful: a binary indicator, if modifications for the use in the automotive domain are required; a binary indicator, if lines of actions are included so that the user knows how to create the proper artifacts during the development; and a binary indicator, if an implementation of the method is included. Many more characteristics are possible and may be helpful, but we identified these three as main choice factors. The type of visualization in the target diagram is depicted in Fig. 6.

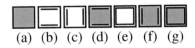

(a) (b) (c) (d) (e) (f) (g)

(a) Modifications   (d) Modifications/Lines of Action   (g) Modifications/
(b) Lines of Action   (e) Lines of Action/Implementation   Lines of Action/
(c) Implementation   (f) Modifications/Implementation   Implementation

Fig. 6 Binary choice indicators [7]

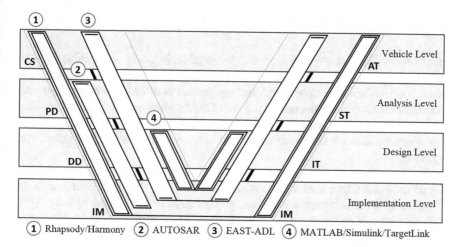

Fig. 7 Method comparison part 1

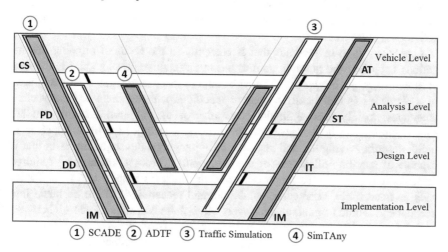

Fig. 8 Method comparison part 2

With the help of the formatting scheme and the previously defined diagram structure, the methods and tools reviewed in Sect. 4 can be sorted in and categorized. The resulting diagrams are shown in Figs. 7 and 8. All traffic simulation methods are represented by one single bar, because they all share the same characteristics and usage scenarios.

These figures can be used as a starting point to choose and evaluate proper candidates for a specific project setup. Because no details about the target project are taken into account, no definite decision guidance can be offered, but the provided diagrams can be used as initial selection method to sort out clearly unfitting candidates and minimize the group of methods that have to be investigated further. This lowers the required resource effort for a proper tool selection.

6 Outlook: The Future of Automotive Development

The overview of the current used development tools in the automotive domain, which was given in Sect. 4, is neither complete nor representative of all automotive manufacturers and their corresponding suppliers. As already mentioned, privacy policies make it difficult to create a full list of the involved tools. Nevertheless, from our point of view, the listing contains the major development methods and tools that are currently used for the specification and implementation of modern automotive systems.

However, the challenges from Sect. 2 are indications for the increasing complexity of next-generation automotive systems and functions. To be able to handle the complexity, the existing tools can be extended or modified, although this can only be done to a certain extent. For a great leap forward in terms of development methods, either completely new methods can be designed and introduced or methods from other domains such as the avionics industry can be migrated and adapted to the specific requirements in the automotive domain. The transfer of tools such as *Esterel SCADE* is a first hint of this process, even though much more of these transfer activities are required.

One major type of methods that is currently in the focus of ongoing investigations regarding the possible application in automotive development is the area of *domain-specific languages*. Although some of these languages are already used (e.g., *Simulink* is technically a *domain-specific language*); there are many other languages, mainly located in the academics or the research domain, that are candidates for use in the development process. The objective in doing so is to build, describe, configure, and deploy a fully integrated development chain that is capable of handling all phases of the development process, manage all required artifacts, and generate the target code and the adjacent models and files. Topics such as traceability, versioning, validation, and reusability have to be taken into account and should be guaranteed. Nevertheless, there is always a trade-off between completeness of the toolchain and usability. *Domain-specific languages* inherently provide consistency and the generation of the required artifacts and are flexible enough to be enriched with wizards, external libraries, *APIs*, and other already available languages to make the process as seamless as possible. For this, there are currently two major development environments for *domain-specific languages* available: *Eclipse Xtext*[29] and *JetBrains MPS*.[30]

Another challenge to overcome is the complexity of developing, testing, and validating new driving functions. They are becoming increasingly complex with each development cycle and have to handle more complex situations. Scenario-based development methods and various kinds of virtual development tools are increasingly used. Traffic scenarios may be used in different ways over the devel-

---

[29]https://eclipse.org/Xtext/ [Accessed: 21-Sept-2018].

[30]https://www.jetbrains.com/mps/ [Accessed: 21-Sept-2018].

opment process, to describe use cases, to help derive requirements, may be used for high-level and detailed system design, and of course are the basis for scenario-based virtual validation methods. Continuous usage of methods for scenario-based system engineering starting at predevelopment phases up to release processes must be achieved to handle the development of future driving systems. Thus, new tools must be developed, existing ones enhanced and then be embedded and used consequently over the whole development process. The growing usage of simulation techniques in driving function development is a first step toward this. But there is much room for further enhancements in developing methods and processes as well as tools and toolchains, to make developing and validating future vehicles efficient, with economically reasonable effort.

With the help of these new concepts and other already existing tools and methods, the development process in the automotive domain can be extended step by step, until the complexity of modern automotive systems can be handled again.

References

1. Ali NM, Hosking J, Grundy J (2013) A taxonomy and mapping of computer-based critiquing tools. IEEE Trans Softw Eng 39(11):1494–1520
2. Azuma M, Coallier F, Garbajosa J (2003) How to apply the bloom taxonomy to software engineering. In: Eleventh annual international workshop on software technology and engineering practice, 2003. IEEE, Piscataway
3. Babar MA, Gorton I (2004) Comparison of scenario-based software architecture evaluation methods. In: 11th Asia-Pacific software engineering conference, 2004. IEEE, Piscataway
4. Baudisch A, Richter K, Sollmann S (2011) Erweiterte Vorgehensmodelle für die Entwicklung echtzeitfähiger, hoch-integrierter, multifunktionaler Steuergeräte-Plattformen
5. Blom H, Hagl F, Papadopoulos Y, Reiser MO, Sjöstedt CJ, Chen DJ, Kolagari R (2012) EAST-ADL - an architecture description language for automotive software-intensive systems. International Standard
6. Blum BI (1994) A taxonomy of software development methods. Commun ACM 37(11):82–94
7. Bock F, Homm D, Siegl S, German R (2016) A taxonomy for tools, processes and languages in automotive software engineering. In: Zizka J, Nagamalai D (eds) Computer science & information technology. AIRCC Publishing Corporation, Chennai
8. Bock F, Sippl C, German R (2017) Fully automated vehicles: challenges, expectations and methods. In: Bargende M, Reuss HC, Wiedemann J (eds) Proceedings of 17th Internationales Stuttgarter Symposium: Automobil- und Motorentechnik, Stuttgart, Germany
9. Bröhl A (1993) The V-model. In: Software - application delevopment - information systems (in German). Oldenbourg, Munich
10. Broy M, Feilkas M, Herrmannsdoerfer M, Merenda S, Ratiu D (2010) Seamless model-based development: from isolated tools to integrated model engineering environments. Proc IEEE 98(4):526–545
11. Caspi P, Pilaud D, Halbwachs N, Plaice JA (1987) LUSTRE: a declarative language for real-time programming. In: Proceedings of the 14th ACM SIGACT-SIGPLAN symposium on principles of programming languages, New York, NY, USA, POPL '87
12. Cheu RL, Tan Y, Lee D (2003) Comparison of paramics and GETRAM/AIMSUN microscopic traffic simulation tools. In: 83rd annual meeting of the transportation research board
13. Dajsuren Y, van den Brand MG, Serebrenik A, Roubtsov S (2013) Simulink models are also software: modularity assessment. In: Proceedings of the 9th international ACM sigsoft conference on quality of software architectures. ACM, New York, pp 99–106

14. De Winter J, Van Leuween P, Happee P (2012) Advantages and disadvantages of driving simulators: a discussion. In: Proceedings of measuring behavior
15. Delgado N, Gates AQ, Roach S (2004) A taxonomy and catalog of runtime software-fault monitoring tools. IEEE Trans Softw Eng 30(12):859–872
16. Di Natale M, Sangiovanni-Vincentelli AL (2010) Moving from federated to integrated architectures in automotive: the role of standards, methods and tools. Proc IEEE 98(4):603–620
17. Djanatliev A, Dulz W, German R, Schneider V (2011) Veritas - a versatile modeling environment for test-driven agile simulation. In: Proceedings of the 2011 winter simulation conference, Phoenix, AZ, USA, WSC 2011
18. Eckhoff D (2016) Simulation of privacy-enhancing technologies in vehicular ad-hoc networks. PhD thesis, University of Erlangen
19. Fürst S, Bechter M (2016) Autosar for connected and autonomous vehicles: the autosar adaptive platform. In: 2016 46th annual IEEE/IFIP international conference on Dependable Systems and Networks Workshop (DSN-W)
20. Fürst S, Mössinger J, Bunzel S, Weber T, Kirschke-Biller F, Heitkämper P, Kinkelin G, Nishikawa K, Lange K (2009) Autosar–a worldwide standard is on the road. In: 14th international VDI congress electronic systems for vehicles, Baden-Baden, vol 62
21. Guo Y, Jones RP (2009) A study of approaches for model based development of an automotive driver information system. In: 3rd annual IEEE systems conference, 2009. IEEE, Piscataway
22. Hoffmann H (2014) Systems engineering best practices with the rational solution for systems and software engineering deskbook release 4.1. Manual
23. International Organization for Standardization (2015) 11898-1: 2015–road vehicles–controller area network (CAN)–part 1: data link layer and physical signalling. International Organization for Standardization
24. IO for Standardization (2009) ISO/DIS 26262-1 - Road vehicles - functional safety - part 1 glossary
25. Kielar PM, Biedermann DH, Borrmann A (2016) MomenTUMv2: a modular, extensible, and generic agent-based pedestrian behavior simulation framework. TUM-I1643, Technische Universität München
26. Klauda M, Hamann R, Kriso S (2013) ISO 26262 – Muss das Rad neu erfunden werden?, Springer Fachmedien Wiesbaden, Wiesbaden, pp 224–227
27. Knauss A, Schröder J, Berger C, Eriksson H (2017) Paving the roadway for safety of automated vehicles: an empirical study on testing challenges. In: 2017 IEEE Intelligent Vehicles symposium (IV)
28. Kornecki AJ, Zalewski J (2003) Design tool assessment for safety-critical software development. In: Conference: software engineering workshop, 2003
29. Kotusevski G, Hawick K (2009) A review of traffic simulation software. Res Lett Inf Math Sci. 13
30. Lachmann R, Schaefer I (2013) Herausforderungen beim Testen von Fahrerassistenzsystemen. In: GI-Jahrestagung
31. Mubasher MM, ul Qounain JSW (2015) Systematic literature review of vehicular traffic flow simulators. In: 2015 international conference on Open Source Software Computing (OSSCOM)
32. Rasshofer RH, Gresser K (2005) Automotive radar and lidar systems for next generation driver assistance functions. Adv Radio Sci 3:205–209
33. Ronaldo A, Ismail T (2012) Comparison of the two micro-simulation software AIMSUN & SUMO for highway traffic modelling, Linköping University, Communications and Transport Systems, The Institute of Technology, p 96. http://www.diva-portal.org/smash/get/diva2: 555913/FULLTEXT01.pdf
34. Schneider V, German R (2013) Integration of test-driven agile simulation approach in service-oriented tool environment. In: Proceedings of the 46th annual simulation symposium, San Diego, CA, USA, ANSS 2013

35. Schneider V, Deitsch A, Dulz W, German R (2016) Combined simulation and testing based on standard uml models. In: Principles of performance and reliability modeling and evaluation. Springer, Cham
36. Stadler C, Gruber T (2016) Functional engineering platform - a continuous approach towards functional development. In: 7th conference on simulation and testing for vehicle technology, Berlin, Germany
37. Toews R (2016) The biggest threat facing connected autonomous vehicles is cybersecurity. https://techcrunch.com/2016/08/25/the-biggest-threat-facing-connected-autonomous-vehicles-is-cybersecurity/
38. Tyndale P (2002) A taxonomy of knowledge management software tools: origins and applications. Eval Program Plann 25(2):183–190
39. Wachenfeld W, Winner H (2015) Die Freigabe des autonomen Fahrens. Springer, Berlin, pp 439–464
40. Wenger J (2005) Automotive radar - status and perspectives. In: IEEE compound semiconductor integrated circuit symposium, 2005. CSIC '05
41. Winner H (2015) Quo vadis, FAS? In: Handbuch Fahrerassistenzsysteme. Springer, Wiesbaden

State-of-the-Art Tools and Methods Used in the Automotive Industry

Harald Altinger

Abstract In recent times, the number of features within a modern-day premium automobile has significantly increased. The majority of them are realized by software, leading to more than 1,000,000 LOC ranging from keeping the vehicle on the track to displaying a movie for rear seat entertainment. The majority of software modules need to be executed on embedded systems, some of them fulfilling mission-critical task, where a failure might lead to a fatal accident. Software development within the automotive industry is different from other industries or open source, as there are more restrictions upon development guidelines and rather strict testing definitions to meet the quality and reliability requirements or even ensure traceability on defect liability. To meet these requirements, various tools and processes have been integrated into the development process, delivering document metadata which can be used for further insights, for example, Software Fault Prediction (SFP).

1 When Reading This Chapter

This chapter represents a short introduction to current tools and development procedures used within the automotive (software) industry. It is a compilation of multiple antecedent publications. Some parts will be extended by knowledge of practitioners to give insights into common processes. As the author is related to one Original Equipment Manufacturer (OEM), some results might be influenced.

H. Altinger (✉)
Audi AG, Ingolstadt, Germany
e-mail: harald.altinger@audi.de

© Springer Nature Switzerland AG 2019
Y. Dajsuren, M. van den Brand (eds.), *Automotive Systems and Software Engineering*, https://doi.org/10.1007/978-3-030-12157-0_4

2   A Short Introduction upon Software within Cars

A modern-day premium car, for example, the 2018 Audi A8 [6], if fully equipped, can contain more than a 100 Electronic Control Units (ECUs), 14 networks and up to 2 SIM cards, and multiple kilometers of cable (compare Fig. 1). According to Charette [13], the software within a modern-day premium car might claim up to 1,000,000 Lines Of Code (LOC). Software modules need to fulfill various tasks ranging from simple comfort features, for example, controlling the stepper motor within an electric seat, up to complex tasks like a predictive chassis. Some features only require a single ECU with a low-power processor, for example, remote garage door control, while others might require a powerful processor with multiple cores and a Graphics Processing Unit (GPU) to encode multimedia content for entertainment. Some features require multiple sensor values connected to different ECUs and influence more than one actuator or even require online and map data, for example, the dynamic matrix headlights (see [6]). The majority of software components need to run under embedded conditions and deliver their result within real time, for example, power steering; others can operate on soft real time, for example, navigation systems. Some features need to meet rather strict development guidelines like Motor Industry Software Reliability Association (MISRA) or need to satisfy safety requirements as Automotive Safety Integrity Level (ASIL). Therefore, one can see a car as a heterogeneous network of multiple distributed computers running a high number of software functions, either

Fig. 1 The cabling topology of the 2018 Audi A8; note the actual ECUs are not shown

stand-alone, distributed asynchronous, or synchronous, which represents a highly complex system.

Some modern-day features might only be realized by software. Most of the required actuators and sensors are already within the car to fulfill various functionalities. Central software components might use these to realize enhanced or additional features, for example, Start/Stop automatic. In this SoftWare Component (SWC) case, it needs to read back the battery capacity level, the climate control state, if the driver is steady on the break, etc. If all these conditions are met, the system instructs the engine control ECU to switch off the engine. Even more complex systems, for example, Advanced Driver Assistance Systems (ADAS), require information from multiple sensors, perform calculation upon more than one ECU, and may control several actuators which sum up to highly complex systems. To visualize this, we are using the Adaptive Cruise Control (ACC) as an example. The system consists of two main sensors, a radar and a camera. Both deliver object information (position, distance, speed, etc.) regarding vehicles ahead of the ego car. Those object information will be calculated on the ECU connected to the sensor; the ego car's motion will be acquired on the Inertial Measurement Unit (IMU) (which is already a part of the car as Electronic Stability Control (ESC) requires precise acceleration information to fulfill its task). The SWC realizing the ACC might be hosted on another ECU, or included on one of the sensors' ECU. It needs to be connected via a bus (CAN, FlexRay) network. This software will calculate acceleration and deceleration upon the driver's presets (distance to vehicle ahead, driving profile, etc.) and send the values to break and engine control ECU. Those components decide upon actual speed and requested (de)acceleration if the engines carrier gas (in case of an Electriv Vehicle (EV) maybe recuperation abilities) is enough or if the break system needs to act. Further details have been released by Duba and Bock [14]; a visualization can be seen in Fig. 2.

Some software are used to compensate mechanical deficits; for example, adding ESC[1] to the 1997 Mercedes A-Class after failing the Elk Test to enhance the car's stability.

These trends lead to an increasing amount of software being indispensable parts of the car. Raising the amount of software might cause the number of software-related bugs to increase too. Figure 3 shows vehicle recalls extracted from the NHTSA database [27]. This database is publicly available and contains all defects which lead to an official recall by the NHTSA on US roads. Similar to Altinger et al. [2], we query the up-to-date database grouping by vehicle models and model years. A software-related recall is counted if the defect or the repair description contains "software," "update," or "program."

One can see that a huge number of vehicles need to be recalled due to software issues. This data is only valid for the USA. The graph uses the term "model year" as a base, meaning the year where the car (or a sub release) has been introduced to the market. This is not necessarily the year where the car has been manufactured or

---

[1]Nowadays ESC is mandatory by law in EU since 11/2014.

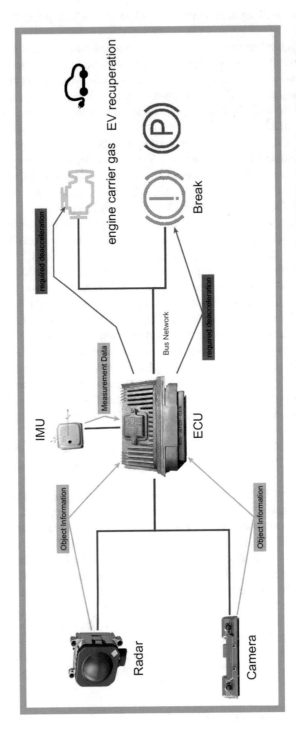

Fig. 2 The ACC workflow composed of three sensors: radar, camera, and IMU. The ECU is symbolic; the ACC SWC might be hosted on one of the sensor ECU. All systems are connected via a bus network

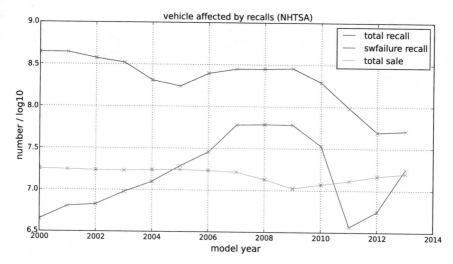

Fig. 3 The software caused recall statistics based on an NHTSA recall database query, [27]. The sale data is originated by Ward's Auto [32]. The graph has been generated using the same settings as Altinger et al. [2]

sold nor the year when the recall has been filed. At high peak, this data shows more recalls than sales which might be caused by vehicles affected by multiple recalls during their lifetime.

Fixing a software bug will be performed by updating the software. If a recall is filed, an OEM needs to pay for contacting the driver, an engineer at a workshop to update the software, maybe a rental car for the customer, etc. There might be lower costs upon using Over The Air (OTA) updates. However, the OEM needs to ensure that all affected vehicles would install the updates. Multiplied with the number of effected vehicles, easily one recall can sum up to millions of euros to fix a simple software bug (compare Capers [12]). This leads even to a high economic interest to prevent bugs in the field and invest a rather high share from the vehicle development costs into testing.

There are no official numbers on the development costs, but Shea [30] did interview John Wolkonowicz, a senior auto analyst for North America at IHS Global, claiming development costs between one (setting up on existing model) and six (new model, new platform, new engines, etc.) billion USD. Broy et al. [11] analyze the development costs for a vehicle electronic system at 300 million EUR, estimating two thirds of this being software development costs.

Developing software to be used within cars is subjected to various restrictions. Coding standards, as the 2004 MISRA-C [23], modeling guidelines as the MAAB [20], limit the usage of programming language features, such as no pointers or no dynamic memory usage or no unlimited depth of function calls, etc. A recent study by Altinger et al. [4] demonstrated the effectiveness of such standard upon preventing common programming bugs. Furthermore, standards like ISO

Table 1 Operators' overview from three automotive projects

Preprocessor		Flow control		Data type		Mathematical		Logic/comp.	
%	Operator	%	Operator	%	Type	%	Type	%	Type
19.41	#define	8.92	break	1.48	int	54.97	+	8.66	!
15.36	#ifndef	6.79	case	96.32	static	13.75	−	3.00	!=
65.24	#include	2.13	default	0.72	struct	1.47	%	7.05	&
		29.24	else	1.48	unsigned	21.85	*	13.06	&&
		0.47	for			7.96	/	49.30	=
		49.46	if					9.08	==
		0.72	return					2.05		
		2.14	switch					4.89	||
		0.02	while					0.84	>=
		0.12	?					0.15	<=
								0.75	<
								0.80	>
								0.32	~
								0.03	^

26262 [17] demand various testing methods addictive to the risk of failure. The ASIL rating represents a software failure impact upon human lives. If a systems failure puts human life in danger, it will be rated *ASIL D*. This level strongly recommends formal verification of the program code. Even in less critical systems, the norm demands testing goals, for example, branch/statement coverage upon test cases. A good overview has been presented by Reactis software [28].

Altinger et al. [4] performed a study on three different automotive software projects, analyzing the usage of operators within those three projects. Table 1 shows an overall summary. All these projects use C as a programming language, limited by the MISRA 2004 coding guidelines. Table 1 demonstrates a strong tendency of using decision-based code elements (if/else, logic operators, etc.) and no usage of pointers or dynamic memory in the three automotive software projects. The whole software has been developed using model-driven approaches realized with Matlab Simulink and TargetLink to generate the actual C code. The workflow is presented within Fig. 6. A more detailed description on the development process has been presented by Altinger et al. [3].

3 Development Process and Available Documents

The VW corporation employed 45,742 people within their Research and Development (RD) according to their annual fiscal report [31]. This amount of engineers working together demands clear development processes and descriptions to handle such huge projects as developing a car. Figure 6 shows an exemplary development process used by three projects we did analyze. A core component is a clear separation between requirements, development, and testing which is realized by using different specialized tools working together with an adopted toolchain as presented by Kiffe et al. [19]. They present their work on linking different tools from various vendors called ENPROVE. This toolchain enables exporting requirements written in DOORS into a Matlab Simulink model. Thus the engineer designing the model can trace if subsystems accord to requirements.

Altinger et al. [2] performed a questionnaire survey receiving 68 responds from Series Development (SD), Pre Development (PD), and Research (Re). The authors presented IBM DOORS as the most common tool to write specifications. Further, they presented a typical work split between dedicated engineers writing specifications and test cases and developers realizing the software. Within their work, they presented Mathworks Matlab Simulink as the most common tool to design models and generate code.

Figure 2 shows the W-development model as a test-enhanced extension to the well-known V-model as presented by Jin-Hua et al. [18]. One can clearly see a related testing stage to every definition and implementation stage. As defined by the MISRA 2004 standard, every module needs to pass code review stages, which are a core part of the W model (Fig. 4).

Bock et al. [10] present the results from a study on various tools used during different development stages. The authors are presenting a rather good overview and short introduction to all tools they present. Their ranking is based on a questionnaire survey asking whether a tool is in use or a tool/method is familiar, targeting engineers working at SD, PD, and Re. Their work developed a taxonomy to guide engineers on the selection of new tools and methods.

Altinger et al. [2] draft commonly used tools for automotive software engineering. The majority of requirements and specifications are written using IBM DOORS. An engineer exports the related subset to his software module and establishes a link between the DOORS database and a Matlab Simulink model. As stated in Altinger et al. [2], software specifier, developer, and tester are dedicated personnel. Thus, another engineer exports the test requirements and generates test cases associated with Software in the Loop (SiL), Hardware in the Loop (HiL), etc. Following this process, every tool will get updates if requirements change. Scripting interfaces are used to ensure requirements are linked with code parts.

In contrast to classic software development, there is a strict milestone plan for every SWC. Figure 5 visualizes them. The automotive industry is dominated by Start Of Production (SOP), which means there is a fixed date where every module, no matter if mechanical, electrical, or software, has to be available to be fit into the new produced model.

Software development has derived various sub-milestones:

- *Interface freeze* – all software interfaces (including network messages on the CAN and FlexRay), similar to an Application Programming Interface (API), have to be defined and are not allowed to be changed afterward.
- *Feature freeze* – all functionality has to be defined and implemented; rapid prototypes are still allowed; code optimization might not be completed.

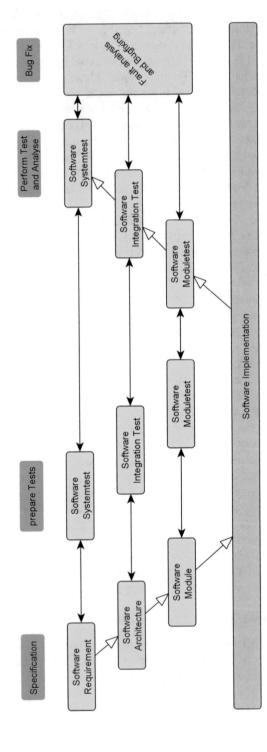

Fig. 4 The W-process as presented by Jin-Hua et al. [18]

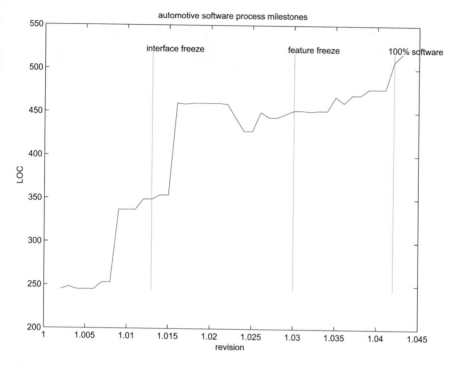

Fig. 5 Sample of software milestones within the automotive domain

- *100 % software* – all implementation shall be done; only bug fixes are allowed after this point.
- *SOP* – Version 1.0 which will be shipped with the first customer cars.

4 Tool Usage

Altinger et al. [2] and Bock et al. [10] yield a list of well-established tools to specify requirements and to perform software tests. Common to both studies, Matlab Simulink is the dominating programming environment.

As outlined by Altinger et al. [3], there exists a wide range of tools supporting the development process. Figure 6 presents their interaction. During the analysis of projects, the following tools served:

- **IBM DOORS**: writes and traces requirements.
- **Matlab Simulink**: develops models and performs basic tests during development.
- **dSpace TargetLink**: generates C code based on the Simulink models.

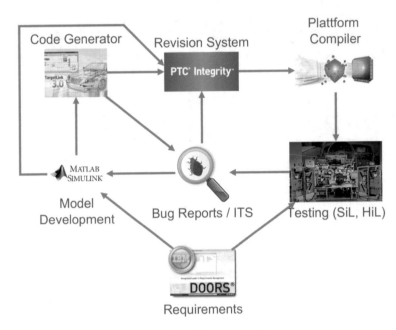

Fig. 6 Development workflow, adopted from Altinger et al. [3]

- **PTC Integrity**: revision system to store each development artifact (model files, configurations, etc.). This tool offers an Integrated Ticketing System (ITS) to handle bug reports and their fixes.
- **Target Compiler**: a compiler specific to the ECUs platform.
- **Testing**: various scripting interfaces using standard automotive tools such as Vector CanOe or CanAp or ADTF, ...

Further and more detailed tool and test strategies are presented by Müller et al. [24].

ISO 26262 requires tools to be qualified before being used for various levels of ASIL-rated software. The EntwicklungsProzess Verbesserung, German: Development Process Improvement (EnProVe) process as described by Kiffe et al. [19] performs such tool qualifications. In addition, this process develops scripting interfaces between the qualified tools to enable automation.

5 Testing Approaches

Traditional testing approaches two decades ago lead to setting up a test car and perform various road tests on a dedicated test track; if the car passes all of them, the phase is over. Modern-day cars consist of a huge amount of components which need to be tested separately. Common approaches are in the loop tests, such HiL,

SiL, Model in the Loop (MiL), etc. As Bergmann et al. [7] state, real test drive will decrease, virtual test drives will increase.

A recent survey by Altinger et al. [2] presents the work distribution upon engineers when developing car software. There is a dedicated group designing test cases which might be executed on HiL or even real test drives. Even within early states of software development, a test case is linked to every stage (compare Jin-Hua et al. [18]) leading to the W-development process (see Fig. 4).

Within computer games, for example, "Need for Speed," one can drive a virtual car on digital reconstructed race tracks. Virtual Test Drive (VTD) (see Dupuis et al. [15]) takes this idea into realistic road simulation. This system uses a simulated world to generate the test stimuli. Based on road network, a car can drive along a virtual street, and other cars might be simulated based on driver model descriptions. The rendering engine can generate an image to be fed into a camera which will generate test input for computer vision algorithm. Sensor model descriptions might extract raw values, for example, LIght Detection And Ranging (LIDAR), based on the virtual world. Nentwig et al. [26] performed analysis upon performance of generated images for lane mark detection considering weather influences. Within their study, they compare generated data with real measurements. Using these toolsets, one might be able to define test cases using real-world assumptions and generate the stimuli for every ECU on a test bench. Müller et al. [24] use this method but extend the data generation. Within their work, they use VTD to calculate sensor input data, for example, video data which is captured by the car's ADAS camera using a video screen. Similar processes are used to generate input data for ultrasonic sensors or even radar. Thus, the complete system can be considered as an HiL.

SiL or MiL tests might be performed with stimuli defined within the test cases, but as systems get more complex, they cannot be tested independently. To cover more aggregated scenarios, one can use, for example, "Virtuelle Probefahrt 2.0." This method is a mixture between real hardware operated on HiL benches and simulated stimuli (sensors, road network, etc.). An in-depth explanation has been given by Miegler et al. [21]. Standardized interfaces to exchange modules, for example, a driver model or road networks, are mandatory. The system is capable of replacing/mixing stimuli with either simulated data (using, e.g., VTD) or recorded real drive data (e.g., captured at road test drives). The modular architecture enables engineers to replace sensor models with different simulation approximations. Using an HiL test bed, the simulated data can be fed into a real ECU or even a network of ECU.

Rather new technologies, such as Vehicle in the Loop (ViL), are using a real vehicle but virtual environment to perform initial test without harming real vehicles. This means a human driver might sit within a real car driving along an empty road. The driver will have to wear virtual reality glasses where he can see the simulated traffic. The simulation will generate sensor inputs to be fed into a System Under Test (SUT) or capture them by real objects on the road. The system will be able to generate real force feedback (via injecting the real vehicle systems) and record the driver's reaction. ADAS development can benefit due to early test chance.

According to Bock et al. [8], the ViL method is usable. Within their work, they performed user study with 36 persons followed by later work [9] where they verified the virtualization assumptions stating that ViL is able to generate realistic driving impressions within a very early stage of development.

Miegler et al. [22] present a new HiL approach where they use existing ECU and real-time rapid prototyping systems to integrate unavailable ECU; for example, vehicle dynamics might be computed on an HiL, and environmental data, for example, traffic participants, is simulated on a PC. Overall, scenario control is part of the simulation; Reset/Replay is possible. Rapid prototyping modules are possible using Automotive Data and Time-Triggered Framework (ADTF) and VTD, where ADTF ensures the connectivity to HiL systems and VTD realizes the environment simulation.

6 Software Fault Prediction (SFP): A New Idea to Be Integrated

As presented in Sect. 2, automotive software development follows restrictive settings in terms of coding guidelines and, for example, commit policies. As outlined by Altinger et al. [3], it is possible to gain advantages. The basic idea in brief is as follows: use code measurements (LOC, cyclomatic complexity by McCabe, etc. as analyzed by Herbold et al. [16]) for every commit; extract bug reports from the ITS and derive a ground truth for bug and fault-free commits; use this data to train a machine learning algorithm, for example, Support Vector Machine (SVM), Näive Bayes (NB), and Random Forest (RF); repeat the measurement step at the time of commit and use the trained machine learning system to derive a probability if the actual commit contains a bug. Figure 7 presents the involved tools and process steps to perform such measurements. Further details and sample projects with measurements are outlined by Altinger et al. [3]. The dataset is publicly available by Altinger [1]; further datasets, including industry grade, are outlined by Sayyad et al. [29].

Altinger et al. [3] could demonstrate a rather high true positive hit rate on fault predictions at more than 90%. They conclude that their data is better than others, for example, Zimmermann et al. [33], on an open source projects; and Nagappan et al. [25], on commercial software, as the changes between commits are smaller and the nature of generated code from model-driven development approaches results within more homogeneous code structure compared to text-based software developers. Further, they could rely on the data in the ITS, as there were policies to enforce correct commit logs by every single developer. Thus, the quality of the measured data is much higher than, for example, on Zimmermann et al.'s [33] analysis on the open source tool Eclipse, which suffers from no clear differentiation between feature commit and a bug fix or needs to deal with blank commit messages.

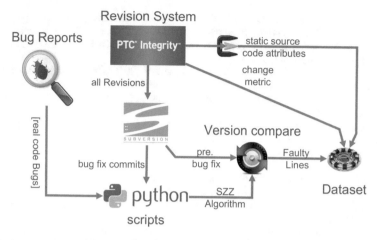

Fig. 7 Workflow to extract metric values for dataset; adopted from Altinger et al. [3]

Due to a low bug rate, as there are only 2.6–15% samples in the bug class, Altinger's [1] datasets suffer from imbalanced class distribution. To overcome this, Altinger et al. [5] suggest to use over- and undersampling to enhance predictive performance. They succeed in showcasing an improved performance using an NB. They state when using up- or downsampling, one has to choose between precision (all reported faults are true bugs) and recall (finding all bugs), both of which cannot be increased at the same time.

The obvious idea, to use a trained prediction model from one project to another, for example, at an early stage where there are too few metric data to train, does not seem to perform well. Altinger et al. [3] performed a cross-project prediction including well-known metric transfer methods but could not archive a prediction performance above the random hit rate. These findings are in line with other industry data as presented by Zimmermann et al. [34].

This method seems to be a promising extension to the W-development process (see Fig. 5), as it could be an early indicator to define the various software tests. As known from Altinger et al. [2], test engineers need to spend free testing budget based on their experience; thus, SFP might aid in these decisions. As of the more statisticñature, SFP cannot replace a full test suite.

References

1. Altinger H (2015) Dataset on automotive software repository. http://www.ist.tugraz.at/_attach/ Publish/AltingerHarald/MSR_2015_dataset_automotive.zip
2. Altinger H, Wotawa F, Schurius M (2014) Testing methods used in the automotive industry: results from a survey. In: Proceedings of JAMAICA. ACM, New York, pp 1–6

3. Altinger H, Herbold S, Grabowski J, Wotawa F (2015) Novel insights on cross project fault prediction applied to automotive software. In: El-Fakih K, Barlas G, Yevtushenko N (eds) Testing software and systems, vol 9447. Springer, Berlin, pp 141–157. http://dx.doi.org/10.1007/978-3-319-25945-1_9

4. Altinger H, Dajsuren Y, Sieg S, Vinju JJ, Wotawa F (2016) On error-class distribution in automotive model-based software. In: 2016 IEEE 23rd international conference on software analysis, evolution, and reengineering, IEEE, Piscataway, pp 688–692. https://doi.org/10.1109/SANER.2016.81

5. Altinger H, Herbold S, Wotawa F, Schneemann F (2017) Performance tuning for automotive software fault prediction. In: 2017 IEEE 24th international conference on software analysis, evolution, and reengineering. IEEE, Piscataway. https://doi.org/10.1109/SANER.2017.7884667

6. Audi AG (2017) Audi A8 (Typ 4N) Selbststudienprogramm 662. Technical manual. AUDI AG

7. Bergmann R, Walesch R (2012) HiL strategie audi. In: 6. dSpace Anwender Konfernz 2012, dSpace. http://www.dspace.com/shared/data/pdf/ankon2013/tag1_pdf/2_audi_walesch_robert_bergmann_richard.pdf

8. Bock T, Maurer M, Farber G (2007) Validation of the vehicle in the loop (VIL); a milestone for the simulation of driver assistance systems. In: 2007 IEEE Intelligent vehicles symposium, pp 612–617

9. Bock T, Maurer M, Meel F, Müller T (2008) Vehicle in the loop. ATZ Automobiltech Z 110(1):10–16. http://dx.doi.org/10.1007/BF03221943

10. Bock F, Homm D, Siegl S, German R (2016) A taxonomy for tools, processes and languages in automotive software engineering abs/1601.03528. http://arxiv.org/abs/1601.03528

11. Broy M, Kruger I, Pretschner A, Salzmann C (2007) Engineering automotive software. Proc IEEE 95(2):356–373. https://doi.org/10.1109/JPROC.2006.888386

12. Capers J (2009) A short history of the cost per defect metric. www.semat.org

13. Charette RN (2009) This car runs on code. IEEE Spectr 46(3):3

14. Duba GP, Bock T (2008) ATZextra Worldw 13:56. https://doi.org/10.1365/s40111-008-0055-0

15. Dupuis M, von Neumann-Cosel K, Weiss C (2010) Virtual test drive vereinheitlichung der simulationsumgebung für SiL-, HiL-, DiL-und ViL-tests bei der entwicklung von fahrerassistenz- und aktiven sicherheitssystemen

16. Herbold S, Grabowski J, Waack S (2011) Calculation and optimization of thresholds for sets of software metrics. Empir Softw Eng 16(6):812–841. https://doi.org/10.1007/s10664-011-9162-z

17. ISO TC 22 SC 3 (2011) ISO 26262:2011:Road vehicles – Functional safety. International. www.iso.org

18. Jin-Hua L, Qiong L, Jing L (2008) The w-model for testing software product lines. In: International Symposium on Computer Science and Computational Technology, 2008. ISCSCT'08, vol 1, pp 690–693

19. Kiffe G, Bock T (2013) Standardisierte entwicklungsumgebung fuer die softwareeigenentwicklung bei audi. In: 7. cSapce User Conference, dSpace, https://www.dspace.com/de/gmb/home/company/events/dspace_events/archive_2013/ankon2013.cfm

20. Mathworks T (2014) Mathworks automotive advisory board checks (MAAB). http://de.mathworks.com/help/slvnv/ref/mathworks-automotive-advisory-board-checks.html

21. Miegler M, Nentwig M (2015) Testing of piloted driving on virtual streets. ATZ Worldw 117(9):16–21. http://dx.doi.org/10.1007/s38311-015-0044-7

22. Miegler M, Schieber R, Kern A, Ganslmeier T, Nentwig M (2009) Hardware-in-the-loop test of advanced driver assistance systems. ATZ Elektron Worldw 4(5):4–9

23. Motor Industry Software Reliability Association (2004) MISRA-C:2004 - Guidelines for the use of the C language in critical systems, 2nd edn. MISRA

24. Müller DIFSO, Brand IM, Wachendorf S, Schröder DIFH, Szot DIFT, Schwab DIS, Kremer B (2009) Integration vernetzter fahrerassistenz-funktionen mit HiL für den VW passat CC 14(4):60–65

25. Nagappan N, Ball T, Zeller A (2006) Mining metrics to predict component failures. In: Proceedings of the 28th international conference on software engineering, pp 452–461
26. Nentwig M, Stamminger M (2011) Hardware-in-the-loop testing of computer vision based driver assistance systems. In: 2011 IEEE Intelligent vehicles symposium (IV), pp 339–344. https://doi.org/10.1109/IVS.2011.5940567
27. NHTSA USDoT {and} USAgov (1997) Office of defects investigation (ODI) recalls database. www-odi.nhtsa.dot.gov
28. Reactis (2015) Achieving ISO 26262 compliance with reactis. http://www.reactive-systems.com/papers/iso-26262.pdf
29. Sayyad Shirabad J, Menzies T (2005) The PROMISE Repository of Software Engineering Databases. http://promise.site.uottawa.ca/SERepository, published: School of Information Technology and Engineering, University of Ottawa, Canada
30. Shea T (2010) Why does it cost so much for automakers to develop new models? http://www.autoblog.com/2010/07/27/why-does-it-cost-so-much-for-automakers-to-develop-new-models/
31. VW Aktiengesellschaft (2014) Geschaeftsbericht 2014. http://geschaeftsbericht2014.volkswagenag.com/konzernlagebericht/nachhaltige-wertsteigerung/forschung-und-entwicklung/f-e-kennzahlen.html
32. WARDS Auto (2013) U.S. car and truck sales, 1931–2013. www.wardsauto.com
33. Zimmermann T, Premraj R, Zeller A (2007) Predicting defects for eclipse. In: International Workshop on Predictor models in software engineering, 2007. PROMISE'07: ICSE Workshops 2007. IEEE, Piscataway, p 9
34. Zimmermann T, Nagappan N, Gall H, Giger E, Murphy B (2009) Cross-project defect prediction: a large scale experiment on data vs. domain vs. process. In: Proceedings of the 7th joint meeting of the European software engineering conference and the ACM SIGSOFT symposium on the foundations of software engineering, pp 91–100

Part III
Automotive Software Reuse

Software Reuse: From Cloned Variants to Managed Software Product Lines

Christoph Seidl, David Wille, and Ina Schaefer

Abstract Many software systems are available in similar, yet different variants to accommodate specific customer requirements. Even though sophisticated techniques exist to manage this variability, industrial practice mainly is to copy and modify existing products to create variants in an ad hoc manner. This clone-and-own practice loses variability information as no explicit connection between the variants is kept. This causes significant cost in the long term with a large set of variants as each software system has to be maintained individually. Software product line (SPL) engineering remedies this problem by allowing to develop and maintain large sets of software systems as a software family.

In this chapter, we give an overview of variability realization mechanisms in the state of practice in the industry and the state of the art in SPL engineering. Furthermore, we describe a procedure for variability mining to retrieve previously unavailable variability information from a set of cloned variants and to generate an SPL from cloned variants. Finally, we demonstrate our tool suite DeltaEcore to manage the resulting SPL and to extend it with new functionality or different realization artifacts. We illustrate the entire procedure and our tool suite with an example from the automotive industry.

1 Introduction

Modern software exists in many similar *variants* that realize slightly different functionality to accommodate specific customer requirements. For example, the automotive industry allows customers of their cars to freely decide whether an optional alarm system should be included or whether the power windows should be manual or automatic. The selected configuration impacts the software required

C. Seidl (✉) · D. Wille · I. Schaefer
Technische Universität Braunschweig, Institute of Software Engineering and Automotive Informatics, Braunschweig, Germany
e-mail: c.seidl@tu-braunschweig.de; d.wille@tu-braunschweig.de; i.schaefer@tu-braunschweig.de

© Springer Nature Switzerland AG 2019
Y. Dajsuren, M. van den Brand (eds.), *Automotive Systems and Software Engineering*, https://doi.org/10.1007/978-3-030-12157-0_5

to operate the car as significant parts of the program logic may differ. Program logic may be specified by source code or, on a more abstract level, by design models such as *function block diagrams (FBDs)* [19], *MATLAB/Simulink*[1] models, or *Rational Rhapsody*[2] statecharts. In consequence, *variability* stemming from different configuration options has to be manifested in various realization artifacts by what is called a *variability realization mechanism* to reflect the differences in program logic.

Variability realization mechanisms used in practice, such as copying and modifying specific software systems through *clone-and-own*, are readily available and do not require additional tools or changes in development process and/or management structure. However, in the long run and with a growing number of variants, these approaches do not support sustainable managed software reuse (see Sect. 3.1). Software product lines (SPLs) [29] offer concepts and facilities for managed software reuse by treating a set of closely related systems as software family. However, there is a gap between the state of practice used in industry and the state of the art for managed software reuse of SPLs in academia with regard to variability realization mechanisms. Furthermore, manually adopting an SPL approach for a grown set of cloned software variants entails significant effort and cost [24].

To remedy these problems, in this chapter, we first give an extensive overview of both the state of practice and the state of the art in variability realization mechanisms. We then describe a procedure we devised to transition from the grown structure of cloned variants to managed reuse of SPLs that analyzes the cloned variants and generates the artifacts for an SPL largely automatically. Finally, we demonstrate our tool suite DeltaEcore to realize the described transition procedure for practical application in order to significantly reduce effort and cost of adopting a managed reuse strategy for grown systems with many variants.

Due to their relevance for industrial practice, we specifically focus on variability realization mechanisms and an associated reverse engineering procedure. To retrieve high-level variability information, such as conceptual *features*, from related product variants, other approaches exist [30]: Feature location techniques analyze natural-language requirements documents [46], product maps describing the software's composition [32, 39], or existing model-based products [50, 51]. However, these approaches do not consider fine-grained variability within realization artifacts or do not generate an SPL as we do.

The structure of this chapter is as follows: Sect. 2 provides an overview of SPL terminology and introduces a running example from the automotive domain. Section 3 explains and contrasts the state of practice and the state of the art in variability realization mechanisms. Section 4 describes our procedure to transition from grown software systems with multiple variants to managed reuse by first analyzing the individual clones and then generating an SPL from the realization artifacts and the

---

[1]http://www.mathworks.com/products/simulink/.

[2]https://www.ibm.com/us-en/marketplace/rational-rhapsody/.

collected information. Section 5 introduces our tool suite DeltaEcore used to create and maintain an SPL. Finally, Sect. 6 closes with a conclusion.

2 Background

In the following section, we provide background on software product lines and provide a running example to illustrate concepts and techniques throughout the chapter.

2.1 Software Product Lines

A software product line (SPL) [8, 29, 43] is an approach for managed reuse where a set of closely related software systems is perceived as a software family consisting of *commonalities* and *variabilities*. Commonalities constitute the functionality contained in all systems of the software family, and variabilities constitute the functionality that sets apart the individual software systems. Individual software systems of the software family are created by combining the commonalities with a selection of the variabilities. However, not all combinations of variabilities are necessarily valid as technical or economical concerns may prohibit certain combinations.

In an SPL, commonalities and variabilities of artifacts as well as the rules governing potential combinations of variabilities are represented on different levels of abstraction. In the *problem space* [9], variabilities are represented on a mere conceptual level with no regard to their technical realization, for example, as names with a description of functionality that can be used to communicate with nontechnical stakeholders, such as customers or management. In the *solution space* [9], variabilities are represented with their effect on realization artifacts, for example, source code or models. Figure 1 depicts an overview of the essential constituents of an SPL and their relations.

Fig. 1 Overview of the essential constituents of a software product line

The main constituent of the problem space is a *variability model*, which governs the valid combinations of variabilities by providing configuration rules. Various notations for variability models exist, such as feature models [9, 20], decision models [28], orthogonal variability models (OVMs) [29], and variability specifications (VSpecs) of the Common Variability Language (CVL) [16]. Despite the wide variety of available notations, feature models are by far the most commonly used in industrial practice [5] so that we elaborate on this notation.

A feature model is a hierarchical decomposition of a variable software system into *features*. A feature is a user visible functionality that, usually, is configurable, that is, may be selected or deselected [6]. Features may be *mandatory*, so that they have to be selected, or *optional*, so that they may be deselected. The root feature is implicitly considered mandatory. Furthermore, features may be grouped into *alternative groups*, so that exactly one of the contained features has to be selected, or *or groups*, so that at least one of the contained features has to be selected. Configuration constraints for a feature only apply if their respective parent feature is selected. For a concrete example of a feature model, see Fig. 2 in Sect. 2.2 for our running example. A configuration defined on the feature model is a selection of features that satisfies all configuration constraints.

The solution space constitutes the realization of all possible software systems of the software family. Realization artifacts may be of a wide variety of languages: Source code in different programming languages may specify program logic. Implementation models, such as statecharts, may define program flow on a higher level of abstraction. Design models, such as class diagrams or component diagrams, may represent parts of the architecture. Documentation material, such as user and developer manuals, may provide information for employing or extending software systems. The selection of variabilities as part of a configuration may have an effect on each of these artifacts as, depending on the choice of variabilities, certain functionality may be present in or missing from the software system. This yields the need to change the design, implementation, and documentation of the respective software system. To perform the respective changes, SPLs employ a *variability realization mechanism*, which takes as input a configuration from a variability model to then collect, adapt, and assemble relevant parts of realization artifacts.

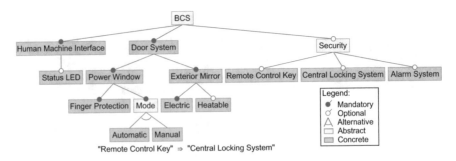

Fig. 2 Excerpt from the *BCS* feature model

The procedure of transforming a configuration into a variant of the SPL is referred to as *variant derivation*. A *variant*[3] or *product* of the SPL is the software system associated with a specific conceptual configuration. The variant derivation procedure differs greatly depending on the concrete type of variability realization mechanism employed by the SPL, which we cover in detail in Sect. 3.

2.2 *Running Example Automotive Body Comfort System*

To illustrate our techniques, we use a running example from an automotive *body comfort system (BCS)* SPL [27] along with its realization as statecharts, which comprises functionalities such as (automatic) power windows and exterior mirror control. In Fig. 2, we show an excerpt of the feature model for the BCS where the full SPL comprises 27 features and 11,616 valid product variants. The depicted feature model comprises different parts of the functionality that are common to all product variants, such as the car's `exterior mirror` or the `human machine interface`. In addition, different optional features exist representing additional functionality, such as `electric` or `heatable` exterior mirrors as well as a `central locking system (CLS)`.

Individual features may be implemented differently depending on the selection of other features. For example, the `CLS` feature exists in alternative implementations depending on different `power window (PW)` modes. In Fig. 3, we show two statechart implementations of the `CLS` feature consisting of the `cls_unlock` and `cls_lock` states with corresponding transitions.

The `ManPW` variant (cf. Fig. 3a) is employed with a manual power window, whereas the `AutoPW` variant (cf. Fig. 3b) is used with an automatic power window.

In terms of implementation, the main difference between the two variants is that, during a transition from `cls_unlock` to `cls_lock`, the `ManPW` variant is only disabled (i.e., `pw_enabled=false`) when the window is closed completely (i.e., `pw_pos==1`). Otherwise, it is still possible to manually close the window. However, the `AutoPW` variant is disabled independent of the position of the window, which is automatically closed by generating a corresponding command (i.e., `GEN(pw_but_up)`). We use these variants of the `CLS` in the remainder of this chapter to illustrate different variability realization mechanisms as well as our variability mining and SPL generation techniques.

---

[3]Note that some publications [29] use a different definition of the term *variant* to describe one concrete option for a specific variation point, for example, a specific value for a configurable parameter.

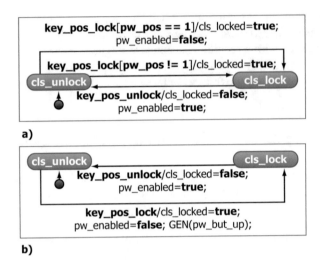

Fig. 3 Two statechart variants associated with the CLS feature. (**a**) ManPW variant of the central locking system (CLS). (**b**) AutoPW variant of the central locking system (CLS)

3 Variability Realization Mechanisms

Different techniques exist to represent changes associated with conceptual features in realization artifacts, such as source code or models. In the following, we describe the state of practice of variability realization in industry and survey the state of the art by explaining in detail different variability realization mechanisms from SPL engineering.

3.1 State of Practice in Variability Realization

In industrial practice, a couple of ad hoc variability realization mechanisms or existing language constructs and modularization concepts (not originally intended to capture variability) are used to encode variability. In the following, we summarize the most widely applied techniques for variability realization in the solution space, while the survey by Berger et al. [5] focuses on variability modeling in the problem space.

- **Clone-and-Own** (or copy-paste-modify) [12, 21]: Developers copy existing models or source code of product variants and modify the respective artifacts until a new variant is obtained. This new variant is then stored under a new name and can be deployed in the same way as existing variants. This process can be repeated for each variant to be developed. The advantage of this approach is that it is very lightweight and saves development effort in the short term. No special modeling notation or tool support is required. However, with an increasing number of variants, the variability in the set of cloned variants becomes difficult to manage as each software system has to be maintained in isolation. In particular, for debugging and maintenance, undocumented variability becomes an obstacle.

- **Conditional Compilation** [26]: In programming languages, conditional compilation techniques allow deriving the implementation for a specific code variant during compilation by appropriately selecting values for preprocessor macros. Conditional compilation is most prominently used within C/C++ where code blocks can be enclosed in `#ifdef` directives that are omitted for compilation if the corresponding constant is set to false. Conditional compilation is a widely used approach within the programming language community and offers very flexible means to obtain custom-tailored code for specific variants. However, it leads to code fragmentation and scattering of variability which is difficult to maintain and debug.

- **Variability Encoding** [44]: Variability in models or source code is encoded by standard programming/modeling language constructs that are originally intended for choice within the control flow during execution. For instance, in programming languages, variability can be encoded using if statements where the if condition is a configuration parameter, such as a specific feature. In MATLAB/Simulink, switch-case statements or variant subsystem blocks are used to capture variability [45]. While variability encoding does not require specific language or tool support to express variability, it is limited to expressing variability in software behavior (in contrast to variability in its structure), and the binding of variability is shifted to runtime which means that the complete code base has to be deployed in all cases which may be disadvantageous for resource-constrained devices or for protecting intellectual property. Additionally, choices due to variability and choices due to program behavior are mixed, which violates the separation of concerns principle and hinders maintenance and debugging.

- **Parametrization**: Variability of a system is captured by setting specific parameter values for system variables. Parametrization in the automotive domain is, for example, used in characteristic curves or maps, such as for engine control, set during calibration phases. Alternatively, electronic control units (ECUs) incorporate a set of behavioral variants that can be configured by parametrization. After the car is readily built, a parametrization string is entered such that the software variant matches the built variant of the car. This requires that all possible variants are already encoded within the ECU. The parametrization strings are often kept in spreadsheets, which complicates analyses such as finding out if the software for a specific car variant is configurable at all. Furthermore, configuration errors may only be detected during system execution, which significantly complicates debugging.

- **Components and Plug-In Frameworks** [41]: Variability on the architectural level can be represented by component or plug-in frameworks. Variants can be obtained by composing different component variants from a component library or by using different plug-ins in plug-in frameworks such as Eclipse. The advantage of component libraries and frameworks is that variability is modularly encapsulated within components. However, variability is subject to the granularity of the components. Hence, fine-grained variability in behavior or structure cannot be captured. Instead, for each (even only fine-grained) change due to variability, a new variant of a component is needed which leads to redundancy and replicated code in the component/plug-in library.

3.2 State of the Art in Variability Realization Mechanisms

Despite ad hoc variability realization mechanisms used in industrial practice and their individual shortcomings, variability realization mechanisms of SPLs support managed reuse. These variability realization mechanisms can be distinguished into three principle groups: *annotative*, *compositional*, and *transformational* [34]. In the following sections, we elaborate on each type.

3.2.1 Annotative Variability Realization Mechanisms

Annotative variability realization mechanisms[4] [22, 34] utilize annotations to denote parts of the realization artifact that belong to a particular feature. As a consequence, with annotative variability realization mechanisms, a single artifact contains all possible variations of one realization asset affected by variability, often referred to as a *150% model*. For example, a C++ class may contain multiple definitions of a method with the same signature (similar name of the method with same number and type of parameters and return value) with different bodies for individual features, where each definition is wrapped in a preprocessor statement (#ifdef) that only enables the respective method when a particular feature is selected. Hence, when used in a disciplined manner, the aforementioned conditional compilation as well as variability encoding may be viewed as annotative variability realization mechanisms (see Sect. 3.1).

For annotative variability realization mechanisms, the connection between a conceptual variability model and the annotations is established through naming conventions, for example, features of a feature model may have the same name as presence variables of annotations. During variant derivation, the presence of elements from the variability model is then resolved to Boolean values for annotations. The realization artifacts are reduced from a 150% model to an artifact of the intended variant by removing those annotated parts of the realization artifact whose conditions in the annotations are not satisfied. The result is a variant containing only the intended functionality. Figure 4 depicts an example of an annotative variability realization mechanism where the two variants of Fig. 2 are represented as a 150% model and the AutoPW variant with automatic power window is derived as example.

---

[4]Annotative variability realization mechanisms are also referred to as *subtractive* or *negative*.

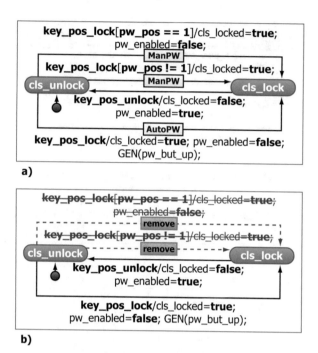

Fig. 4 Example of an annotative variability realization mechanism. (**a**) 150% model. (**b**) Variant derivation by removing parts of the 150% model

Annotative variability realization mechanisms pose certain requirements to be applicable: If annotations are internal to realization artifacts, constructs for adequate annotation have to either be included in the realization language (e.g., if statements) or in another embeddable language (e.g., the C++ preprocessor). Furthermore, if annotations are external to realization artifacts (e.g., through a specific annotation model), elements need to be referenceable from outside of the realization artifact. Finally, the specification of the 150% model requires full knowledge of all possible variations at design time.

A variety of tools for managing SPLs is based on an annotative variability realization mechanism: BigLever's Gears[5] [25] and pure-system's pure::variants[6] [7] are industrial tools for managing SPLs. FeatureIDE[7] [42], Clafer[8] [3], and FeatureMapper[9] [17] are tools for managing SPLs stemming from academia.

---

[5] http://biglever.com/solution/product.html.

[6] http://pure-systems.com/Products.html.

[7] http://wwwiti.cs.uni-magdeburg.de/iti_db/research/featureide.

[8] http://clafer.org.

[9] http://featuremapper.org.

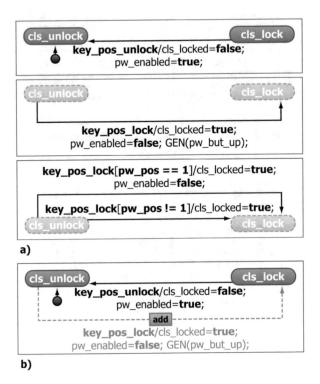

Fig. 5 Example of a compositional variability realization mechanism. (**a**) Core model and two units of composition. (**b**) Variant derivation by adding units of composition to the core model

3.2.2 Compositional Variability Realization Mechanisms

Compositional variability realization mechanisms[10] [4, 22, 34] represent a variable software system as a common *core model* and multiple *units of composition*. The core model comprises the realization of functionality common to all variants. The units of composition contain the realization of individual configuration options, usually on the granularity level of a single feature, but it is principally possible to define finer-grained units of composition. During variant derivation, relevant units of composition are collected and combined with the core model to form a variant containing valid realization artifacts. Figure 5 depicts an example of a compositional variability realization mechanism where the two variants of Fig. 2 are represented as a common core model with two units of composition and the AutoPW variant with automatic power window is derived as example.

It is possible that, in isolation, neither the core model nor the individual units of composition are valid artifacts with regard to the syntax of the language in

---

[10]Compositional variability realization mechanisms are also referred to as *additive* or *positive*.

which they are specified as they may define partial and incomplete information. For example, a statechart may define a core model only containing the states where the units of composition contain various transitions that will connect the states in different variants.

Compositional variability realization mechanisms require certain conditions: The functionality affected by variability has to be accessible for modification through composition, which makes certain structures of realization artifacts more favorable. For example, in source code, it may be complicated to replace fragments of methods through composition so that smaller methods are beneficial for composition. Furthermore, a component-based software architecture [41] or a plug-in framework allows for easier alignment of features with units of composition but poses restrictions on the architecture (see Sect. 3.1).

Compositional variability realization mechanisms do not necessarily have to know all possible variation in advance as new units of composition may be added later on. However, in contrast to 150% models of annotative variability realization mechanisms, which store all variations of a realization artifact in a single element, the (possibly many) units of composition of compositional variability realization mechanisms lead to an increased scattering, which may increase maintenance effort.

Compositional variability realization mechanisms are used in different approaches and tools: *feature-oriented programming (FOP)* [4] captures modifications to artifacts resulting from a different feature in a *feature module*. A feature module uses superimposition to express changes to a realization artifact by either adding new or overriding parts of existing realization artifacts or, potentially, utilizing the previous content. As an example, a feature module for a C++ class may provide an alternative implementation for an existing method and, as part of the new implementation, call the previous definition of the method. During variant derivation, the features selected in a configuration are resolved to the respective feature modules via name matching, which are then used to compose the core model of the SPL with the units of composition. FOP is implemented in various tools, such as the AHEAD Tool Suite[11] [4] or FeatureHouse[12] [2]. Furthermore, FeatureIDE [42] may be configured to use a compositional variability realization mechanism.

Aspect-oriented programming (AOP) [23] may be perceived as a compositional variability realization mechanism when employed within an SPL. An *aspect* captures (potentially cross-cutting) concerns of a software system as additions to various significant locations of the targeted realization artifact called *join points*. Individual features may be realized as aspects that are then combined with the core model of the SPL by *weaving* the additions of relevant aspects into the respective join points. AOP is utilized for the realization of variability in SPLs in various different approaches [1, 13, 15].

---

[11] http://cs.utexas.edu/~schwartz/ATS/fopdocs.

[12] http://infosun.fim.uni-passau.de/spl/apel/fh.

3.2.3 Transformational Variability Realization Mechanisms

Transformational variability realization mechanisms represent variabilities as transformations that restructure a *base variant* of an SPL to a specific *target variant* that constitutes the functionality associated with the selected features of one particular configuration. Sequences of calls to transformation operations may be grouped into *transformation modules* if they have a sufficiently high level of cohesion. A transformation module may realize variability of an entire feature or parts thereof. Transformations may have different complexity: On an atomic level, transformations may be perceived as addition, modification, and removal of elements of the addressed realization artifact. On a more complex level, these atomic operations may be synthesized to form compound operations, for example, to remove an element and all depending references.

During variant derivation, the base variant is copied and relevant transformation modules are collected and sequentially applied to transform the base variant to the intended target variant. In the process of transformation, newly required functionality is added, and functionality of the base variant that is rendered redundant due to the selected configuration is removed. Figure 6 depicts an example of a transformational variability realization mechanism where the ManPW variant with the manual power window of Fig. 2 is used as base variant with transformations to create additional variants and the AutoPW variant with automatic power window is derived as example.

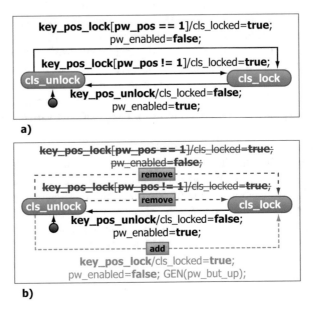

Fig. 6 Example of a transformational variability realization mechanism. (**a**) Base variant. (**b**) Variant derivation by adding, modifying, and removing elements

The base variant of an SPL may, in principal, be selected arbitrarily from the set of products. However, the choice of the base variant greatly influences the number of transformations to create the other variants and, thus, the complexity of the SPL. Hence, a number of considerations may influence which product is selected as base variant of the SPL depending on the intended use: Selecting the product most commonly purchased by customers aligns with business practices when functionality of other products is viewed as deviation from the common product. Selecting the product created first aligns with the development process as subsequent changes may directly be represented as delta modules. Selecting the product with the most features aligns with product configuration where undesired functionality may be deselected.

There are multiple approaches that can be perceived as transformational variability realization mechanisms if applied in the context of an SPL: Model transformation may be employed as transformational variability realization mechanism if all realization artifacts can be perceived as models, for example, as instances of meta-models (see Sect. 4). In this case, model transformation operations are employed to alter realization artifacts of the SPL. Model transformation operations may either be provided as *general purpose transformation operations* [35, 38] that are agnostic of the realization language or as *domain-specific transformation operations* [31] that are tailored to the respective realization language.

The *Common Variability Language (CVL)* [16] is an attempt at adding standardized variability to arbitrary modeling notations by overlaying variability information over realization assets of a software family. CVL provides a set of standard operations to manipulate realization assets of a software family in order to manifest changes associated with variability so that it can be regarded as providing a transformational variability realization mechanism. For example, the approach allows binding variability at one variation point, for example, by choice from a fixed set of options or by setting the value of a variable. Variant derivation facilities may be employed to manifest the effects of changes associated with variability in implementation artifacts.

Delta modeling [33] is the most prominent transformational variability realization mechanism for SPLs. In delta modeling, transformations are performed by invoking specific *delta operations*. Delta operations are provided by a *delta language*, which is a domain-specific variability language tied to the *source language* employed by realization artifacts it modifies, for example, *DeltaStateCharts* for statecharts. The delta language allows fine-grained control over which operations are available for realizing variability while, at the same time, allowing complex transformation operations specific to the source language of the artifacts that should be modified.

Delta modules encapsulate sequences of calls to delta operations with particularly strong cohesion, for example, because they realize a particular feature. Usually, multiple delta modules are required to realize a specific variant of an SPL. The order in which the delta modules are applied may influence the created variant. Hence, not necessarily all application orders of delta modules are valid as, for example, an element cannot be modified before it was added. To express these restrictions, *application-order constraints* may be specified, which state that a specific delta module can only be applied after another delta module was applied.

During variant derivation, first, the relevant delta modules are collected. This may be a manual selection of delta modules by a user of the SPL, or, if the SPL also employs a feature model, the selection of features of a configuration may be resolved to the set of associated delta modules. Then, the application-order constraints of delta modules are evaluated and used as input to a topological sorting algorithm, which creates one valid sequence of application. Finally, the base variant of the SPL is copied and the transformations are applied sequentially in the order determined for the delta modules to create the intended target variant of the SPL.

Transformational variability realization mechanisms have both as input and output a valid variant of the SPL. This is in contrast to annotative and compositional variability realization mechanisms, which depend on a specific software family representation as input with either a 150% model or a core model, which both potentially are not considered valid realization artifacts with regard to syntax and static semantics of the realization language. Hence, transformational variability realization mechanisms have various benefits: For one, provided that the configuration knowledge is sound, both the input and output of transformational variability realization mechanisms may be inspected with standard tools that depend on valid syntax and static semantics of the respective artifacts. In addition, the variability realization process of creating variants by starting out with one concrete software system and transforming it aligns closely with the common practice of companies to first develop an individual software solution on request and, later on, to transform it into an SPL with multiple products to sell off the shelf. Moreover, using transformations is flexible as not necessarily all features have to be known in advance and new features can be added seamlessly later on.

Finally, delta modeling as transformational realization mechanism can emulate the behavior of both annotative and compositional variability realization mechanisms when restricting the delta operations employed in delta modules to only remove or add elements, respectively. This well-formedness property is referred to as *monotonicity* of delta modules [10], and there are procedures to transform any delta-oriented SPL to a monotonous form [11]. Due to these beneficial properties, we employ delta modeling for our work and the explanations in the remainder of this chapter.

4   From Cloned Variants to Managed Software Product Lines

In this section, we explain a procedure to migrate from cloned variants in grown systems to managed reuse in SPLs. The presented approach is applicable to models of different block-based modeling languages, such as MATLAB/Simulink models, FBDs, and statecharts [18, 47, 49].

A common way to describe the notation of a modeling language is through a *metamodel* [14, 40]. In a metamodel, *metaclasses* describe the elements of the modeling language, *metareferences* the relations of these elements, and *metaattributes* the properties of each element. Concrete artifacts of the modeling language are then perceived as instances of the metamodel. The Eclipse Modeling Framework (EMF) provides *Ecore* as a notation to specify metamodels and offers a wide variety of tools on that basis: With Xtext[13] or EMFText,[14] models can be specified in a textual language, for example, source code. With GEF,[15] GMF,[16] or Graphiti,[17] it is possible to create visual editors so that models can be specified in a graphical language, for example, statecharts. With these tools, it is possible to perceive any realization language (e.g., the language for statecharts) as a metamodel and its artifacts (e.g., concrete statecharts) as models instantiating the metamodel. This has the benefit that artifacts may be defined in different representations and may stem from different sources but can still be handled uniformly as models of an explicitly defined metamodel. Predefined metamodels, grammars, and parsers exist for many popular languages to treat their source code as instances of Ecore models, such as *JaMoPP*[18] for Java or *srcML*[19] for C/C++. In addition, it is possible to define metamodels for further languages.

In Fig. 7, we show the metamodel for statecharts we utilize for the BCS running example. The metaclasses State and Transition represent the respective elements of the statechart notation, where elements of both are uniquely identified by the value of the metaattribute id. States may carry a name and transitions may specify events upon whose occurrence they prompt certain actions if the respective guards are satisfied. Metareferences define potential compositions of the respective elements. A StateChart consists of a number of states, where the initialState is designated by the specialized class InitialState. Furthermore, a state contains multiple transitions of which each references a sourceState and a targetState. Additionally, a state may be a compound in the sense that it may contain further states and, indirectly, transitions that define its behavior. Depending on the number of elements instances may reference, metareferences may be distinguished into *single-valued references* (e.g., initialState of StateChart) and *multivalued references* (e.g., states of StateChart), which is of relevance for our approach. The remaining elements of the metamodel in Fig. 7 capture variability as will be explained in Sect. 4.1. Finally, the *metamodel URI* is the unique identifier of the metamodel which can be utilized to retrieve the metamodel from a central registry, for example, when checking models for conformance with the respective metamodel. For the BCS running example, the metamodel URI is http://www.tu-braunschweig.de/isf/states.

---

[13] http://eclipse.org/Xtext.

[14] http://emftext.org/.

[15] http://eclipse.org/gef.

[16] http://eclipse.org/modeling/gmp.

[17] http://eclipse.org/graphiti.

[18] http://jamopp.org/index.php/JaMoPP.

[19] http://www.srcml.org/.

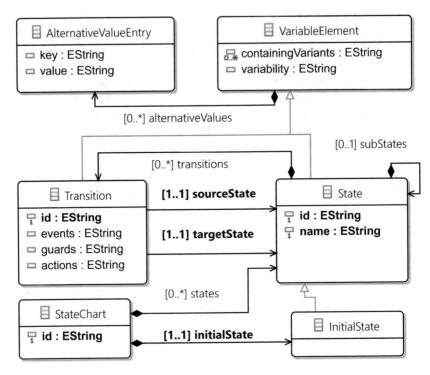

Fig. 7 Metamodel for the statechart notation used in the running example

Due to the benefits of employing metamodels, we heavily utilize model-based
development. However, we do not require users of our mining and generation
technology to use it as well because the model-based character is completely
transparent and required inputs may be provided in textual languages. In Sect. 4.1,
we explain how a set of cloned model variants can automatically be analyzed
to extract variability information (i.e., common and varying parts between the
variants). In Sect. 4.2, we show how to generate delta modules of a delta-oriented
SPL from the cloned variants.

4.1 Mining Variability from Cloned Variants

To support transition from a set of cloned model variants to a managed reuse
strategy, it is essential to identify variability relations between the variants. In Fig. 8,
we show our family mining process, an approach to semiautomatically reverse-
engineer variability information from a set of block-based model variants [18, 47,
49]. The approach relies on metamodeling techniques and first translates the input
models in an instance of a metamodel specifically tailored to the modeling language

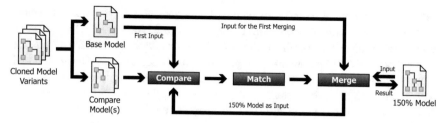

Fig. 8 Workflow for variability mining from cloned model variants

employed to realize variants, for example, for statecharts, the metamodel of Fig. 7. Furthermore, we created metamodels capable of handling *MATLAB/Simulink* models and *FBDs* [18, 47, 49]. The family mining algorithm consists of the following three phases. During the *Compare Phase* (cf. Sect. 4.1.1), models are compared and possible relations are identified. In the *Match Phase*, unambiguous one-to-one relations are selected from these comparisons. In the *Merge Phase*, the resulting relations are used to create a *150% model*.

In the metamodel for statecharts presented in Fig. 7, we provide classes `VariableElement` and `AlternativeValueEntry` to store the determined variability information: In `VariableElement`, we allow to store the identified variability for compared elements (i.e., whether they are contained in all compared variants or represent variability only present in certain variants). In addition, we store the model variants containing the corresponding elements. In `AlternativeValueEntry`, we allow to define mappings from model names to alternative values (e.g., when we identify state names with minor deviations in two compared variants).

The following sections explain each of the steps for mining variability from cloned variants in details.

4.1.1 Compare Phase

Our family mining approach is realized in a pairwise manner and, thus, iteratively compares two models at a time. The algorithm starts comparing the models by selecting a *base model* (e.g., the smallest variant) and regards all remaining models as *compare models*. Next, the algorithm compares the selected base model with one of the compare models by analyzing the dataflow in the model. For our running example, we compare the two statechart implementation variants of the CLS feature (cf. Fig. 3). Starting from the *start elements* on the highest hierarchy level where data is introduced to the model or where the execution is started (i.e., the `cls_unlock` initial states from the variants with manual and automatic central locking system in Fig. 3), the algorithm separates the currently analyzed model hierarchy for both models into *stages*. These stages are created by analyzing the dataflow and contain only elements that are separated by the same number of edges (e.g., transitions in statecharts) compared to the start elements. For instance, the compare algorithm creates stage S0 with state `cls_unlock` and stage S1 with state `cls_lock` for both compared variants in Fig. 3. Depending on the employed modeling language, not only the analyzed model *nodes* (e.g., states for statecharts) are relevant but their connecting *edges* (e.g., transitions for statecharts) contain additional information worth considering during comparison. For example, as transitions in statecharts contain important execution information, they should be analyzed during family mining. Thus, two additional stages are created for both variants from Fig. 3 containing the outgoing transitions from the states `cls_unlock` and `cls_lock`, respectively.

Next, each stage from the base model is compared with its counterpart from the compare model by iterating and inspecting all possible combinations of the contained elements. For each comparison between two model elements, a so-called *compare element* is created storing the compared elements together with a *similarity value* calculated according to a user-defined *metric* [48, 49]. Such a metric allows to assign different weights to the properties of compared elements and, thus, allows to rank their influence on the model functionality. For instance, when comparing two transitions from Fig. 3, we could assign a higher weight to *actions* triggered by the transitions than to their *events* (i.e., the events triggering the transition's execution) and *guards* (i.e., conditions that have to be fulfilled in order to execute the transition) because actions trigger new events and the execution of other transitions. Consequently, these actions have a high influence on the semantics of a particular statechart. As the metric is adjustable to different settings, users can easily modify the weights to different needs. During the comparison of two elements, the similarity value is calculated by summing up the metric's weights according to the elements' similarity. To allow comparison of the calculated similarity values, we normalize the metric's values to the interval [0..1]. A concrete example for such a metric can be found in [48]. In cases where no counterpart exists for the comparison of elements in a stage, we create comparisons with *null*, which indicates that the respective element is optional as it is not present in some of the variants.

Our procedure natively supports comparison of hierarchical models by recursively starting the algorithm when comparing two hierarchical elements (e.g., two hierarchical states in statecharts or two subsystems in *MATLAB/Simulink*) [49]. Using this approach, we are able to calculate the overall similarity value of compared hierarchical elements by averaging the similarity value of their sub-elements.

4.1.2 Match Phase

The resulting list of compare elements might contain ambiguous relations between the compared model elements, because, during the comparison of stages, multiple combinations with the same model element might be created. For instance, during the comparison of the stages containing the transitions going from the `cls_unlock` state to the `cls_lock` state, the algorithm creates two compare elements. Both elements contain the corresponding transition from Fig. 3b comparing it with both possible variants from Fig. 3a.

As these ambiguous relations hinder to identify distinct one-to-one variability relations between the compared model elements, the algorithm traverses the list of all possible compare elements. For each contained compare element, it identifies all other compare elements sharing at least one of the compared elements (i.e., the element from the base model or the compare model). Afterward, the algorithm identifies a distinct match for these compare elements by selecting the compare element with the highest similarity value. All ruled-out compare elements are deleted and the algorithm continues until no unmatched elements are left. Thus, in our example, the algorithm matches the transition between the states `cls_unlock` and `cls_lock` in Fig. 3b with the transition containing the `pw_pos==1` guard in Fig. 3a because they have a higher similarity compared to the other possible transition containing the `pw_pos!=1` guard (i.e., a higher number of statements match). In case the algorithm cannot identify distinct relations for a compare element, it first sorts the conflicting elements to the end of the list. Using this approach, the algorithm tries to implicitly solve ambiguous relations automatically by first matching other elements. However, in some cases user intervention may be necessary to resolve the conflict. Elements that were ruled out completely from all compare elements (i.e., they have no matching partner) are regarded as optional elements and are added to the final list of matches in compare elements. For example, the ruled-out transition between the states `cls_unlock` and `cls_lock` in Fig. 3a has to be added as such an optional compare element to not lose information contained in the compared variants.

4.1.3   Merge Phase

The resulting list of distinctively matched compare elements can now be used to create a 150% model storing all implementation artifacts from the compared variants together with their identified variability (i.e., how the elements are related and in which models they are contained). The algorithm processes the list of distinct matches and creates the 150% model by merging the compare elements with a copy of the base model. For the merging process, we define the following mapping function to classify variability of compared elements:

$$rel(A, B) = \begin{pmatrix} similarity >= 0.95 & mandatory \\ 0 < similarity < 0.95 & alternative \\ similarity = 0 & optional \end{pmatrix} \tag{1}$$

This mapping function is adjustable to user preferences and defines default thresholds that were identified during an impact analysis of differing properties in compared elements with additional interviews on how similar these elements are regarded according to domain experts [48, 49]. The threshold of 95% categorizes two compared elements as *mandatory* (i.e., they are regarded as equal). However, we allow minor deviations between the elements because of the 5% interval up to 100% equality. For example, this allows us to regard elements as mandatory despite minor differences in their names. Mandatory elements do not have to be merged into the 150% model as they are already contained in the base model copy. However, we have to annotate differing values for the properties where they are not equal (e.g., the changed name). Otherwise, we lose information when creating the 150% model. Compare elements with a similarity value of 0% are regarded as *optional* (i.e., they are only contained in some of the variants) as they have no counterpart. Depending on whether the element was already contained in the base model copy, we have to copy the corresponding elements to the 150% model. All elements with a similarity value between the mandatory and optional threshold are regarded as *alternatives* (i.e., all variants contain exactly one of the possible alternative elements). Here, only the element that was not contained previously in the base model copy has to be copied to the 150% model. For all elements in the 150% model, we compare the model containing the corresponding element and explicitly store the variability identified according to the thresholds in Eq. 1. The resulting 150% model is used as an input for the iterative comparison with the next model.

In Fig. 9, we present the 150% model for our running example. As we can see, the algorithm correctly identified and annotated the models containing the different transitions (i.e., either the ManPW variant or the AutoPW variant). For readability reasons, we neglected the explicit variability annotations in Fig. 9. However, the algorithm correctly identifies that the annotated transition from the AutoPW variant is an alternative to the annotated transition with the pw_pos==1 guard from the ManPW variant. The remaining annotated transition with the pw_pos!=1 guard is identified as an optional element. All elements without annotations are regarded as mandatory as they are contained in both variants.

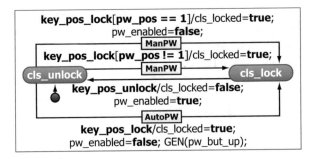

Fig. 9 Part of the 150% model showing the variability of the compared CLS variants

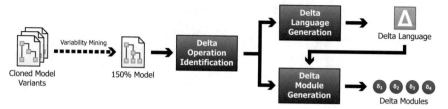

Fig. 10 Workflow for the generation of a delta-oriented SPL consisting of a delta language and multiple delta modules

4.2   Generating a Delta-Oriented Software Product Line

After creating a 150% model, we are now able to generate an SPL for managed reuse. In particular, we generate a delta-oriented SPL consisting of a delta language tailored specifically to the modeling language used to realize implementation artifacts and a set of delta modules using this delta language to specify the transformations to create the original set of variants from the base variant. Figure 10 depicts the workflow for SPL generation.

The workflow consists of three steps. First, the algorithm decides on which transformation operations are required by processing annotations of the 150% model (cf. Sect. 4.2.1). Then, the algorithm uses these operations to generate a delta language for the modeling language of the inspected variants (cf. Sect. 4.2.2).[20] Finally, using this delta language, delta modules are generated that describe the transformations from the base variant to all of the initially analyzed variants (cf. Sect. 4.2.3).

4.2.1   Delta Operation Identification

As we apply delta modeling, we intend to transform a *base variant (BV)* to a *target variant (TV)*. Our process allows users to freely select any variant contained in the input 150% model as BV. To determine appropriate delta operation calls for transformation to the respective TV, we have to decide for each element from the generated 150% model whether it has to be *added*, *removed*, or *modified* by calling the respective delta operation.

To identify appropriate operations to call, we use a decision process: The algorithm analyzes the annotations in the 150% model to identify in which variants the currently considered element is contained. In case it *is not* contained in BV and TV or it *is* contained in both variants but is regarded as *identical*, the process decides to not generate a delta operation. Two elements are regarded as identical

---

[20]To be exact, the algorithm generates a delta dialect, which can be used to generate the delta language (cf. Sect. 5).

Table 1 Decisions for Fig. 9 with BV=ManPW and TV=AutoPW

Element	Decision process
Transition 4	Decide $\xrightarrow{!in\ BV}$!in BV $\xrightarrow{in\ TV}$ Add
Source state for transition 4	Decide $\xrightarrow{!in\ BV}$!in BV $\xrightarrow{in\ TV}$ Set
Transition 3	Decide $\xrightarrow{in\ BV}$ in BV $\xrightarrow{in\ TV\ \&\ identical}$ Nothing
Source state for transition 2	Decide $\xrightarrow{in\ BV}$ in BV $\xrightarrow{!in\ TV}$ Unset
Transition 2	Decide $\xrightarrow{in\ BV}$ in BV $\xrightarrow{!in\ TV}$ Remove

if the family mining process marked them as mandatory without any alternative properties (e.g., for elements with minor name differences). In all other cases, we have to generate a delta operation for the transformation between both variants. If the analyzed element is only contained in BV and does not have an annotation for TV, it has to be *unset/removed* to generate the final variant. Similarly, the element has to be *set/added* when it is only contained in TV and not in BV. For elements that are contained in both variants but are *not similar*, the original element has to be *modified*.

In Table 1, we present the identification of delta operation calls for the 150% model in Fig. 9 with BV=ManPW (i.e., the variant in Fig. 3a) and TV=AutoPW (i.e., the variant in Fig. 3b). For space reasons, we limit the table to demonstrate an example for each type of identified delta operation (i.e., *add*, *set*, *remove*, *unset*, and *nothing*). However, for complete transformations, all elements have to be analyzed by the delta operation identification process. To allow easier reasoning on the decisions in Table 1, we have numbered the transitions of Fig. 9 consecutively from top (i.e., the transition with the pw_pos==1 guard) to bottom (i.e., the transition with the AutoPW annotation). As we can see, the algorithm identifies correctly that transition 4 is not contained in ManPW and has to be *added* during a transformation to AutoPW. Similarly, transition 2 has to be *removed* as it is not contained in AutoPW. In addition, the algorithm identifies that the source states for transition 4 and 2 have to be *set* and *unset*, respectively. These operations are required to update the corresponding references because of the metamodel structure used to store the statechart variants in Fig. 7 (i.e., the sourceState reference is used to store the corresponding state). For transition 3, the delta operation identification process correctly determines that the element does not have to be transformed as it is contained in both variants.

At present, we analyze atomic differences and identify atomic delta operations that modify a single value or reference. However, in the future we plan on identifying semantically richer delta operations by utilizing domain knowledge or knowledge of the semantics of the realization language, for example, a delta operation for statecharts that can remove a state as well as all its incoming and outgoing transitions to preserve well-formedness of the statechart.

4.2.2 Delta Language Generation

Using the described decision process, it is possible to determine which delta operations have to be called to perform the transformations to retrieve target variants TVs for all inspected cloned variants from the selected base variant BV. Depending on the selected BV, different delta languages are generated as only delta operations are considered that are needed to transform BV to the analyzed TVs. For example, when generating a delta language for the comparison of BV=ManPW with another TV that only contains additional states and transitions, no *remove* operation for transitions is generated. For each decision returned by the algorithm, we generate a corresponding delta operation and store it in a set to prevent generation of redundant operations. For instance, in our running example, this prevents the generation of multiple delta operations to *add* transitions to states.

In Listing 1, we present an excerpt from the delta language generated for the 150% model in Fig. 9 with BV=ManPW and TV=AutoPW. The excerpt contains the delta operations generated for the decisions in Table 1. This delta language was generated with our tool suite DeltaEcore (cf. Sect. 5) using information on the required type of delta operation, the transformed reference or attribute, and the containing class. The automatically generated names describe the functionality of the corresponding operation in a unique and descriptive way but they may be changed manually without impacting further generation (cf. Sect. 4.2.3)

4.2.3 Delta Module Generation

Using the generated delta language, it is now possible to define delta modules that contain transformations from one variant to another. Our algorithm automatically generates appropriate delta modules for the variability identified using the family mining algorithm (cf. Sect. 4.1). To start the delta module generation, our algorithm expects the selected BV and TVs from the 150% model generated during family

```
 1  deltaDialect {
 2    configuration:
 3      metaModel: <http://www.tu-braunschweig.de/isf/states>;
 4
 5    deltaOperations:
 6      addOperation addTransitionToTransitionsOfState(
 7        Transition value, State [transitions] element);
 8      setOperation setSourceStateOfTransition(State value,
 9        Transition [sourceState] element);
10      removeOperation removeTransitionFromTransitionsOfState(
11        Transition value, State [transitions] element);
12      unsetOperation unsetSourceStateOfTransition(
13        Transition [sourceState] element);
14      // ...
15  }
```

Listing 1 Excerpt from the delta language generated for the running example

mining as well as a delta language providing operations to transform to the inspected variants. The used delta language could either be generated automatically using the approach described in Sect. 4.2.2 or could be specified manually.

Using the selected BV, the algorithm analyzes the decision of each element from the currently considered TV to identify the appropriate delta operation to apply. For this procedure, the adequate delta operation is determined from the provided delta language by looking up its type (e.g., *set, modify*) and the metamodel element it addresses as parameter. First, the number of possible delta operations to apply is reduced by filtering out all delta operations whose type does not conform with the decision. For example, the *set* decision in Table 1 reduces the number of possible operations for the delta language in Listing 1 to exactly one remaining element (i.e., the `setSourceStateOfTransition` operation). Then, the algorithm compares the references or attributes transformed by the remaining delta operations and the modified types with the needed transformation according to the generated decision (i.e., in our example setting the *sourceState* reference in the `Transition` class). Only after these additional checks, a corresponding delta operation call is generated to realize the needed transformation. It is worth noting that determining the respective delta operation to call is independent of the name of the operation so that the latter may be chosen freely before generation.

All generated delta operation calls for a transformation from a selected BV to a TV are stored in delta modules. In case of *add* operation calls, the corresponding constructor calls to create the element to be added are generated automatically and are also stored in the delta modules. In Listing 2, we show an excerpt for the delta module to transform BV=ManPW to TV=AutoPW. The excerpt contains all delta operation calls with corresponding constructor calls for the decisions from Table 1 (cf. Sect. 5.2).

Using the generated delta modules, it is now possible to derive variants from the defined BV that correspond to all originally inspected cloned variants. In the following Sect. 5, we explain our implementation of the tool suite DeltaEcore, which is used for the generation process and which allows automatic variant derivation from the specified delta modules.

```
1  delta "CLS-ManPW->CLS-AutoPW"
2    dialect <http://www.tu-braunschweig.de/isf/states>
3    modifies <CLS-ManPW.statechart> {
4      unsetSourceStateOfTransition(<trans3>);
5      removeTransitionFromTransitionsOfState(<trans3>,
6        <cls_unlock>);
7      Transition t = new Transition(id: "trans1",
8        events: "key_pos_lock", actions: "cls_locked = true;
9        pw_enabled = false; GEN(pw_but_up);");
10     addTransitionToTransitionsOfState(t, <unlock>);
11     setSourceStateOfTransition(<unlock>, <trans1>);
12     // ...
13 }
```

Listing 2 Excerpt from the delta module generated for the running example

5 Realization as Tool Suite DeltaEcore

DeltaEcore[21] [36, 37] is a tool suite for variability management using the transformational variability realization mechanism delta modeling. It employs a model-based development process, which allows the tool suite to be tailored to specific implementation languages with low effort. DeltaEcore has three major application areas: *delta language creation* to adapt the tool suite to work with specific realization languages, *software product line definition* to apply a managed reuse strategy to a family of related software systems, and *variant derivation* to generate individual software products from the software product line.[22] Figure 11 provides an overview and the following sections elaborate on each of these application areas in detail.

5.1 Delta Language Creation

Before creating an SPL with DeltaEcore, suitable delta languages for all *source languages* used for the implementation have to be created, for example, a delta language for statecharts. A delta language provides dedicated operations to alter artifacts of the source language and, thereby, governs the level of control to these artifacts, for example, by providing operations to change the transitions of a state but not its id.

DeltaEcore assumes source languages to be available as an Ecore-based meta-model where the source language's elements are represented as metaclasses with references and attributes, which is feasible for both textual and graphical languages. For example, the statechart notation used throughout the running example is defined by the metamodel depicted in Fig. 7.

A delta language in DeltaEcore consists of two parts: the *common base delta language* and a *delta dialect*. The common base delta language is agnostic of the source language and defines language constructs shared by all delta languages, such as the definition of variables or requiring other delta modules. A delta dialect is specific to a source language as it references elements of the source language's metamodel to define delta operations for the respective source language. When specifying a delta module to alter an artifact of a specific source language, DeltaEcore combines the common base delta language with the respective delta dialect to provide an appropriate delta language.

The common base delta language is provided entirely by DeltaEcore. However, the delta dialect has to be defined once for each source language, for example, a

---

[21] http://deltaecore.org.

[22] In this chapter, we focus on functionality of DeltaEcore regarding delta modeling. In addition, DeltaEcore also allows for a seamless integration of feature models, provides a graphical editor and configurator, supports integrated management of SPL evolution, and may be interfaced with other tools, such as FeatureIDE [42].

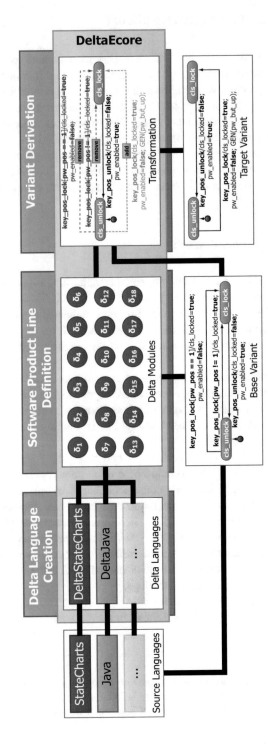

Fig. 11 Overview of DeltaEcore's major application areas for delta modeling

delta dialect for statecharts by specifying the signatures of the delta operations that should be provided. As the principle nature of altering implementation artifacts in delta modules is similar even across different languages, DeltaEcore provides seven types of *standard delta operations*:

- *Set* and *unset* operations alter values of single-valued references, such as the reference `initialState` of `StateChart`, by supplying a new value or resetting the reference to its default value, respectively.
- *Add* and *remove* operations alter values of multivalued references, such as the reference `states` of `StateChart`, by adding or removing a value to the set of values, respectively. *Insert* operations alter values of multivalued references with ordered values by adding a value to the ordered set of values at a specific position.
- *Modify* operations alter values of attributes, such as the attribute `name` of `State`, by supplying a new value, which, in contrast to altering values of references, is guaranteed to be free of side effects.
- *Detach* operations remove an element from its container so that it can be deleted from the model upon save if no other references to the element exist.[23]

Due to their uniform definition, the standard delta operations of DeltaEcore have defined semantics in terms of how they affect the artifacts of the source language. In addition to standard delta operations, it is also possible to supply *custom delta operations* with user-defined semantics, for example, to realize complex operations specific to the source language, such as removing a state along with all its incoming and outgoing transitions. As an example, we have already presented the delta dialect for the statechart notation used throughout the chapter in Listing 1 in Sect. 4.2.2.

A delta dialect may be defined in multiple ways: First, it may be defined entirely manually. Second, it may be generated by DeltaEcore by analyzing the structure of the source language's metamodel for suitable delta operations [36]. Third, it may be generated as part of mining variability of cloned variants as presented in Sect. 4.2.2 for statecharts. Furthermore, combinations of these approaches are also possible so that a generated delta dialect may be refined manually to supply specific user-defined operations. In combination with the common base delta language, the definition of a delta dialect suffices to create a delta language tailored to a specific source language.

---

[23] We refrained from defining a delete operation due to its potentially cross-cutting effects when resetting all references to the deleted element. However, a delete operation can be supplied manually with minimal effort if suitable for the source language.

5.2 Software Product Line Definition

To apply a managed reuse strategy to a family of related software systems, DeltaEcore supports the definition of an SPL based on delta modeling. For this purpose, one of the products of the family of software systems is designated as base variant, and all other variants are described in terms of transformations realizing the differences of the implementation artifacts to those of the base variant, which are captured in delta modules.

A delta module modifies a realization artifact (e.g., a statechart) through a sequence of calls to delta operations to add, modify, and remove elements. In DeltaEcore, the delta operations available for altering a realization artifact are provided by the delta dialect as part of the delta language for the respective source language (e.g., the delta dialect for statecharts).

As an example, we have already presented a delta module in Listing 2 in Sect. 4.2.3. In line 2, the delta dialect for the source language of the artifact to be altered is set by providing the URI of the language's metamodel, which DeltaEcore resolves to the appropriate delta dialect. In line 3, the statechart realizing the manual power window, which serves as base variant, is referenced to be modified by the delta module. In lines 4–12, a sequence of calls to delta operations alters the statechart to contain functionality related to the automatic power window.

To create complex variants, multiple delta modules may be used where each one encapsulates strongly coherent changes, for example, to realize one feature of the SPL. However, delta modules may not be completely independent of one another, for example, when one delta module alters an element that is created by another delta module. For this purpose, DeltaEcore allows for delta modules to specify *delta module dependencies*, which state that a delta module requires another delta module to be applied before its transformations can be invoked.

DeltaEcore also allows for specifying *application-order constraints*, which state that a certain delta module may only be applied after another delta module was applied. In contrast to delta module dependencies, application-order constraints do not entail that the referenced delta module is inevitably necessary so that application-order constraints are only evaluated should both delta modules be selected explicitly.

The set of all delta modules and their application-order constraints comprise the SPL, which allows for managed reuse and creation of individual products through variant derivation.

5.3 Variant Derivation

To create a concrete software product of the SPL defined in DeltaEcore, a variant derivation has to be performed. For this purpose, a set of delta modules has to be selected, which is then applied in a suitable order to transform a base variant to a target variant containing the intended functionality.

The initial set of delta modules is supplied by a user through selecting those delta modules that are associated with the functionality for the software product that differs from the base variant, for example, to enable the automatic power window.[24] DeltaEcore automatically completes the initial set of delta modules by adding all (transitively) required delta modules.

To ensure deterministic variant derivation, a valid application sequence for the relevant delta modules has to be determined. For this purpose, DeltaEcore performs a topological sorting, which takes into account delta module dependencies as well as application-order constraints posed upon those delta modules to determine an application sequence that satisfies all constraints.

DeltaEcore then copies the base variant of the SPL and applies the delta modules in the determined sequence to perform the transformations described by calls to delta operations within each delta module. The result is the target variant of the SPL containing the functionality of the intended software product. In DeltaEcore, the variant derivation procedure is fully automated so that users of the SPL only need to supply the initial set of delta modules intended to create a specific software product. Furthermore, it is also principally possible to use delta modules to perform changes specific to individual customers, for example, to customize a previously generated variant.

With the capacities of DeltaEcore for delta language creation, software product line definition, and variant derivation, it is possible to develop a set of closely related software systems as delta-oriented SPL. Combined with the variability mining and SPL generation approaches we presented in Sect. 4, it is possible to seamlessly migrate from the industrial practice of clone-and-own to a managed reuse strategy using an SPL.

6　Conclusion

In this chapter, we have demonstrated an approach for seamless transition from the industrial practice of creating software variants through clone-and-own to a managed reuse strategy with a delta-oriented SPL. We first reviewed the state of practice and state of the art in variability realization mechanisms. We then introduced our procedure for variability mining from variants created through clone-and-own to retrieve previously unavailable variability information. We demonstrated how our procedure generates a delta-oriented SPL from the mined variability information. With the resulting delta-oriented SPL, it is possible to maintain and create a large set of variants from a single set of managed variable artifacts.

In the future, we plan on incorporating domain knowledge to generate semantically richer delta operations to be used within the generated delta language and

---

[24]If a feature model is supplied, it is further possible to select a configuration from the feature model and have DeltaEcore resolve the selected features to the respective associated delta modules.

associated delta modules. Furthermore, we will also investigate how to generate an initial feature model to be used with the delta modules to also represent the problem space of the SPL in order to further increase usefulness for the industry.

References

1. Alves V, Matos P, Cole L, Vasconcelos A, Borba P, Ramalho G (2007) Extracting and evolving code in product lines with aspect-oriented programming. In: Transactions on aspect-oriented software development IV. Springer, Berlin, pp 117–142
2. Apel S, Kästner C (2009) An overview of feature-oriented software development. J Object Technol 8(5):49–84
3. Bąk K, Czarnecki K, Wąsowski A (2011) Feature and meta-models in Clafer: mixed, specialized, and coupled. In: Proceedings of the international conference on software language engineering (SLE), SLE '11. Springer, Berlin, pp 102–122
4. Batory D (2004) Feature-oriented programming and the AHEAD tool suite. In: Proceedings of the international conference on software engineering (ICSE), ICSE '04. IEEE, Piscataway, pp 702–703
5. Berger T, Rublack R, Nair D, Atlee JM, Becker M, Czarnecki K, Wąsowski A (2013) A survey of variability modeling in industrial practice. In: Proceedings of the international workshop on variability modeling in software-intensive systems (VaMoS), VaMoS '13. ACM, New York, pp 7:1–7:8
6. Berger T, Lettner D, Rubin J, Grünbacher P, Silva A, Becker M, Chechik M, Czarnecki K (2015) What is a feature?: a qualitative study of features in industrial software product lines. In: Proceedings of the international software product line conference (SPLC), SPLC '15. ACM, New York, pp 16–25
7. Beuche D (2012) Modeling and building software product lines with pure::variants. In: Proceedings of the international software product line conference (SPLC), SPLC '12. ACM, New York, pp 255–255
8. Clements PC, Northrop LM (2001) Software product lines: practices and patterns. Addison-Wesley, Boston
9. Czarnecki K, Eisenecker UW (2000) Generative programming: methods, tools, and applications. Addison-Wesley, Boston
10. Damiani F, Lienhardt M (2016) On type checking delta-oriented product lines. In: Proceedings of the international conference on integrated formal methods (iFM), iFM '16. Springer, Berlin, pp 47–62
11. Damiani F, Lienhardt M (2016) Refactoring delta oriented product lines to enforce guidelines for efficient type-checking. In: Proceedings of the international symposium on leveraging applications of formal methods, verification and validation (ISoLA), ISoLA'16. Springer, Berlin
12. Dubinsky Y, Rubin J, Berger T, Duszynski S, Becker M, Czarnecki K (2013) An exploratory study of cloning in industrial software product lines. In: Proceedings of the European conference on software maintenance and reengineering (CSMR), CSMR '13. IEEE, Piscataway, pp 25–34
13. Figueiredo E, Cacho N, Sant'Anna C, Monteiro M, Kulesza U, Garcia A, Soares S, Ferrari F, Khan S, Dantas F (2008) Evolving software product lines with aspects. In: Proceedings of the international conference on software engineering (ICSE), ICSE '08. IEEE, Piscataway, pp 261–270
14. Greenfield J, Short K (2003) Software factories: assembling applications with patterns, models, frameworks and tools. In: Proceedings of the international conference on object-oriented programming, systems, languages and applications (OOPSLA), OOPSLA '03. ACM, New York, pp 16–27

15. Groher I, Voelter M (2009) Aspect-oriented model-driven software product line engineering. In: Transactions on aspect-oriented software development VI. Springer, Berlin, pp 111–152
16. Haugen Ø, Møller-Pedersen B, Oldevik J, Olsen GK, Svendsen A (2008) Adding standardized variability to domain specific languages. In: Proceedings of the international software product line conference (SPLC), SPLC '08. IEEE, Piscataway, pp 139–148
17. Heidenreich F, Kopcsek J, Wende C (2008) FeatureMapper: mapping features to models. In: Proceedings of the international conference on software engineering (ICSE), ICSE '08. ACM, New York
18. Holthusen S, Wille D, Legat C, Beddig S, Schaefer I, Vogel-Heuser B (2014) Family model mining for function block diagrams in automation software. In: Proceedings of the international workshop on reverse variability engineering (REVE), SPLC '14. ACM, New York, pp 36–43
19. International Electrotechnical Commission (2009) Programmable logic controllers – part 3: programming languages. IEC61131-3 Standard
20. Kang KC, Cohen SG, Hess JA, Novak WE, Peterson AS (1990) Feature-oriented domain analysis (FODA) feasibility study. Tech. Rep. CMU/SEI-90-TR-021, Carnegie-Mellon University Software Engineering Institute
21. Kapser C, Godfrey MW (2006) "Cloning Considered Harmful" considered harmful. In: Proceedings of the working conference on reverse engineering (WCRE), WCRE '06. IEEE, Piscataway, pp 19–28
22. Kästner C, Apel S, Kuhlemann M (2008) Granularity in software product lines. In: Proceedings of the international conference on software engineering (ICSE), ICSE '08. ACM, New York, pp 311–320
23. Kiczales G, Lamping J, Mendhekar A, Maeda C, Lopes C, Loingtier JM, Irwin J (1997) Aspect-oriented programming. ECOOP '97. Springer, Berlin
24. Krueger C (2002) Variation management for software production lines. In: Software product lines. Springer, Berlin, pp 37–48
25. Krueger CW (2008) The Biglever software gears unified software product line engineering framework. In: Proceedings of the international software product line conference (SPLC), SPLC '08. IEEE, Piscataway, pp 353–353
26. Liebig J, Apel S, Lengauer C, Kästner C, Schulze M (2010) An analysis of the variability in forty preprocessor-based software product lines. In: Proceedings of the international conference on software engineering (ICSE). ACM, New York, pp 105–114
27. Lity S, Lachmann R, Lochau M, Schaefer I (2012) Delta-oriented software product line test models – the body comfort system case study. Tech. Rep. 2012-07, Technische Universität Braunschweig, Braunschweig
28. Muthig D, Atkinson C (2002) Model-driven product line architectures. In: Software product lines. Springer, Berlin, pp 110–129
29. Pohl K, Böckle G, van der Linden FJ (2005) Software product line engineering: foundations, principles and techniques. Springer, Berlin
30. Rubin J, Chechik M (2013) A survey of feature location techniques. In: Domain engineering: product lines, languages, and conceptual models. Springer, Berlin, pp 29–58
31. Rumpe B, Weisemöller I (2011) A domain specific transformation language. In: Proceedings of the international workshop on models and evolution (ME), ME '11
32. Ryssel U, Ploennigs J, Kabitzsch K (2011) Extraction of feature models from formal contexts. In: Proceedings of the international software product line conference (SPLC), SPLC '11. ACM, New York, pp 4:1–4:8
33. Schaefer I, Bettini L, Bono V, Damiani F, Tanzarella N (2010) Delta-oriented programming of software product lines. In: Software product lines: going beyond. Lecture notes in computer science, vol 6287. Springer, Berlin, pp 77–91
34. Schaefer I, Rabiser R, Clarke D, Bettini L, Benavides D, Botterweck G, Pathak A, Trujillo S, Villela K (2012) Software diversity: state of the art and perspectives. Int J Softw Tools Technol Transfer 14(5):477–495
35. Schmidt DC (2006) Model-driven engineering. Computer 39(2):25

36. Seidl C, Schaefer I, Aßmann U (2014) DeltaEcore – a model-based delta language generation framework. In: Modellierung, Modellierung'14, pp 81–96
37. Seidl C, Schaefer I, Aßmann U (2014) Integrated management of variability in space and time in software families. In: Proceedings of the international software product line conference (SPLC), SPLC '14. ACM, New York
38. Sendall S, Kozaczynski W (2003) Model transformation the heart and soul of model-driven software development. Tech. rep., Microsoft
39. She S, Lotufo R, Berger T, Wasowski A, Czarnecki K (2011) Reverse engineering feature models. In: Proceedings of the international conference on software engineering (ICSE), ICSE '11. IEEE, Piscataway, pp 461–470
40. Steinberg D, Budinsky F, Paternostro M, Merks E (2008) Eclipse modeling framework, 2nd edn. Addison-Wesley, Boston
41. Szyperski CA (1998) Component software - beyond object-oriented programming. Addison-Wesley, Boston
42. Thüm T, Kästner C, Benduhn F, Meinicke J, Saake G, Leich T (2014) FeatureIDE: an extensible framework for feature-oriented software development. Sci Comput Program 79:70–85
43. van der Linden F, Schmid K, Rommes E (2010) Software product lines in action: the best industrial practice in product line engineering. Springer, Berlin
44. von Rhein A, Thüm T, Schaefer I, Liebig J, Apel S (2016) Variability encoding: from compile-time to load-time variability. J Log Algebr Methods Program 85(1):125–145
45. Weiland J, Manhart P (2014) A classification of modeling variability in Simulink. In: Proceedings of the international workshop on variability modeling in software-intensive systems (VaMoS), VaMoS '14. ACM, New York, pp 7:1–7:8
46. Weston N, Chitchyan R, Rashid A (2009) A framework for constructing semantically composable feature models from natural language requirements. In: Proceedings of the international software product line conference (SPLC), SPLC '09. ACM, New York, pp 211–220
47. Wille D (2014) Managing lots of models: the FaMine approach. In: Proceedings of the international symposium on the foundations of software engineering (FSE), FSE '14. ACM, New York, pp 817–819
48. Wille D, Holthusen S, Schulze S, Schaefer I (2013) Interface variability in family model mining. In: Proceedings of the international workshop on model-driven approaches in software product line engineering (MAPLE), SPLC '13. ACM, New York, pp 44–51
49. Wille D, Schulze S, Seidl C, Schaefer I (2016) Custom-tailored variability mining for block-based languages. In: Proceedings of the international conference on software analysis, evolution, and reengineering (SANER), SANER '16, vol 1. IEEE, Piscataway, pp 271–282
50. Zhang X, Haugen Ø, Møller-Pedersen B (2011) Model comparison to synthesize a model-driven software product line. In: Proceedings of the international software product line conference (SPLC), SPLC '11. IEEE, Piscataway, pp 90–99
51. Zhang X, Haugen Ø, Møller-Pedersen B (2012) Augmenting product lines. In: Proceedings of the Asia-Pacific software engineering conference (APSEC), vol 1. IEEE, Piscataway, pp 766–771

Variability Identification and Representation for Automotive Simulink Models

Manar H. Alalfi, Eric J. Rapos, Andrew Stevenson, Matthew Stephan, Thomas R. Dean, and James R. Cordy

Abstract This chapter presents an automated framework for identifying and representing different types of variability in Simulink models. The framework is based on the observed variants found in similar subsystem patterns inferred using Simone, a model clone detection tool, and an empirically derived set of variability operators for Simulink models. We demonstrate the application of these operators to six example systems, including automotive systems, using two alternative variation analysis techniques, one text-based and one graph-based, and show how we can represent the variation in each of the similar subsystem patterns as a single subsystem template directly in the Simulink environment. The product of our framework is a single consolidated subsystem model capable of expressing the observed variability across all instances of each inferred pattern. The process of pattern inference and variability analysis is largely automated and can be easily applied to other collections of Simulink models. We provide tool support for the variability identification and representation using the graph-based approach.

M. H. Alalfi (✉)
Department of Computer Science, Ryerson University, Toronto, ON, Canada
e-mail: manar.alalfi@ryerson.ca

E. J. Rapos · M. Stephan
Department of Computer Science and Software Engineering, Miami University, Oxford, OH, USA
e-mail: rapose@miamioh.edu; stephamd@miamioh.edu

A. Stevenson · J. R. Cordy
School of Computing, Queen's University, Kingston, ON, Canada
e-mail: andrews@cs.queensu.ca; cordy@cs.queensu.ca

T. R. Dean
Electrical and Computer Engineering Department, Queen's University, Kingston, ON, Canada
e-mail: dean@cs.queensu.ca

© Springer Nature Switzerland AG 2019
Y. Dajsuren, M. van den Brand (eds.), *Automotive Systems and Software Engineering*, https://doi.org/10.1007/978-3-030-12157-0_6

1 Introduction

Software variability management (SVM) research has gained a lot of interest in the last two decades, especially for its vital role in developing reusable software product line (SPL) assets [5]. SVM is a complex, multifaceted problem that intersects with several traditional software engineering topics, including software configuration management, run-time dynamism, domain-specific languages, model-driven engineering, and software architecture. SVM offers a powerful toolbox to help manage complexity in these fields and is rapidly evolving into an independent research area that is of vital importance for systems that include configuration and run-time dynamism of components, in addition to software product lines.

One facet of SVM is variability modeling, an enabling technology for delivering a variety of related software systems in a fast, consistent, and comprehensive way. The key is to build a common base from which to efficiently express and manage variations. SVM is often closely associated with SPLs, which are mainly aimed at creating and maintaining a collection of similar software systems derived from a shared set of software assets. Variability can be expressed as stand-alone models, such as feature models in SPLs, or as annotations on a base model, by means of extensions to the base modeling language, such as UML profiles with stereotypes [15].

Variability modeling continues to gain interest from industry, and variability support in modeling tools, including Mathworks' Simulink and IBM's Rhapsody, is one of their most desirable features. Several industrial standards, such as SysML and AUTOSAR, are actively working to create extensions that help to express variability.

Understanding variability in existing systems and the variation points of their artifacts is the first and most important step toward enabling variability modeling. Many methods have been proposed for analyzing commonality and variability from a requirement's point of view, as well as connecting the analysis to the implementation [11, 21]. However, there remains a need for techniques that analyze existing system requirements and implementations for commonality and variability in an automated way.

In this chapter we present a framework for identifying variability candidates from existing software-intensive systems modeled using Simulink [25], the most popular modeling languages for hybrid hardware/software systems. Automotive Simulink models are particularly prone to cloning due to the copy-paste authoring paradigm of the Simulink IDE and the inherent similarity of elements and tasks in automotive applications. Our framework, shown in Fig. 1, uses an efficient model clone detection technique to automatically identify subsystem variants from a large pool of existing Simulink models. Once all potential variants are identified, the framework classifies and represents those variants using a set of empirically derived variability operators.

The framework is aimed at providing tool support to automatically represent model subsystem variability directly in the Simulink environment and thus provide practical assistance to engineers to identify, understand, and visualize patterns of

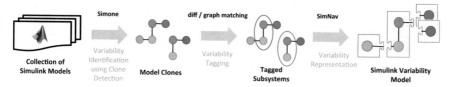

Fig. 1 Variability identification and representation framework

similar subsystems in a large model set. This understanding may help, among other things, in reducing maintenance effort and bug identification at an early stage of software development, both on the model level and before the model semantics are transformed into actual code. We demonstrate our framework on six systems from the Mathworks demonstration set and describe the stages of our framework using a running example.

In a previous short paper, we proposed a set of empirically derived variability operators for Simulink models [2] and provided evidence of the soundness of our operators based on the analysis of six Simulink systems representing a range of diverse applications. In this chapter, we use those proposed variability operators as the basis of an automated framework for the identification and representation of system variability in Simulink models.

The contributions of this chapter are as follows:

- Detailed description of an automated framework for the identification and representation of variability in Simulink models
- Demonstration of a text-based tagging approach to identify and mark variability in Simulink models using our previously proposed set of variability operators
- Demonstration of a graph-based approach for the identification and representation of variability in Simulink models
- Tool support for the graph-based approach that works directly in the Simulink IDE environment

The following sections provide a detailed description of the proposed framework and our experience with it.

2 Variability Identification and Representation Framework

The stages of framework as illustrated in Fig. 1 are explained below.

Stage 1: Variability Identification This stage uses model clone detection to identify groups of similar subsystems in a repository of Simulink models. We used Simone [1], a hybrid text-based model clone detection tool, to identify common Simulink subsystem patterns and variability candidates. In this framework we have configured Simone to identify subsystems that are at least 80%

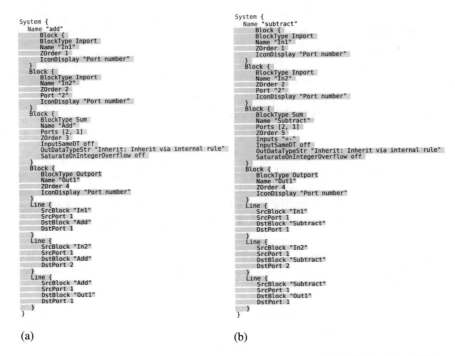

Fig. 2 Textual representations of example subsystem Variants A and B. (**a**) Variant A—addition.
(**b**) Variant B—subtraction

similar to each other as a first approximation. Similar subsystems are clustered
into clone classes, sets of subsystems that are similar to one another. Figure 2
shows an example of two instances of the same subsystem pattern as identified
by Simone. Variant A takes two numbers and adds them together, while Variant
B subtracts one number from the other—both models have two inports and one
outport. Section 3 provides a more detailed discussion of this first stage.

Stage 2: Variability Tagging   This stage uses two techniques, #ifdef preprocessor text tagging and subgraph similarity algorithms, to identify and tag subsystem
variability between the subsystems in each clone class identified in Stage 1.
This identification and tagging process is based on a set of proposed variability
operators that we have empirically inferred from a large set of observations
of variance in pattern candidates identified by Simone. Section 4 presents
our proposed variability operators for Simulink. The example in Fig. 2 shows
portions of the model that are similar among all instances (highlighted in green),
while the elements in different colors represent the variation between the two
models. This stage is aimed at automating the identification of commonality and
variability in this way and at marking the type of variability according to our
variability operators. A detailed description of this stage is presented in Sect. 5.

Stage 3: Variability Representation   This stage presents our approach to representing the identified Simulink subsystem variability using Simulink's built-in

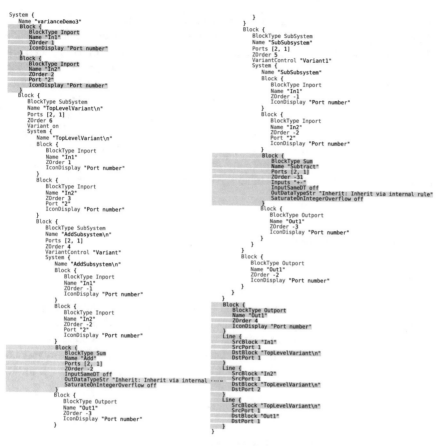

Fig. 3 Textual representation of the variance model for Variants A and B of Fig. 2

Variant Subsystem Block capability. Referring to our running example in Fig. 2, portions of the model that are similar among all instances (highlighted in green) are placed directly into the new variance model, with some renaming for sources and destinations pertaining to Variant Subsystem Blocks. The variability [e.g., the "Add" block of Variant A (highlighted in pink), and the "Subtract" block of Variant B (highlighted in cyan)] must be encapsulated in Variant Subsystem Blocks for the corresponding variability operators, with additional source lines to implement the variants. Figure 3 presents the textual representation of the created variant model representing both Variant A and Variant B. The sections highlighted in green are the common elements from Variant A to Variant B, the pink highlighting represents the "Add" block from Variant A, and the cyan highlighting represents the "Subtract" block from Variant B. All of the other necessary information to construct the textual representation of the variance model is available from the text of Variants A and B. Section 6 discusses variability representation in Simulink based on our proposed operators.

We believe that automating the application of variability operators will streamline the process of representing subsystem variability in Simulink Models and reduce the risk of error introduced through manual representation. In the following sections, we discuss the stages of our approach in more detail, beginning with a discussion of the automated similarity identification inferred by Simone. We then introduce two options for tagging variation in the identified sets of similar subsystems, one based on *diff* and *#ifdef* on the normalized textual representation of the models and one using graph matching algorithms to determine identical subgraphs. Finally, we discuss the final step of translating the tagged variations to Simulink Variant Subsystem Blocks, completing an end-to-end automated process for the identification and representation of subsystem variability in Simulink models.

3   Variability Identification

To determine an appropriate set of Simulink subsystem variability operators, we used the set of models in the six diverse Simulink systems of the Mathworks Simulink demonstration set as a starting point. These systems include models for a range of applications in industrial, automotive, aerospace, and other domains and are intended to demonstrate the range of ways to represent model features in these applications using Simulink. They include a range of model versions and variants for each application and represent a rich source of examples of Simulink model variation.

3.1   Simone: An Initial Approximation

To begin, we first required some indication of which subsystems in the models of each system were similar enough to be considered variants of each other. For this we used the Simone sub-model clone detector [1]. Simone is a hybrid model clone detection technique that uses a normalized text representation of graphical models to efficiently identify near-miss subsystem clones, that is, those that are similar up to a given threshold of difference (in this experiment, up to 20% different). Simone is based on the NICAD code clone detector [17], extended to handle graphical models.

To identify and categorize subsystem variations, we applied Simone to the entire set of models in each of the six Simulink demonstration systems. From each set of models, Simone generated a database of near-miss subsystem clone pairs, representing pairs of model subsystems which are largely similar but may vary up to 20% in components, connections, inputs, outputs, or other attributes. Simone automatically groups these clone pairs into "clone classes," which are sets of subsystems that are nearly similar to one another. It inherits from NICAD an efficient exemplar-based algorithm to achieve this clustering, choosing a particular cloned subsystem and then gathering all those other cloned subsystems that are similar to it within the difference threshold. By beginning with the largest exemplars, it automatically identifies the most inclusive set of variants of each cloned subsystem.

In practice even the raw clone classes resulting from this analysis can already be used by Simulink model engineers to understand variations in their systems directly from the examples in each class. In our previous work, we have integrated the results of Simone directly into the Simulink IDE using a Simulink plugin called SimNav [8, 16] that directly presents similar subsystems in the Simulink model editor (Fig. 4).

Table 1 presents the initial clustering results provided by Simone for the set of models in each of the six Simulink demonstration systems. Each subsystem in each clone class has at least 80% common elements with the others in the class. A particular element of each clone class is chosen by our framework as an exemplar, from which the others in the class are considered to be variants. We then classified the nature of these variants to empirically derive the variability operators presented in the next section.

4 Variability Operators

Using a manual inspection of the Simone results for the six systems using SimNav, and investigating the variants in each Simone-reported clone class, we identified the following types of variability in similar Simulink subsystems:

Block Variability Changes at the block level, such as added or removed blocks, or one block replaced with another. An example of this type of variability is shown in Fig. 5 (encircled in red).

Input/Output Variability Changes in the input/output ports for a specific block. These can be changes to the number of ports or the signatures of the ports. A changed signal falls into this category as well. This type of variability is shown in Fig. 6 (encircled in red).

Function Variability Changes to the contained function of a specific block or set of blocks, such as constant values, data parameters, or the entire function. This type of variability is shown in Fig. 7 (note the different functions and constants in the corresponding blocks of the two subsystems).

Layout Variability Changes to the layout information of the model elements, such as block position. This type of variability is shown in Fig. 8 (note the mirroring of parts of the model).

Subsystem Name Variability Changes to the names of similar subsystems. This type of variability is shown in Fig. 9 (encircled in red).

For each of the six systems, we determined the number of instances of each type of variability and ensured that all observed variations could be covered by the set of variability operators. The results of this categorization can be found in Table 2. The most common types of variability we observed were block variability and input/output variability, with the others occurring less frequently. There were no instances of variability that did not fall into one of these five categories.

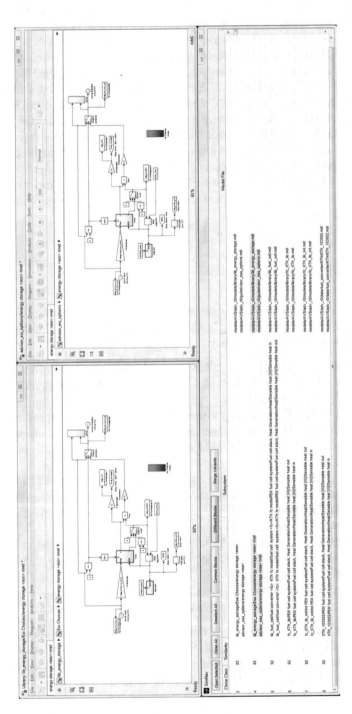

Fig. 4 The SimNav subsystem clone class exploration interface in the Simulink IDE (from [8])

Table 1 Simone clone detection results at a difference threshold of 20%

System name	# subsystems	# clone pairs	# clone classes
Automotive	357	189	24
Aerospace	188	62	15
Industrial	16	4	2
Features	935	85	25
General	146	11	7
Others	28	6	4

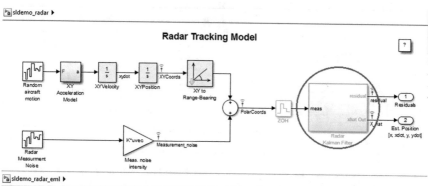

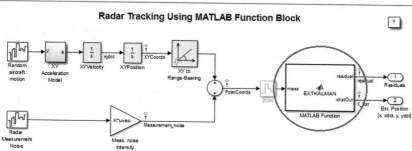

Fig. 5 Block variability

5   Tagging Subsystem Variability

To model the variability across the instances of a given subsystem pattern, we must first determine the common components of the subsystem across all of the instances in the clone class. Once we determine the commonalities between all instances, the remaining components of the subsystem represent the variations we wish to model using the variability operators.

f14 ▶

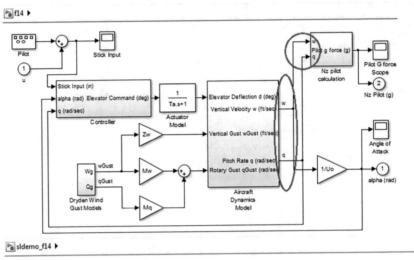

sldemo_f14 ▶

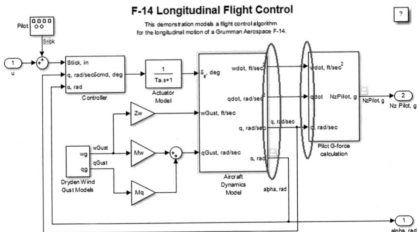

Fig. 6 Input/output variability

5.1 Tagging Using #ifdef

Since Simone computes the clone classes based on a normalized textual difference between the subsystems, one straightforward way to tag the variability between models is in their textual representation. In this section we describe how to use the unix *diff* command and source transformation to tag the variation in similar subsystems.

We begin with a single difference file generated using the command *diff -DFIRST model1.mdl model2.mdl*. This command merges the two files using C-style *#ifdef* statements to characterize the differences. For example, Fig. 10 shows two *Outport*

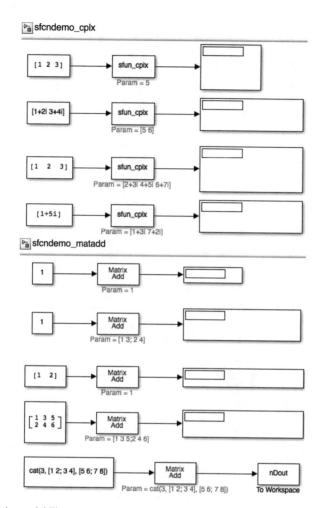

Fig. 7 Function variability

blocks from two models taken from the Simulink example model set. Figure 11 shows the difference between the two outport blocks.

In this example, the differences between the corresponding outport blocks are three different attributes: *Name*, *SID*, and *Position*. The *Name* attribute identifies the name of the block, while the *Position* attribute is part of the layout of the model. The *SID* attribute is the unique identifier given to each element of a Simulink model. We manually split the difference into three differences and append a tag to each difference condition to indicate which of the variability operators of Sect. 4 applies.

The resulting difference file is shown in Fig. 12. The first difference uses the condition *FIRST_Name* to indicate that this is a subsystem name variability. The third difference appends the variability tag *_Layout* to indicate a layout difference.

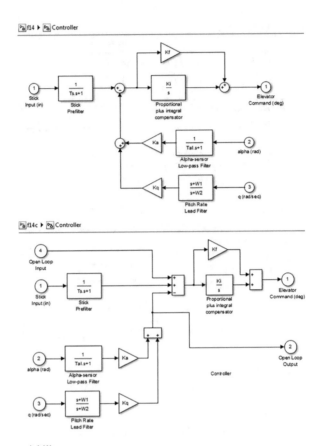

Fig. 8 Layout variability

The second difference is a difference internal to the representation of the model, and we use the variability tag _Other for this case.

In general, the variability can be categorized based solely on the entity attributes involved in the differences. In the Simulink demonstration systems, the attributes that reflect the names of entities are the *Name* and *Text* attributes. Some of the attributes associated with layout are *Position*, *Location*, *ZoomFactor*, *Points*, *FontName*, and *FontSize*. The value attributes are much more varied as they specify the options for each of the function blocks in the models. Some examples of value attributes in the demonstration models are *Value*, *DataFormat*, *TimeRange*, *YMin*, and *YMax*.

Variability in structure appears in the output of diff in two different ways. First, they may appear as differences in attributes that express the connectivity between elements of the model. These are attributes such as *DstBlock*, *SrcBlock*, *DstPort*, *SrcPort*, *PortNumber*, and *Port*. The second way these differences can appear is as additions and deletions of structural elements such as blocks and

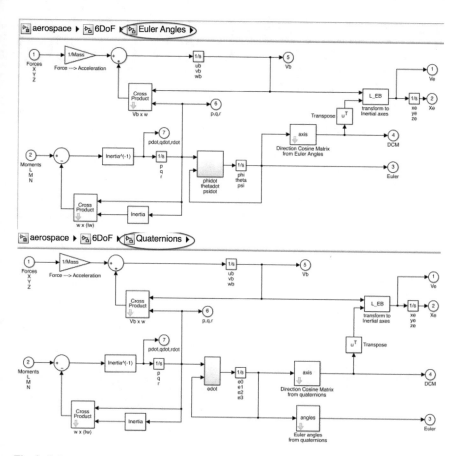

Fig. 9 Subsystem name variability

Table 2 Observed instances of the identified variability operators

System name	Block	Input/output	Function	Layout	Subsystem name
Automotive	10	6	1	3	8
Aerospace	5	17	2	4	13
Industrial	5	2	0	0	0
Features	22	22	17	2	4
General	5	3	1	1	1
Others	14	24	4	3	5

lines. Figure 13 shows the diffs resulting from the addition of a block in the one of the models, *m_SimulinkDemo Models_aerospace_0.3__2_123_so.mdl*. The *diff* algorithm triggers on the first difference, and since the element following an additional block is often another block, the first line of that block (i.e., *Block {*) often matches and appears outside of the difference at the beginning and inside

```
File m_SimulinkDemoModels_aerospace_0.3__2_171_so.mdl:
    Block {
        BlockType Outport
        Name "alpha, rad"
        SID "71"
        Position [675, 357, 705, 373]
        IconDisplay "Port number"
        InitialOutput "0"
    }

File m_SimulinkDemoModels_aerospace_0.3__2_123_so.mdl:
    Block {
        BlockType Outport
        Name "alpha (rad)"
        SID "60"
        Position [630, 235, 650, 255]
        IconDisplay "Port number"
        InitialOutput "0"
    }
```

Fig. 10 Example subsystem difference

```
Block {
    BlockType Outport
#ifndef FIRST
    Name "alpha (rad)"
    SID "60"
    Position [630, 235, 650, 255]
#else /* FIRST */
    Name "alpha, rad"
    SID "71"
    Position [675, 357, 705, 373]
#endif /* FIRST */
    IconDisplay "Port number"
    InitialOutput "0"
}
```

Fig. 11 Example subsystem difference

```
  Block {
    BlockType Outport
#ifndef FIRST_Name
    Name "alpha (rad)"
#else /* FIRST */
    Name "alpha, rad"
#endif /* FIRST */
#ifndef FIRST_Other
    SID "60"
#else /* FIRST */
    SID "71"
#endif /* FIRST */
#ifndef FIRST_Layout
    Position [630, 235, 650, 255]
#else /* FIRST */
    Position [675, 357, 705, 373]
#endif /* FIRST */
    IconDisplay "Port number"
    InitialOutput "0"
  }
```

Fig. 12 Example subsystem difference split into tags

```
  Block {
#ifndef FIRST
    BlockType Gain
    Name "Gain5"
    SID "42"
    Position [530, 222, 580, 268]
    ShowName off
    Gain "1/Uo"
  }
  Block {
#endif /* ! FIRST */
```

Fig. 13 Example additional block difference

```
#ifndef FIRST_Structure
  Block {
    BlockType Gain
    Name "Gain5"
    SID "42"
    Position [530, 222, 580, 268]
    ShowName off
    Gain "1/Uo"
  }
#endif /* ! FIRST */
  Block {
```

Fig. 14 Example transformed additional block difference

```
  Annotation {
#ifndef FIRST
    Name "F-14 Flight Control(an updated
      version of this demo is available
      by running 'sldemo_f14')"
    Position [328, 377]
#else /* FIRST */
    Name "F-14 Longitudinal Flight Control"
    Position [368, 17]
    FontName "Arial"
    FontSize 18
    FontWeight "bold"
  }
  Annotation {
  Name "This demonstration models a flight
      control algorithm for the
      longitudinal motion of a Grumman
      Aer" "ospace F-14."
    Position [367, 47]
#endif /* FIRST */
  }
```

Fig. 15 Example complex difference

the difference at the end (Fig. 13). This is easily handled by moving the difference markers slightly earlier as shown in Fig. 14. We use _Structure_ when tagging both structure attribute differences and added/deleted structural elements, also shown in Fig. 14.

The differences can interact in interesting ways, but they can always be broken down into either additional elements or changes to attributes. Thus complex differences such as the diffs shown in Fig. 15 can be separated into multiple diffs,

```
 Annotation {
#ifndef FIRST_Name
    Name "F-14 Flight Control(an updated
        version of this demo is available
        by running 'sldemo_f14')"
#else /* FIRST */
    Name "F-14 Longitudinal Flight Control"
#endif
#ifndef FIRST_Layout
    Position [328, 377]
#else /* FIRST */
    Position [368, 17]
    FontName "Arial"
    FontSize 18
    FontWeight "bold"
#endif
    }
#ifdef FIRST_Layout
  Annotation {
    Name "This demonstration models a flight
        control algorithm for the
        longitudinal motion of a Grumman
        Aer" "ospace F-14."
    Position [367, 47]
    }
#endif /* FIRST */
```

Fig. 16 Example transformed complex difference

as shown in Fig. 16. This one difference encompasses all of a name change for an annotation, layout changes for the annotation, and an additional annotation.

These transformations, which we explored manually, can be implemented in a straightforward manner as a source transformation in TXL [7], following the approach taken by Malton et al. [14]. Malton et al. used a trace-based approach to expand preprocessor statements in conventional programming languages, handling overlaps between macros as expansions in the scope of the replacement.

Figure 17 shows the beginning of three-way diff (using the command *diff3*) of three elements of a clone class from the Matlab demonstration model set. As can be seen from the figure, the differences are the names of the systems, the names of several blocks, the location of one block, and the internal identifier *SID*. All of the remaining differences in these three files are to the internal identifiers of the blocks. All of the remaining block types, values, and other attributes are identical.

Figure 18 shows the final result; we have merged the contents of the diff with the original files. The highlights show the line common to all three files. Each of the *ifdef* lines has also been annotated with the type of the difference based on the attribute within.

The remaining issue is that blocks in the model are occasionally in different orders in different model files. Simone performs a canonical sort of elements on the subsystems extracted from the models before making comparisons when identifying clone pairs. We can apply this same sorting algorithm to the original model files before performing the diff for tagging.

```
====
1:2c
    Name "m_SimulinkDemoModels_automotive_10_13_so"
2:2c
    Name "m_SimulinkDemoModels_automotive_10_17_so"
3:2c
    Name "m_SimulinkDemoModels_automotive_10_33_so"
====2
1:4c
3:4c
        Name "validate_driver"
2:4c
        Name "validate_passenger"
====2
1:8c
3:8c
            Name "validate_driver"
2:8c
            Name "validate_passenger"
====1
1:12c
            Location [65, 299, 664, 654]
2:12c
3:12c
            Location [69, 319, 668, 674]
====2
1:15c
3:15c
            Name "validate_driver"
2:15c
            Name "validate_passenger"
====
1:29c
            SID "74"
2:29c
            SID "110"
3:29c
            SID "83"
====
...
```

Fig. 17 Example three-way diff

5.2   Tagging via Graph Algorithms

An alternate approach to discover and tag variability in Simulink model clones is to treat a Simulink system as a directed graph and apply subgraph matching techniques. In this approach, Simulink blocks represent graph nodes, and the connections between blocks represent directed graph edges. This graph-based abstraction makes it immune to changes in layout, which is beneficial for finding a set of common blocks between clones but does not help to discover layout-based variability.

The first step in this approach is to discover a set of common blocks between the system clones. Our current algorithm supports an arbitrary number of clones, but we describe it with two clones for simplicity. The goal is to map a subset of blocks in clone 1 to a subset of blocks in clone 2. This mapping is accomplished

```
Model {
#ifdef FIRST_Name
      Name "m_SimulinkDemoModels_automotive_10_13_so"
#elif SECOND_Name
      Name "m_SimulinkDemoModels_automotive_10_17_so"
#else
      Name "m_SimulinkDemoModels_automotive_10_33_so"
#endif
   System {
#ifdef FIRST_Name || THIRD_Name
         Name "validate_driver"
#elif
         Name "validate_passenger"
#endif
         Location [428, 407, 944, 880]
         Block {
            BlockType SubSystem
#ifdef FIRST_Name || THIRD_Name
            Name "validate_driver"
#elif
            Name "validate_passenger"
#endif
            Ports [1
            Position [115, 123, 300, 177]
            System {
#ifdef THIRD_Layout
               Location [65, 299, 664, 654]
#else
               Location [69, 319, 668, 674]
#endif
...
```

Fig. 18 Example transformed three way diff

by first mapping a single block from clone 1 to clone 2 known as the root and then recursively matching each roots' neighbors as well as possible.

This algorithm incorporates two types of block matches: strong match (block type and name must both match) and weak match (block type must match but name can differ). The root blocks are chosen by selecting the strongly matched block pair (one from each clone) with the most connections. Since only one connected subgraph is produced from this algorithm, more connections on the root block increases the chances of a larger resulting subgraph. As block matching grows outward from the root blocks, strong matches are prioritized over weak matches to help disambiguate potential match candidates. It is possible for strong matches to exist in the clones that are not found by this algorithm, for example, if they are separated from the root block by an unmatchable region.

The end result is a connected subgraph G_1 from clone 1 and a connected subgraph G_2 from clone 2, where each node in G_1 is mapped to a corresponding node in G_2. These subgraphs represent the set of common blocks between two clones, as shown in Fig. 19.

Once the common set of blocks is established, the remaining blocks in each clone represent some form of variation. In a merged model file, such as that shown in Sect. 6.7, the common blocks and their connections remain untagged, but the other blocks can be tagged with their clone variant. This can be accomplished by using

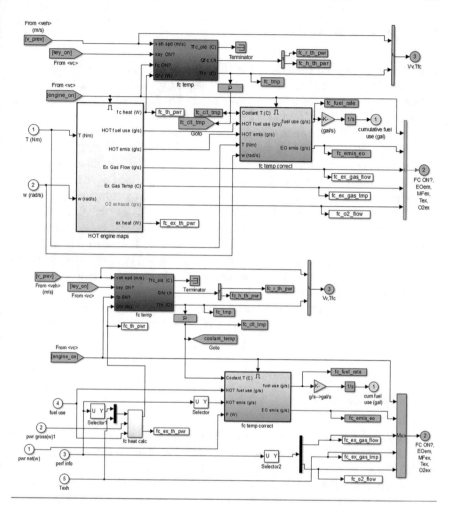

Fig. 19 Common blocks computed by the graph matching algorithm. The root block (red) is determined, then neighboring blocks are recursively included first by strong match (blue) and then by weak match (yellow)

the #ifdef approach from Sect. 5.1 or by simply adding a new Simulink parameter such as "Variant clone1" to each appropriate variant block. When extending this algorithm to find variation in three or more clones, a tag will specify each clone where the block exists.

6   Representing Variability

Once the variability has been tagged in all instances in a clone class, our goal is to produce a single subsystem model capable of serving for all the instance subsystems of that clone class. To do this, we make use of the Simulink Variant Subsystem Block, a built-in feature of Simulink designed to offer developers the choice between any number of different options for a particular subsystem.

A Variant Subsystem Block can contain any number of different subsystems, as long as they all have the same number of inports. The contained subsystems represent alternatives for the variant subsystem, and only one of them may be active at any given time. The active subsystem is determined by a logical expression, often making use of a Simulink mode variable. While on the face of it the Variant Subsystem Block seems limited in its expressiveness, being restricted to replacement of entire subsystems, in our work we have leveraged this feature to represent not the subsystem alternatives of the model itself, but rather our variability operators as Variant Subsystem Blocks, allowing us to expose the individual points of variation explicitly in the Simulink environment.

The following subsections outline how we use the Variant Subsystem Block to represent each of our variability operators. We refer back to the example figures in Sect. 4 as examples of each type of variability.

6.1   Block Variability

Block variability is perhaps the most intuitive operator to model using the Variant Subsystem Block, especially in the instance of a block being replaced by another similar block. To model this using a Variant Subsystem Block, we simply place each of the alternative blocks in its own subsystem and place all subsystems in a Variant Subsystem Block. In instances where a block, or group of blocks, is added (or removed, since there is no concept of directionality associated with the block variability operators), the variability is modeled by having the added blocks contained within a subsystem and placed in the Variant Subsystem Block. To represent those instance(s) without those blocks, an empty subsystem, where the inports connect directly to the outports (or sometimes a terminator), is placed in the Variant Subsystem Block.

To illustrate this operator, recall the example presented in Fig. 5. This operator is represented on the main level of the newly created *variability1* model by inserting a Variant Subsystem Block in its place, which then contains two subsystems, one for each of the original options. This can be seen in Fig. 20, which shows the top-level model with the Variant Subsystem Block, as well as the two options inside of it (outlined in red).

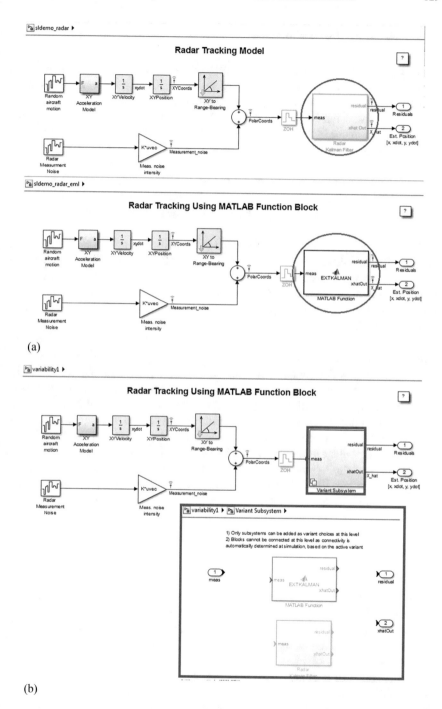

(a)

(b)

Fig. 20 Representing block variability. (**a**) Clone pair with block variation. (**b**) Corresponding variability model

6.2   Input/Output Variability

Modeling input/output variability using Variant Subsystem Blocks is somewhat less intuitive. In order to represent this type of variability, the top-level subsystem must contain the greatest number of inputs and outputs across all instances. Extra inputs are then dealt with inside the options of the Variant Subsystem, typically by sending them to a terminator in the variants where they are not used. Instances where there are extra outputs are even more difficult to represent, as anything that follows is affected, and will need to be represented inside another Variant Subsystem Block. We can consider the outputs as inputs to the conceptual block "remainder of the model." The extra outputs are then sent to terminators in the instances that do not contain them and are used as they normally would be in the instances that do contain them.

To illustrate representation of this operator, recall the example of Fig. 6, and more specifically the additional outputs from the Aircraft Dynamics Model subsystem in the *sldemo_f14* model, and thus the additional input to the block to the right, which also is in an instance of block variability, as our example. At the top level, there must be a Variant Subsystem Block (Variant Subsystem 1—outlined in red) to model the different options for the Aircraft Dynamics Model, which will have four outputs, as this is the maximum number from all of the options. To handle the extra outputs, a second Variant Subsystem Block (Variant Subsystem 2—outlined in blue) is used. Variant Subsystem 2 also handles the changed subsystem block by offering two options—note one option uses three inputs, and the other uses two. In the instance where only two inputs are required, it is contained in a Container Subsystem (outlined in green), and the third input is sent to a terminator. This can all be seen Fig. 21, which shows the top-level model (top), as well as the contents of each of the Subsystem Variant Blocks [Variant Subsystem 1 (bottom left) and Variant Subsystem 2 (bottom center)] and the contents of the created Container Subsystem (bottom right). Note that the *variability2* model only accounts for the discussed input/output variability.

6.3   Function Variability

While function variability is its own type of operator, it can be modeled in the same manner as block variability. Consider that the two blocks with different functions can be thought of as different blocks entirely. Just as with block variability, we represent this with a Variant Subsystem Block, with an option for each of the original blocks.

To illustrate this operator, recall the example from Fig. 7. Each block is replaced with a Variant Subsystem Block, thus allowing a choice between the two options. Each Variant Subsystem Block can use its own mode variable, thus allowing combinations of options, or a common mode variable, thus only representing the

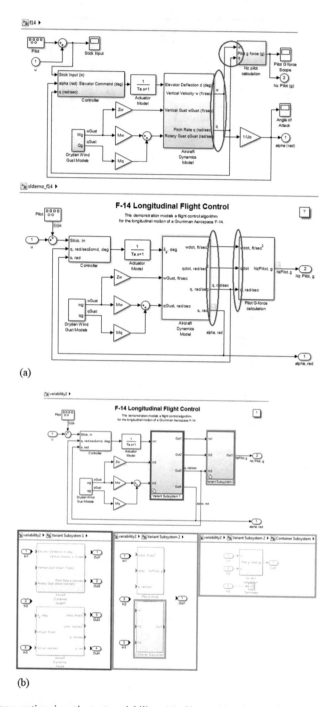

(a)

(b)

Fig. 21 Representing input/output variability. (**a**) Clone pair with input/output variation. (**b**) Corresponding variability model

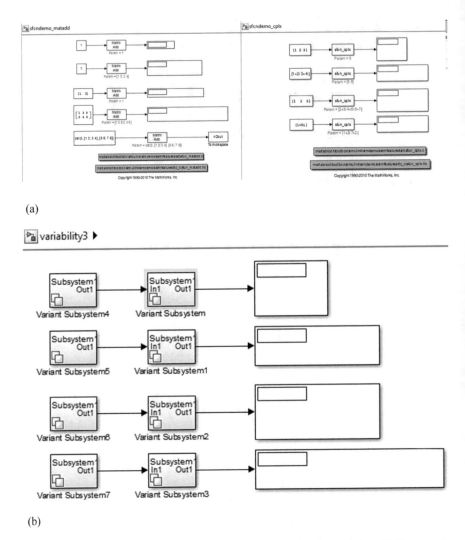

(a)

(b)

Fig. 22 Representing function variability. (**a**) Clone pair with function variation. (**b**) Corresponding variability model

two original observed variants. Figure 22 shows the top level of the *variability3* model, with the eight blocks replaced with Variant Subsystem Blocks.

6.4 Layout Variability

Because the layout of the model has no effect on its behavior, we have chosen to not represent changes in layout in the resulting variability model. Representing the other types of variability in a class that also has layout variability, one layout instance is arbitrarily chosen to represent all models in that class, regardless of their initial layout.

6.5 Subsystem Name Variability

If the contents of a subsystem have not changed, and only the name has changed, there is no behavioral change to the model; however we still wish to represent this variability as it does have meaning to the developers. This representation would be handled in the exact same way that block variability would be; we can just consider the two differently named blocks different versions of a block and use them as options in the Variant Subsystem. This would also account for instances where the name has changed, and the actual contents vary slightly, as is the case with our example in Fig. 9. Since the implementation for block variability has already been demonstrated, there is no need to explicitly illustrate it here.

6.6 Combinations of Operators

Through observation of the studied systems, it is evident that each subsystem pattern may require more than one type of variability, and as such, more than one variability operator may need to be applied. Rather than defining combinations of operators as their own unique operator, we have determined that applying any individual operators in succession is sufficient in representing the variability. For example, in an instance where there exists both function variability and block variability, each is handled individually following their respective process.

6.7 Creating Variability Models Directly in Simulink

Currently the process of creating variability models has been automated directly within SimNav for pairs of models. Given two models, the similar blocks and different blocks can be tagged (using an extension of the algorithm described above), and the different blocks are then merged using the Subsystem Variant Block.

In Sect. 5.2 we discuss an algorithm to find the common blocks among models in a class of near-miss clones. We use this algorithm to automatically construct a variability model representing the clone class. For this procedure, consider a clone class with n clones $(C_1, C_2, C_3, \ldots, C_n)$ where each clone contains a set of blocks $(C_i = \{block_1, block_2, block_3, \ldots\})$.

1. Compute the blocks in common between all clones in the clone class using the algorithm from Sect. 5.2:

$$C_{common} = \bigcap_{i=1}^{n} C_i$$

2. Take the complement of the common blocks to find the blocks that vary between clones, one variant set per clone: $V_i = C_i - C_{common}$
3. For each variant set V_i, place its blocks into a subsystem S_i
4. Create a variant system containing subsystems $S_1, S_2, \ldots, S_n$.
5. The final variability model contains the common blocks C_{common} and the variant subsystem from the previous step.

The connections between blocks are kept wherever possible (i.e., within the common blocks and within each individual set of variant blocks). Connections that traverse the boundary between a variant set and the common set are severed and replaced by input/output ports in the resulting subsystems. The blocks within each subsystem are then connected to these ports such that the meaning of the original clone is restored when its corresponding Variant Subsystem Block is activated. Figure 23 shows a clone pair both before and after the variability procedure has been applied.

Due to the scalability limitations of graph algorithms, we also plan to continue exploring other possibilities for automating the creation of variability models. We plan on extending the work from Sect. 5.1 to allow us to directly manipulate the textual representation of subsystem variants into a single Simulink model capable of expressing variability in a similar manner to that described in this section.

7 Related Work

Model variability is a richly researched area. There have been a number of techniques developed for many different domains [9]. Typically, variability is looked at from a management perspective [5], in that it is an essential property of projects that needs to managed. There have also been steps taken to semiautomatically extract variability in code-based projects [13] and model-based projects [20] in order to manage it. The difference between our work and the latter is we use model clone detection, via Simone, as the starting point for finding variability among, and grouping into classes/patterns, sets of models, whereas they compare systems recursively by mapping similar components of the same type based on different criteria. like name similarity, number of identical parameter values, connections, and more, in order to get a weighted similarity sum between zero and one. Similar to one of our two presented approaches for tagging variability, they identify variation points using a graph-based approach. It is our contention that using classes, patterns, and clustering provides a better basis for representing variability and more useful

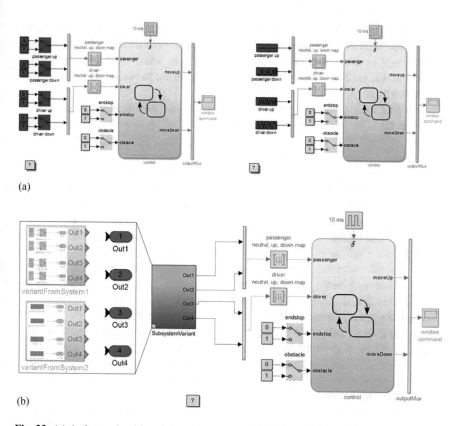

Fig. 23 **(a)** A clone pair with variation, i.e., common blocks in yellow and different blocks in red and **(b)** the corresponding variability model

information than basic name and property similarity. In addition, graph-based approaches for matching typically do not perform well on larger model sets due to subgraph isomorphism [6]. Our approach avoids this issue because our model clone detection algorithm uses the textual representations of the models [1]. While one of our proposed tagging approaches uses graph algorithms, it is applied only to sets of single models/graphs that have already been identified as similar, so the subgraph isomorphism complexity will not be an issue on this small scale.

Albeit a relatively new subarea, there is some existing work on variability in Simulink models. Weiland and Manhart [26] argue the necessity for modeling variability in Simulink. They introduce a classification of possible concepts that can be employed in order to represent Simulink variability: model elements for model adaptation, conditional model elements, and model elements for data variability. In this chapter, we realize the first of these concepts proposed by them in order to explicate variability among Simulink clones. While Weiland and Manhart note that using the variant subsystem block does not perform well in regard to their binding time, we have found, from speaking to our industrial partners and engineers, that this solution is most preferable for them. They want the variability to be demonstrated and usable within their native Simulink environments. In addition, we have yet to witness any binding time limitations or concerns in using the variant subsystem block, but this is something we will continue to monitor as we employ this solution in our methods.

One Simulink-specific approach to encoding variability is accomplished by Haber et al. [11], who note that functional modeling approaches for representing Simulink variability are often complex and do not scale well to larger systems. Thus, they propose Delta Simulink, which is a first-class language that includes single-step operations like add, remove, modify, and replace. While it is an operational approach, it is also graphical in that users can illustrate their deltas in a separate, non-Simulink, viewer. Because their approach is operational, they note that "some modification operations have to be split up into several deltas to be applied in a sequence." Our representation avoids this in that it is declarative. The sets of models belonging to a related cluster indicate, all at once, exactly how the models differ. This declarative representation is understandably simpler, while still being precise, and has been received well thus far by our industrial partners. Another concrete deficiency is that Delta Simulink is another modeling language, and the tools they develop are external from Simulink. During our conversations with the industrial engineers, it was made clear from the beginning that a key priority was to have an approach that works within Simulink and can be as least disruptive as possible to their processes. Seeing as Delta Simulink is a new language, albeit an extension to Simulink, and exists outside the Simulink editor environment, their solution was not ideal for our purposes.

Steiner et al. [21] manage Simulink variability by using and contrasting Pure::variants, which has a Simulink connector that uses "point of change" information, and Hephaestus, which has a graphical interface that allows developers to select system elements to be used to generate specific product line instances. Their approach uses conditional model elements in order to represent Simulink variability, which, as we discussed previously in this chapter, would not be ideal for Simulink clone variants. In addition, the learning curve for using their technique is quite high as engineers would have to familiarize themselves with Pure::variants and Hephaestus. It uses the Hephaestus graphical interface, which is external to the Simulink editor native environment and is another reason this approach was not well suited for us nor our partners.

Managing clones in product lines involves cases where systems using product lines or feature models have exact duplicates or similar segments of a related product line. Rubin et al. [18, 19] provide a framework for handling such systems that includes abstract operators that allow engineers to reason and manage clones detected in these systems. Their work is focused on the product line, higher level of abstraction, level, while our work is intended explicitly for Simulink models. In addition, our approach is declarative, while theirs is operational.

While variability involves looking at how systems differ at a somewhat larger scale, model mutations focus on stepwise changes to a model in order to perform various types of analysis. Recently, we proposed and validated a taxonomy of

Simulink model mutations for the purposes of injecting various types of Simulink model clones [24]. There is also work on Simulink model mutations that describe mutation instances that explicitly try to mutate a model's run-time properties [4, 12, 27]. While this mutation analysis work was helpful in guiding how we viewed Simulink variability, we essentially were focused on a higher and more feature-oriented level.

Basit and Dajsuren [3] use a constraint language in order to model mutations among Simulink clones with the purpose of allowing clone management that is entirely separate from the models. Their work is concerned with a different, but related, task. Their work looks at the model clones as simply clones, while we focus on the (sub)system level in order to identify candidates for (sub)system variants. Our approach is more geared toward tool support and working directly in the Simulink environment to assist engineers. In addition, they do not have a tool at this point, and engineers would have to use their constraint language.

Calculating and representing variability in models is analogous to calculation and representation phases of model comparison [23]. The first phase, calculation, involves discovering what is the same and what is different, while the second phase of model comparison, representation, addresses the form that the differences and similarities among models take. There are many different ways of achieving model comparison calculation [22], but nothing was specifically suited to identifying variability among a set of Simulink models as outlined in our framework. As such, we presented both a Unix diff and graph-based approach in this chapter. Model comparison representation can be realized in an operational fashion, such as through the use of edit scripts, or in a more declarative fashion like those that represent the differences in a model-based or abstract syntax-like form. The representation we present in this chapter falls more in the declarative category as we are representing the variability in model form by having multiple model implementation options linked to a specific variation point.

For our graph-based variation analysis, where we explicate the similarities and differences within a set of already identified "similar" models, we based our approach on what Deissenboeck et al. [10] did in their ConQAT graph-based model clone algorithm. The difference here is that we are using a graph-based approach on only the microlevel in order to compare and contrast a small set of models. The algorithm ConQAT uses cannot detect near-miss clones, while our model clone detection approach can [1]. Using an approach that does not identify near-miss clones from the graph-based variation analysis perspective is sufficient, as we need only to identify variation points.

8   Conclusion

Based on the six example systems of the Simulink demonstration set, we have empirically derived five variability operators for Simulink models. These five operators encompass all of the different types of variability observed from the initial analysis of similar subsystem variance provided by Simone, a hybrid sub-model clone detector. We have presented two methods for tagging variability across a set of similar Simulink models, one based on text differencing and one on graph matching. Both of these processes have been automated for pairs of similar subsystems, and we are currently extending them to handle N-way differencing. We have shown how each of the five variability operators can be represented directly in the Simulink environment through a novel use of the Variant Subsystem Block by extending our SimNav tool to support this feature. While the variability representation using the graph matching approach showed good results in representing variability for small subsystems, we are still experimenting our tool for larger subsystems and expect to face some scalability issues related to our graph matching algorithm. For that reason, we continue to explore the alternative text-based implementation carrying on our tagging approach using #ifdef. This alternative approach has the advantages that it automatically tags variations based with the types of variability operators and is likely to scale better than the graph-based approach.

Acknowledgements This work is supported in part by the Natural Sciences and Engineering Research Council (NSERC) of Canada as part of the NECSIS Automotive Partnership with General Motors, IBM Canada, and Malina Software Corp. and by an Ontario Research Fund Research Excellence grant.

References

1. Alalfi MH, Cordy JR, Dean TR, Stephan M, Stevenson A (2012) Models are code too: near-miss clone detection for Simulink models. In: ICSM'12 - 28th international conference on software maintenance, pp 295–304
2. Alalfi MH, Rapos EJ, Stevenson A, Stephan M, Dean TR, Cordy JR (2014) Semi-automatic identification and representation of subsystem variability in Simulink models. In: ICSME'14 - 30th international conference on software maintenance and evolution, pp 486–490
3. Basit HA, Dajsuren Y (2014) Handling clone mutations in Simulink models with VCL. In: IWSC'14 - 8th international workshop on software clones, pp 1–8
4. Binh NT, et al (2012) Mutation operators for Simulink models. In: KSE'12 - 4th international conference on knowledge and systems engineering, pp 54–59
5. Capilla R, Bosch J, Kang KC (2013) Systems and software variability management. Springer, Berlin
6. Cook SA (1971) The complexity of theorem-proving procedures. In: 3rd ACM symposium on the theory of computing. ACM, New York, pp 151–158
7. Cordy JR (2006) The TXL source transformation language. Sci Comput Program 61(3):190–210
8. Cordy JR (2013) Submodel pattern extraction for Simulink models. In: SPLC'13 - 17th international conference on software product lines, pp 7–10
9. Czarnecki K, Grunbacher P, Rabiser R, Schmid K, Wąsowski A (2012) Cool features and tough decisions: a comparison of variability modeling approaches. In: VaMoS'12 - 6th international workshop on variability modelling of software-intensive systems, pp 173–182
10. Deissenboeck F, Hummel B, Jürgens E, Schätz B, Wagner S, Girard JF, Teuchert S (2008) Clone detection in automotive model-based development. In: Proceedings of the 30th international conference on software engineering. ACM, New York, pp 603–612

11. Haber A, Kolassa C, Manhart P, Nazari PMS, Rumpe B, Schaefer I (2013) First-class variability modeling in Matlab/Simulink. In: VaMoS'13 - 7th international workshop on variability modelling of software-intensive systems, pp 11–18
12. He N, Rümmer P, Kroening D (2011) Test-case generation for embedded Simulink via formal concept analysis. In: DAC'11 - 48th design automation conference, pp 224–229
13. Kastner C, Dreiling A, Ostermann K (2013) Variability mining: consistent semiautomatic detection of product-line features. IEEE Trans Softw Eng 40(1):67–82
14. Malton A, Schneider K, Cordy J, Dean T, Cousineau D, Reynolds J (2001) Processing software source text in automated design recovery and transformation. In: IWPC'01 - 9th international workshop on program comprehension, pp 127–134. https://doi.org/10.1109/WPC.2001.921724
15. Object Management Group (2009) Variability modeling. http://www.omgwiki.org/vari-ability/doku.php?id=introduction_to_variability_modeling
16. Rapos EJ, Stevenson A, Alalfi MH, Cordy JR (2015) SimNav: Simulink navigation of model clone classes. In: 2015 IEEE 15th international working conference on source code analysis and manipulation (SCAM), pp 241–246. https://doi.org/10.1109/SCAM.2015.7335420
17. Roy CK, Cordy JR (2008) NICAD: accurate detection of near-miss intentional clones using flexible pretty-printing and code normalization. In: ICPC'08, 16th IEEE international conference on program comprehension, pp 172–181
18. Rubin J, Chechik M (2013) A framework for managing cloned product variants. In: ICSE'13 - 35th international conference on software engineering, pp 1233–1236
19. Rubin J, Czarnecki K, Chechik M (2013) Managing cloned variants: a framework and experience. In: SPLC'13 - 17th international conference on software product lines, pp 101–110
20. Ryssel U, Ploennigs J, Kabitzsch K (2010) Automatic variation-point identification in function-block-based models. In: GPCE'10 - 9th international conference on generative programming and component engineering. ACM, New York, pp 23–32
21. Steiner E, Masiero P, Bonifácio R (2013) Managing SPL variabilities in UAV Simulink models with pure: variants and hephaestus. CLEI Electron J 16(1):1–7
22. Stephan M, Cordy JR (2012) A survey of methods and applications of model comparison. Tech. Rep. 2011–582, Queen's University, revision 3
23. Stephan M, Cordy JR (2013) A survey of model comparison approaches and applications. In: International conference on model-driven engineering and software development (Modelsward), SCITEPRESS, pp 265–277
24. Stephan M, Alalfi MH, Cordy JR (2014) Towards a taxonomy for Simulink model mutations. In: International conference on software testing, verification and validation workshops (ICSTVVW), pp 206–215. https://doi.org/10.1109/ICSTW.2014.17
25. The Mathworks Inc (2014) Simulink version 8. http://www.mathworks.com/products/simulink/
26. Weiland J, Manhart P (2014) A classification of modeling variability in Simulink. In: VaMoS'14 - 8th international workshop on variability modelling of software-intensive systems, pp 1–7
27. Zhan Y, Clark J (2005) Search-based mutation testing for Simulink models. In: GECCO'05 - genetic and evolutionary computation conference, pp 1061–1068

Defining Architecture Framework
for Automotive Systems

Yanja Dajsuren

Abstract Although architecture frameworks have not been standardized in the automotive industry, different types of architecture viewpoints and views have been introduced recently as part of automotive architecture frameworks. In this chapter, we first present a literature review which has been carried out to discover the existing architecture frameworks and architecture description languages for the automotive industry as well as their benefits and gaps. We propose an architecture framework for automotive systems (AFAS) based on the extracted viewpoints from existing automotive architecture description mechanisms.

1 Introduction

An architecture description language (ADL) is considered a viable solution to manage multidisciplinary engineering information in an effective way [7, 24, 34]. According to the ISO 42010 international standard [18], an ADL provides one or more *model kinds* (data flow diagrams, class diagrams, state diagrams, etc.) as a means to frame some *concerns* for its *stakeholders*. Model kinds can be organized into *architecture views*, which are governed by *architecture viewpoints*.

Recognizing the importance of ADLs, automotive companies have been actively involved in their development over the last decade. These include BMW who have been involved in developing AML [6, 31], as well as Volvo, Fiat, and VW/Carmeq who have been involved in developing the EAST-ADL (Embedded Architectures and Software Technologies-Architecture Description Language) [9] and TADL [39]. EAST-ADL is being extended to model the fully electric vehicle in the scope of the ICT MAENAD project, where many automotive manufacturers and suppliers are participating [23]. Besides the automotive ADLs, SysML [27] and

Y. Dajsuren (✉)
Department of Mathematics and Computer Science, Eindhoven University of Technology, Eindhoven, The Netherlands
e-mail: y.dajsuren@tue.nl

© Springer Nature Switzerland AG 2019
Y. Dajsuren, M. van den Brand (eds.), *Automotive Systems and Software Engineering*, https://doi.org/10.1007/978-3-030-12157-0_7

MARTE [26] are also attracting considerable attention of automotive companies [1, 2, 30].

According to the ISO 42010 international standard [18], in addition to an ADL, an architecture framework is another key mechanism used to describe architectures. An architecture framework provides conventions, principles, and practices for the description of architectures within a specific domain and/or community of stakeholders [18]. The benefits of existing architecture frameworks such as Kruchten's 4+1 view model [20], Ministry of Defense Architecture Framework (MODAF) [5], The Open Group Architecture Framework (TOGAF) [38], and ISO Reference Model for Open Distributed Processing (RM-ODP) [17] drive the creation of architecture frameworks for other industries.

Having a standardized architectural foundation and specifically automotive-specific architecture frameworks is very important for the automotive industry. The key elements of this proposed architecture framework was first introduced in the scope of the automotive architecture framework (AAF) [8]. The AAF aimed to describe the entire vehicle system across all functional and engineering domains and drive the thought process within the automotive industry [8]. Only in recent years, automotive companies have started to take initiative in defining an architecture framework for automotive systems, for example, architecture design framework (ADF) by Renault [14].

Automotive embedded systems are categorized into vehicle-centric functional domains (including powertrain control, chassis control, and active/passive safety systems) and passenger-centric functional domains [covering multimedia/telematics, body/comfort, and human machine interface (HMI)] [24]. Each functional domain needs to tackle different system concerns. For example, the powertrain control enables the longitudinal propulsion of the vehicle, and body domain supports the functioning of the airbag, wiper, and lighting and other functions for the vehicle users. However, all the integrated functionalities must not jeopardize the key vehicle requirements of safety and efficiency.

The automotive industry is vertically organized [7], which facilitates independent development of vehicle parts. An automobile manufacturer (called an "original equipment manufacturer", or OEM) creates the functional architecture and distributes the development of the functional components to the suppliers, who implement and deliver the software models and/or hardware [7]. Software models for each functional component or subsystem can be developed in different ADLs or programming languages, which may make the integration process at the OEM more cumbersome. This process requires common architecture frameworks between OEMs and suppliers or at least better formalization of architecture views and consistency between them.

Therefore, there needs to be a common definition of an ADL and architecture framework, and these should be applicable for all functional domains. However, architecture description elements of an automotive-related ADL and architecture frameworks (i.e., architecture viewpoints, views, and correspondences) are not systematically defined. Figure 1 shows the timeline of the automotive architecture description mechanisms.

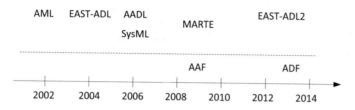

Fig. 1 Timeline of the automotive architecture description mechanisms

This chapter extracts architecture elements (viewpoints, views) from automotive ADLs, compares the extracted elements with the existing automotive architecture frameworks, and proposes an architecture framework for automotive systems (AFAS) with a coherent set of architecture views.

1.1 Chapter Outline

Section 2 presents the automotive architecture frameworks and describes the architecture viewpoints defined in the automotive frameworks. Section 3 introduces automotive-related ADLs and presents the extracted architecture viewpoints from the ADLs. Section 4 presents an architecture framework for automotive systems (AFAS), which contains architecture viewpoints and views consistent with the automotive AFs and ADLs. Section 5 summarizes the chapter.

2 Automotive AFs and Viewpoints

An architecture framework establishes a common practice for creating, interpreting, analyzing, and using architecture descriptions within a particular domain of application or stakeholder community [18]. While an architecture description language (ADL) is used to describe or represent an architecture, an architecture framework enables the efficient use of an ADL for a particular domain. Therefore, a standard architecture framework in the automotive industry can enable an efficient architecture description for system stakeholders. In the ISO 42010 international standard, a conceptual model of an architecture framework as shown in Fig. 2 is almost identical to the conceptual model of an ADL as shown in Fig. 3. The differences are as follows:

- An architecture framework should provide at least a single *architecture viewpoint*, which is used to organize the *model kinds*.
- An ADL should define at least a single *model kind* without necessarily providing a *architecture viewpoint*.

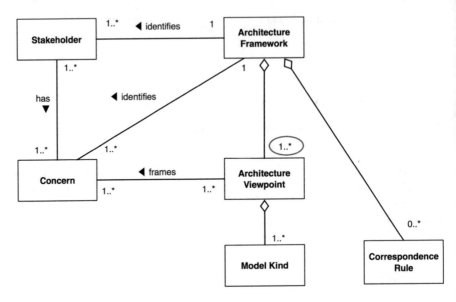

Fig. 2 A conceptual model of an architecture framework [18]

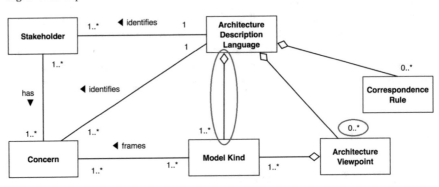

Fig. 3 A conceptual model of an architecture description language [18]

In this section, we present the automotive architecture frameworks, extract common architecture viewpoints, and summarize other architecture viewpoints that exist only in one of the architecture frameworks.

2.1 Automotive Architecture Frameworks

Automotive architecture framework (AAF) [8] is the first architecture framework for the automotive industry to pave the way for a standardized architecture description. The AAF was defined to describe the entire vehicle system across all functional and engineering domains. Since the AAF conforms to the ISO 42010 international standard [8], a set of viewpoints and views are explicitly defined. The AAF proposes two sets of architecture viewpoints: mandatory or general viewpoints and optional viewpoints. Mandatory viewpoints and their respective views include *functional viewpoint, technical viewpoint, information viewpoint, driver/vehicle operations viewpoint*, and *value net viewpoint*. Optional viewpoints suggested by the AAF are *safety, security, quality and RAS* (reliability, availability, serviceability), *energy, cost, NVH* (noise, vibration, harshness), and *weight*. The general viewpoints are intended to be closer to the already proven frameworks in other manufacturing industries, for example, RASDS [37] and RM-ODP [17]. Since the introduction of the concepts in the first draft of the AAF, further research is needed to identify automotive-specific architectural elements.

Architectural design framework (ADF) [14] is developed by an OEM to support the construction of an architecture framework for the automotive industry. The ADF includes *operational, functional, constructional*, and *requirements* viewpoints. Although the AAF and ADF are constructed to provide the basis for the architecture framework for the automotive industry, architecture viewpoints and views are extracted from architecture frameworks from other industries. Furthermore, in these frameworks, the definition of architectural elements including architecture viewpoints, views, and correspondences have not been addressed consistently with automotive ADLs.

2.2 Extracting Viewpoints from Automotive AFs

An architecture framework may include one or more architecture viewpoints, which consist of a set of model kinds [18]. We discussed above the architecture viewpoints and views of AAF and ADF frameworks. The viewpoints are described in a similar way to the viewpoint catalog [32]. Below we extract the common viewpoints of AAF and ADF according to the following template:

- Definition: Definition of the viewpoint is presented.
- Stakeholders: Although the stakeholders are not explicitly identified for the viewpoints in the AAF and ADF, we list the stakeholders.
- Concerns: Stakeholder concerns are defined.
- Views: The views governed by the viewpoints are presented.
- Model kinds: The model kinds used in the viewpoint are presented.

Functional Viewpoint Table 1 summarizes the functional viewpoint, which is defined both in the AAF and ADF frameworks. A *function* realizes a feature in a set of interacting and interdependent software and/or hardware components.

The functional viewpoint extracted from automotive ADLs as discussed in Sect. 3.2 generally matches the description of the functional viewpoint in AAF and ADF frameworks.

Table 1 Functional viewpoint definition extracted from AAF and ADF frameworks

Functional viewpoint	
Definition	It describes the vehicle functions and their interactions
Stakeholders	AAF: OEMs, suppliers, tool vendors, and research institutes
	ADF: Undefined
Concerns	Functional composition and interfaces
Architecture views	AAF: Functional view
	ADF: Functional breakdown structure view, functional architecture view, allocation on functions view
Model kinds	AAF: Functional architecture (Functional composition of a vehicle, its functional entities, interfaces, interactions, interdependencies, behavior, and constraints)
	ADF: AD, BDD, IBD for the functional breakdown structure view; AD, BDD, IBD for the functional architecture; *allocation* concept for the requirements allocation on functions views
Correspondence rules	AAF: Correspondences to technical and optional viewpoints, for example, energy
	ADF: Refinement and conformance correspondence to the operational viewpoint

In AAF, the *functional viewpoint* describes vehicles in terms of vehicle functions and their logical interactions. The AAF functional viewpoint governs a *functional view*, which describes the functional composition of a vehicle, its functional entities, interfaces, interactions, interdependencies, behavior, and constraints [8]. Although AAF does not specify a particular model kind for the functional viewpoint, it defines the *functional architecture*. The functional architecture describes the system from the black-box perspective by describing the system's functionality that is presented to the outside world [8]. The stakeholders of the AAF are defined as OEMs, suppliers, tool vendors, and research institutes. Stakeholder concerns are not explicitly defined for the AAF functional viewpoint. Based on the description of the functional viewpoint, we defined them as functional composition and interfaces. The functional viewpoint corresponds to the technical and optional viewpoints.

In ADF, the *functional viewpoint* supports three main views: *functional breakdown structure*, *functional architecture*, and *allocation on functions* [14]. ADF defines SysML model kinds for each functional views. SysML activity diagram (AD), block definition diagram (BDD), and internal block diagram (IBD) are defined for the functional breakdown structure view. In the activity diagram, the *system functions* are defined by regrouping or refining *activities* (actions) identified in the operational scenario views and allocating them to SysML *blocks*. In the BDD and IBD, *ports* and *connectors* conform to a flow type (e.g., energy, information) of external interfaces and object flows specified in ADs [14]. Although it is not explicitly mentioned in the ADF, an *allocation* concept is plausibly used for allocating requirements to functions (blocks). Stakeholders, their concerns, and correspondence rules are not explicitly determined in the ADF. We expect the same stakeholders and concerns for the AAF are applicable to the ADF. Regarding correspondence, the functional viewpoint conforms or refines the operational viewpoint.

Technical/Constructional Viewpoint Table 2 presents the technical/constructional viewpoint, which looks at a vehicle in terms of its physical components, their relationships, and constraints. AAF refers to it as a technical viewpoint and ADF refers to it as a constructional viewpoint.

In AAF, the *technical viewpoint* addresses a vehicle from the perspective of its physical components. This includes electronic control units (ECUs), their geometry, and composition within superordinate geometric structures as well as their relationships. It also includes the vehicle's behavior such as physical aspects like thermodynamics, acoustics, vibrations, mechanical deformation, as well as dependencies and constraints [8].

The AAF technical viewpoint governs a *technical view*, which consists of *runtime model* view, *hardware topology* view, and *allocation* view. As in the AAF functional view, the technical view does not specify the model kinds for its constituent views; instead the definitions of what they should represent are provided. The *technical architecture* describes how the system can be *realized* into a given hardware platform [8]. It consists of the runtime model, the hardware topology, and the allocation model. The *runtime model* describes the behavior of the system from a physical/technical perspective. The *hardware topology* model describes the structure of the hardware platform using *physical units*, which represent hardware components (ECUs, sensors, mechanical components, etc.) and their connections (buses, wires, etc.) [8]. The *allocation* model maps the elements of the runtime

Table 2 Technical/constructional viewpoint

Technical/constructional viewpoint	
Definition	It describes vehicle physical components, their relationships, constraints, and allocation
Stakeholders	AAF: OEMs, suppliers, tool vendors, and research institutes
	ADF: Undefined
Concerns	Physical component composition and their relationships
Architecture views	AAF: Technical architecture view consisting of runtime model view, hardware topology view, and allocation view
	ADF: Product breakdown structure view, organic architecture view, requirements and function allocation on components view
Model kinds	AAF: Technical architecture for the technical view consisting of runtime model (for the runtime view), hardware topology (for the hardware topology view), allocation model (for the allocation view)
	ADF: BDD, IBD for a product breakdown structure view; BDD, IBD for an organic architecture view, requirements and function allocation on components
Correspondence rules	AAF: Correspondences to the functional viewpoint and optional viewpoints, for example, energy viewpoint
	ADF: Conformance correspondence to the functional viewpoint

model to the elements of the hardware topology model [8]. As in the functional viewpoint, all stakeholders are considered relevant to the technical viewpoint. AAF determines that the technical viewpoint has strong correspondences to the functional viewpoint and optional viewpoints, for example, energy viewpoint.

The ADF *constructional viewpoint* supports the *product breakdown structure*, *organic architecture*, and *allocation on components* views. ADF also defines SysML the model kinds for each constructional view. SysML BDD and IBD model kinds are selected for the product breakdown structure and organic architecture views. The *allocation* concept is used for allocating requirements and function to components [14]. The *product breakdown structure* identifies and allocates the system functions to physical components. The *organic architecture* defines the components of the system, their interfaces, and connections, which satisfy the system's technical requirements (e.g., cost, weight, size, authorized/forbidden use of materials) and other criteria (e.g., performance, effectiveness) [14]. Architecture models for the *allocation on components* view captures the allocation and structuring of the system requirements and functions to physical components to achieve an optimal allocation. The flows between functions are associated with the interfaces/connectors (e.g., mechanic, electric, network) between components [14].

As in the functional viewpoint, all stakeholders are considered relevant to the constructional viewpoint. ADF does not specify the concerns and correspondences explicitly. However, we identified the same concerns as AAF. The *conformance* correspondence is detected according to the implicit description of architecture views of the ADF constructional viewpoints.

Requirements Viewpoint Table 3 presents the requirements viewpoint, which looks at the vehicle from the perspective of the vehicle stakeholders including end users (drivers and passengers) and vehicle environment. We map the AAF *driver/vehicle operations* mandatory viewpoint, *value net* mandatory viewpoint, and all the *optional viewpoints*, that is, safety, security, quality, RAS (reliability, availability, serviceability), energy, cost, NVH (noise, vibration, harshness), and weight viewpoints to the ADF *requirements viewpoint*.

In ADF, requirements viewpoint captures elicitation of stakeholder requirements and elaboration of system technical requirements. ADF *requirements viewpoint* supports the stakeholder requirements view, high-level requirements view, and system technical requirements view. The ADF requirements viewpoint is in alignment with the AAF mandatory viewpoints *driver/vehicle operations* and *value net* viewpoints. The AAF *driver/vehicle operations* viewpoint looks at the interactions, interfaces, and interdependencies between vehicle and its end user (driver and passengers) as well as the surrounding environment (e.g., road, other vehicles, and traffic control systems) [8]. In addition, it describes the related behavior, constraints, and priorities. The *driver/vehicle operations* viewpoint governs *driver/vehicle operations* view.

Actors and system boundary are also captured as part of the ADF *stakeholder requirements view*. The AAF *value net* viewpoint is used to optimize the efficiency of the value creation process [8]. It can also be captured by the ADF *stakeholder requirements* view. High-level requirements are identified after the stakeholder

Table 3 Requirements viewpoint definition extracted from AAF and ADF frameworks

Requirements viewpoint	
Definition	It captures the vehicle from the perspective of the vehicle driver and the world around the vehicle
Stakeholders	AAF: All stakeholders (end users, OEMs, suppliers, tool vendors, and research institutes)
	ADF: Undefined
Concerns	Interactions between vehicle, end user, environment
Architecture views	AAF: Driver/vehicle view, value net view, optional views (safety, security, quality, RAS, energy, cost, NVH, and weight views)
	ADF: Stakeholder requirements view, high-level requirements view, system technical requirements view
Model kinds	AAF: Driver/vehicle operations model, value net model, models for safety, security, quality, RAS, energy, cost, NVH, and weight views
	ADF: Requirements diagram for the stakeholder requirements, high-level requirements, and system technical requirements views
Correspondence rules	AAF: Correspondences to other mandatory viewpoints
	ADF: Correspondence to the operational, functional, constructional viewpoints

requirements are elicited. An example high-level requirement can define measures of effectiveness or key performance parameters (KPP) [14]. The technical requirements are built after the operational models are defined, for example, by defining functional requirements from operations identified in sequence diagrams in the operational view [14]. Technical requirements capture functional, performance, and interface requirements or constraints [14]. What is captured in the AAF *optional viewpoints* depends on the vehicle system. However, ADF *requirements viewpoint* can support viewpoints such as safety, security, quality, RAS, energy, cost, NVH, and weight viewpoints.

In AAF, no specific model kind is defined for requirements-related viewpoints. In ADF, SysML requirements diagram type is selected for the requirements viewpoint [8].

Formalization of stakeholder and high-level requirements and elaboration of system technical requirements are captured by the SysML requirements diagram for all these views. All stakeholders, including vehicle end users (drivers and passengers), are defined for this viewpoint. Interactions, interfaces, and interdependencies between vehicle, end users, and the surrounding environment are key concerns. This viewpoint corresponds to other viewpoints to enable the requirements traceability of each viewpoint.

Other Viewpoints AAF *information viewpoint* is mandatory but does not have a similar viewpoint in the ADF. The information viewpoint looks at the vehicle from the perspective of information or data objects used to define and manage a vehicle [8]. It governs the *information view*, which describes information or data objects, their metadata, properties, relationships, configurations, and configuration constraints [8].

ADF *operational viewpoint* is the most abstract viewpoint of the ADF framework. The operational viewpoint governs structural and behavioral operational views. The *structural operational view* consists of the *maximal system scope, system environment, operational context, external interfaces,* and *use case* views [14]. The actors, system scope, system environment, and high-level interactions are identified in these structural views. The *behavioral operational view* consists of *operational scenarios* and *system working mode* views. These views are built from the structural operational views [14]. System use cases are used to identify actors, the system boundary and high-level interactions, which are refined in SysML sequence diagrams. *Operational scenarios* view addresses detailed interactions between the system and external systems/user/environment to realize the use cases. *System working states* view uses state machines to describe alternative conditions for operational scenarios [14]. SysML diagram types are mapped to the operational viewpoint as follows: SysML internal block diagram type is selected for the maximal system scope, system environment, operational context, and external interfaces views. SysML use case diagram type is selected for the use case view. SysML sequence and activity diagram types are selected for the operational scenarios view. SysML state machine diagram type is selected for the system working modes view.

Although these viewpoints exist only in one of the architecture frameworks, we address these viewpoints in the definition of the architecture framework for automotive systems (AFAS) in Sect. 4, for example, the information viewpoint of the AAF is included in the AFAS framework.

2.3 Discussion

An Architecture framework for the automotive systems has not been standardized in the automotive industry. Automotive architecture framework (AAF) and architecture design framework (ADF) aim to define a complete and integrated architecture framework for the automotive industry. We have identified common architecture viewpoints of these frameworks and summarized those that exist only in one of the frameworks. In the following section, we present the automotive ADLs and extract the viewpoints defined in the scope of the automotive ADLs. In Sect. 4, we then integrate the common architecture viewpoints of architecture frameworks and ADLs. Other viewpoints are also considered in the definition of the architecture framework.

3   Automotive ADLs and Viewpoints

According to the ISO 42010 international standard for systems and software engineering [18], an *architecture description language* (ADL) is any form of expression used to describe an architecture. As illustrated in Fig. 4, an ADL provides

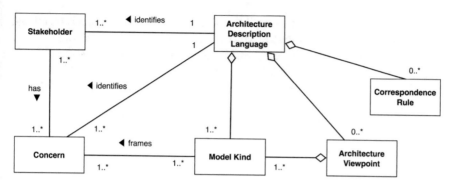

Fig. 4 ADL conceptual model [18]

one or more *model kinds* (data flow diagrams, class diagrams, state diagrams etc.) as a means to frame some *concerns* for its *stakeholders* [18]. In the case of several model kinds provided by an ADL to capture complex architectural representations, *architecture viewpoints* can be used to organize them. *Correspondence rules* can be used to express and enforce architecture relations, for example, refinement, composition, and traceability.

In this section, we present the automotive architecture ADLs, extract common architecture viewpoints, and summarize other architecture viewpoints that exist only in one of the ADLs. We apply the same template followed in Sect. 2.2, when describing the architecture viewpoints.

3.1 Automotive ADLs

EAST-ADL [9] (Embedded Architectures and Software Technologies-Architecture Description Language) is an architecture description language for automotive domain. It has been defined in the scope of an European research initiative, ITEA project EAST-EEA since 2001 [9]. The EAST-EEA project aimed to reduce automotive software's dependency on hardware, allowing more flexibility regarding the allocation of software [24]. The EAST-ADL has been refined in the ATESST project to EAST-ADL2 [36], which was extended further to support modeling of fully electric vehicles in the scope of the MAENAD project to EAST-ADL2.1.12 [22]. In the remainder of the chapter, EAST-ADL refers to the EAST-ADL2.1.12. The main purpose of EAST-ADL is to capture engineering information of automotive electrical/electronic (E/E) systems to enable modeling of the entire system development life cycle. The language consists of four main abstraction levels, which can be considered architecture viewpoints of the ISO 42010 standard. The highest level is called a *vehicle level*, where the basic vehicle features, requirements, and use cases are captured. The abstract functionalities based on the requirements and features are further defined in the *analysis level* and further refined as the

concrete functionalities in the *design level*. The design level also contains functional
definitions of application software, hardware components, and middleware. It also
covers function to hardware (e.g., ECU) allocations. The lowest abstract level,
implementation level, uses AUTOSAR [3] concepts to realize the higher-level
models. Requirements, variability, timing, dependability, and environment models
are captured in parallel with these abstraction levels.

TADL [39] (timing-augmented description language) is originated from EAST-
ADL, AUTOSAR, and MARTE. It was developed by the TIMMO project. TADL
addresses timing issues early in the development cycle by standardizing specification, analysis, and verification of timing constraints in all levels of abstraction of
EAST-ADL2.

AADL [12] (architecture analysis and design language) was developed to model
software, hardware, and system architecture of real-time embedded systems such
as aircraft, motorized vehicles, and medical devices. The Society of Automotive
Engineers (SAE) defined the AADL as SAE AS5506 Standard based on the MetaH
ADL [40]. Initially AADL was known as the Avionics Architecture Description
Language. In AADL, a system is constructed as a composite component consisting
of application software and execution platform. AADL enables a system designer to
perform analyses of the composed components such as system schedulability, sizing
analysis, and safety analysis. The focus of AADL is on task structure and interaction
topology, although generalization to more abstract entities is possible. It supports
the definition of mode handling, error handling, and interprocess communication
mechanisms. As such, it acts as a specification of the embedded software, which
can be used for automatic generation of an application framework where the actual
code can be integrated smoothly. The language supports different types of analysis
mechanisms, for example, for safety and timing analysis. Further, a behavioral
annex is proposed, to allow a common behavioral semantics for AADL descriptions.

AML [31] (automotive modeling language) is developed in the scope of the
FORSOFT project, which defined an architecture centric language to analyze
and synthesize automotive embedded systems. Similar to other ADLs, it offers
commonly accepted modeling constructs to specify the software and hardware parts
of the system architecture. The architecture is described by using components, in-
and outports, and connectors. The abstract syntax of the AML provides a conceptual
and methodological framework as a prerequisite for well-defined semantics of the
offered modeling constructs. The usage of different kinds of textual, graphical, or
tabular notations for a concrete model representation is supported. AML models
can be represented by various notational elements offered by widespread modeling
languages and tools such as ASCET-SD,[1] UML 1.4/2.0, and UML-RT.

SysML [27] (Systems Modeling Language) of OMG is a general purpose graphical modeling language to support specification, analysis, design, and verification
of complex systems. It is sponsored by INCOSE/OMG with broad industry and
vendor participation and adopted by the OMG in 2006 as OMG SysML. The

[1]ETAS ASCET-S http://www.etas.com/.

SysML adjusts UML2 [28] to system engineering by excluding unrelated diagrams and including new modeling concepts and diagrams for systems engineering. The SysML concepts concern requirements, structural modeling, and behavioral constructs. New diagrams include a requirement diagram and a parametric diagram and adjustments of UML activity, class, and composite structure diagrams. Tabular representations of requirements or allocations, for example, are also included as an alternative notation. Multiple vendors support SysML tools such as MagicDraw (No Magic) [25], Enterprise Architect (Sparx Systems) [35], Sirius (Eclipse) [11], Rational Rhapsody (IBM) [16], and PolarSys (Former TOPCASED) (Eclipse) [10]. One of the drawbacks of SysML is that SysML, as in UML, does not have a well-defined semantics.

Figure 5 illustrates the SysML structure, which consists of the following diagram types:

- The **requirement diagram** provides cross-cutting relationships between requirements and system models.
- The **structure diagrams** are *block definition diagrams* (BDD), *internal block diagrams* (IBD), *package diagrams*, and *parametric diagram*. UML class and composite structure diagrams are the basis of the BDD and IBD. A parametric diagram is a new diagram type, which can define quantitative constraints like maximum acceleration, minimum curb weight, and total air conditioning capacity.
- The **behavior diagrams** are *use case, state machine, activity diagrams*, and *sequence diagrams*. Activity diagram is modified from UML 2.0 activity diagram.

Tabular representations of requirements or allocations, for example, are also included as an alternative notation. SysML can be used to model hardware, software, information, processes, etc.

MARTE [26] (Modeling and Analysis of Real Time and Embedded) profile is an OMG standard for modeling real-time and embedded applications in UML2. It provides fundamental concepts of modeling and analyzing concerns of the real-time and embedded systems such as performance and schedulability issues. MARTE design model supports real-time embedded models of computation and communication and software and hardware resource modeling, while analysis model

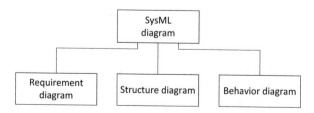

Fig. 5 SysML structure

enables generic quantitative analysis, schedulability, and performance analysis and refinement [13]. Both hardware and software aspects are supported.

3.2 Extracting Viewpoints from Automotive ADLs

The relationship between the architecture description elements (i.e., stakeholders, concerns, viewpoints, views, and model kinds) is presented in IEEE 1471-2000 standard and subsequently in ISO 42010 international standard [18]. Correspondences and correspondence rules are used to express and enforce architecture relations (e.g., composition, refinement, consistency, traceability, and dependency) within or between architecture description elements [18]. However, architecture description elements remain vague in automotive ADLs. Therefore, in this section, we identify the viewpoints together with other architecture elements, namely, stakeholders, concerns, viewpoints, and respective model kinds from automotive ADLs introduced in Sect. 3.1. The summary of the viewpoints extracted from the automotive ADLs is presented in Table 4.

Feature Viewpoint Product line engineering is one of the software engineering approaches to reduce software development costs. It is used by some automotive suppliers, but it is not used by the OEMs [7]. A *feature* is an end-user visible characteristic of a system [19], and it is captured in the feature viewpoint. The feature viewpoint is absent in the extracted viewpoints from automotive architecture frameworks as discussed in Sect. 2.2. However, EAST-ADL is the only automotive ADL to support product lines in the architecture description. Table 5 summarizes

Table 4 Automotive ADLs and viewpoints

Viewpoint	EAST-ADL	AADL	AML	SysML	MARTE
Feature	Technical feature				
Functional	Functional analysis	Layered system modeling	Functional network	Functional viewpoint (from ADF)	System configuration, generic component
Logical	Functional design (functional design architecture)	Composite system	Logical architecture	A subset of functional viewpoint (from ADF)	High-level application
Implementation	AUTOSAR software representation, hardware design architecture	Application software, execution platform	Technical architecture	Constructional viewpoint (from ADF)	Allocation

Table 5 Feature viewpoint definition extracted automotive ADLs

Feature viewpoint	
Definition	It captures the vehicle from the perspective of the vehicle features and the world around the vehicle
Stakeholders	End user, system architect, tier-x designer, safety engineer, tester or maintenance engineer
Concerns	Vehicle features, interactions between vehicle features, end user, environment
Architecture views	Vehicle view
Model kinds	Vehicle feature diagram
Correspondence rules	Correspondences to environment, requirements, and functional viewpoints

the feature viewpoint, which is extracted from the EAST-ADL. As discussed in Sect. 3.1, the highest abstraction level of EAST-ADL is called a *vehicle level*, where the basic vehicle features, requirements, and use cases are captured [36]. The vehicle level can be interpreted as a *vehicle view*, which contains a *vehicle feature model*. The vehicle feature model is used to describe a product line in terms of available features and their dependencies. The feature model can be used as a starting point to related requirements, use cases, and other constructs [9]. It can be used by all the stakeholders. Feature viewpoints have a correspondence with the environment, requirements, and functional viewpoints.

From other automotive-related ADLs, MARTE has mechanisms that can be used for the product line engineering. For example, *CombinedFragments*, *abstract class*, *inheritance*, *interface implementation*, *variables* can be used for analyzing software product line models [4]. However it is not considered a feature viewpoint, given that the MARTE is not a profile for software product line engineering.

Functional Viewpoint The functional viewpoint describes the vehicle from the abstract functions and their interaction point of view. Table 6 presents the functional viewpoint, which is defined in all automotive ADLs. The definition and purpose of the functional viewpoint of automotive ADLs is the same as the functional viewpoint of the automotive architecture frameworks as discussed in Sect. 2.2. However, the architecture views and model kinds differ among automotive ADLs.

In EAST-ADL, the vehicle features are realized by *abstract functions* in the *Functional Analysis Architecture* (FAA) at the functional *analysis view*. The FAA specifies what the system will do by specifying the main structure, interfaces, and behavior to realize the features and requirements from the vehicle view [9]. The FAA does not provide detailed design or implementation decisions. There is an *n-to-m* mapping between vehicle feature entities and FAA entities i.e., one or several functions may realize one or several features [36]. EAST-ADL provides the concepts for function component modeling to define the logical functionality and decomposition in the FAA [36]. *Functions* interact with each other via *ports* that are

Table 6 Functional viewpoint definition extracted from automotive ADLs

Functional viewpoint	
Definition	It describes the vehicle functions and their interactions
Stakeholders	End user, system or functional architect, tier-x designer, safety engineer, tester or maintenance engineer
Concerns	Functional composition and interfaces
Architecture views	EAST-ADL: Analysis view
	AADL: Layered architecture modeling view
	AML: Functional network view
	SysML: Functional breakdown structure view, functional architecture view, allocation on functions view
	MARTE: Breakdown structure view, functional architecture view, allocation on functions view
Model kinds	EAST-ADL: Functional analysis architecture (Function component modeling conepts)
	AADL: Core AADL language
	AML: AML metamodel and semantics for the functions and functional network
	SysML: AD, BDD, IBD for the functional breakdown structure view; AD, BDD, IBD for the functional architecture; *allocation* concept for the requirements allocation on functions views
	MARTE: AD, BDD, IBD for the functional breakdown structure view; AD, BDD, IBD for the functional architecture; *allocation* concept for the requirements allocation on functions views
Correspondence rules	EAST-ADL: Correspondences to feature, environment, and requirements viewpoints (an n-to-m mapping between vehicle feature entities and FAA entities (i.e., one or several functions may realize one or several features)
	AADL: Refinement correspondence to the composite system view
	AML: Refinement and allocation correspondence to logical architecture
	SysML, MARTE: Refinement and conformance correspondence to the logical viewpoint

linked by *connectors*. The *system boundary, environment model*, and *abstract safety analysis* can be carried out in the analysis view [36].

AADL introduced the *layered architecture modeling* to support hierarchical containment of components, layered use of threads for processing and services, and layered virtual-machine abstraction of the execution platform [33]. In AADL, a system is constructed as a *composite system* consisting of *application software, execution platform*, or *system components*, which are all considered specific type of components. AADL defines components by *type* and *implementation* declarations [12]. A component's interface and external attributes (e.g., interaction points, flow specifications, and internal property values) are defined in a component type declaration [12]. A component's internal structure (e.g., its subcomponents, subcomponent connections, flow implementations, and properties) are defined in a component implementation declaration [12]. The AADL core modeling language for the component-based representation enables modeling of components, interactions, and properties [12]. It has graphical and textual representations. The layered architecture and composite system models are further refined in the composite system. Because the functional viewpoint describes the system's functionality in black-box perspective, we map the layered system modeling to the functional viewpoint.

We map the *functional network* abstraction level to the functional viewpoint, because a network of functions, that is, generic and reusable building blocks, are defined at this level. A function has an interface, which specifies the *required* and *provided signals* [31]. Local signals of a function are not accessible to enable reusability [31]. Regarding correspondence to other viewpoints/views, functions can be refined and deployed on different control units of the lower level logical architecture view.

For SysML, we reuse the architectural elements of the functional viewpoint in the ADF framework in Sect. 2.2. In the ADF *breakdown structure view*, *functional architecture view*, and *allocation on functions view* are defined for the functional viewpoint. SysML activity diagram, block definition diagram, and internal block diagrams are selected for the *breakdown structure view* and *functional architecture view*.

Logical Viewpoint We generalized a more concrete viewpoint that refines the functional viewpoint as a *logical viewpoint*. Table 7 presents the logical viewpoint, which is (implicitly) defined in all automotive ADLs except AML. Note that the AAF defines logical viewpoint as a white-box representation of a system, but it does not define it as an architecture viewpoint [8]. Thus the logical viewpoint is not listed as one of the architecture viewpoints of the automotive architecture frameworks in Sect. 2.2. However, it is a viewpoint that is common among automotive ADLs.

In EAST-ADL, the functional analysis architecture (FAA) of the analysis view (governed by the functional viewpoint) is refined by the functional design architecture (FDA) and hardware design architecture (HDA) at the design level/view [22]. We exclude the HDA from the design viewpoint, because the logical architecture needs to be independent from the underlying hardware. The FDA decomposes the functions defined in the FAA by adding a behavioral description and a detailed interface definition to meet constraints regarding nonfunctional requirements such as efficiency, reuse, or supplier concerns [22]. There are *n-to-m* mappings by realization relationships between entities in the FDA and entities in the FAA [22].

In AADL, the internal structure of a system is constructed as a *composite system* consisting of *application software*, *execution platform*, or *system components* [12], which are all considered specific type of components as described in the functional viewpoint section. Therefore, we map the composite system to the logical viewpoint. The AADL core modeling language for the component-based representation is also applied for the composite system representation. The composite system models are further refined in the application software view.

Table 7 Logical viewpoint definition extracted automotive ADLs

Logical viewpoint	
Definition	It refines the functional architecture into logical components, which are independent from implementation details and underlying hardware
Stakeholders	End user, system architect, tier-x designer, safety engineer, tester or maintenance engineer
Concerns	Internal structure of the vehicle functions, detailed interactions between and inside vehicle functions
Architecture views	EAST-ADL: Functional design view
	AADL: Composite system view
	AML: Logical architecture view
	SysML: Functional breakdown structure view
Model kinds	EAST-ADL: Functional design architecture
	AADL: Core AADL language
	AML: AML metamodel and semantics for the logical architecture
	SysML: AD, BDD, IBD for the functional breakdown structure view
Correspondence rules	EAST-ADL: Refinement correspondence to functional viewpoint (*n-to-m* mappings by realization relationships between entities in the FDA and entities in the FAA)
	AADL: Refinement correspondence to the application software, physical platform
	AML: Refinement correspondence to the functional network view
	SysML: Refinement and conformance correspondence to the functional viewpoint

In AML, logical architecture model refines the functional network models [31]. The logical architecture model describes the logical control units, actors, and sensors of the environment [31]. The functions defined in the functional network are deployed on different logical control units. However, implementation details like the system is clocked (not event driven), and communication between/within logical control units are synchronous and are specified at this stage.

For SysML, we reuse the part of the architectural elements of the functional viewpoint in the ADF framework in Sect. 2.2. The ADF *breakdown structure view* is defined to capture function identification and decomposition. SysML activity diagram, block definition diagram, and internal block diagrams are selected for the *breakdown structure view*.

Implementation Viewpoint The implementation viewpoint describes the software architecture of the electrical/electronic (E/E) system in the vehicle [22]. Table 8 summarizes the implementation viewpoint elements extracted from the automotive ADLs.

In EAST-ADL, the implementation viewpoint is supported by the system architecture and software architecture of AUTOSAR [22]. AUTOSAR serves as a basic infrastructure for the management of functions within both future applications

Table 8 Implementation viewpoint definition extracted automotive ADLs

Implementation viewpoint	
Definition	It realizes the logical architecture into software and hardware components
Stakeholders	End user, system architect, tier-x designer, safety engineer, tester or maintenance engineer
Concerns	Implementation of logical components into software and hardware components, optimal resource utilization, allocation, performance estimation, etc.
Architecture views	EAST-ADL: Implementation view, design view (hardware design)
	AADL: Application software view, execution platform view
	AML: Technical architecture view
	SysML: Product breakdown structure view, organic architecture view, requirements and function allocation on components view
Model kinds	EAST-ADL: AUTOSAR application software, AUTOSAR basic software (using AUTOSAR software component template, ECU resource template, and system template), hardware design architecture from the design level
	AADL: Core AADL language
	AML: AML metamodel and semantics for the technical architecture
	SysML: BDD, IBD for a product breakdown structure view; BDD, IBD for an organic architecture view, requirements and function allocation on components
Correspondence rules	EAST-ADL: Realization correspondence to logical viewpoint (*n-to-m* mappings by realization relationships between entities in the implementation view and entities in the design view)
	AADL, AML, SysML: Realization correspondence to the logical viewpoint

and standard software modules [3]. In EAST-ADL, AUTOSAR software components realize the functional design architecture, and AUTOSAR basic software components realize the hardware design architecture using the AUTOSAR software component, ECU resource, and system templates. Regarding the correspondence, traceability is supported from implementation level elements (AUTOSAR) to upper level elements by *realization* relationships [22].

In addition to the AUTOSAR system and software architectures, the EAST-ADL hardware design architecture (HDA) is also mapped in this viewpoint. HDA is then refined further by ECU specifications and topology.

In AADL, a system instance consists of *application software components* and *execution platform components* [12].

In AML, the technical architecture enriches the logical architecture with concrete technical information, for example, concrete bus, control unit, and operating system specifications [31]. The performance estimation can be carried out in this architecture modeling [31]. AML language is used for this viewpoint.

We consider the *constructional viewpoint* discussed in the ADF in Sect. 2.2 as part of the implementation viewpoint, because it decomposes a vehicle into physical components and defines their relationships and constraints. Then the implementation viewpoint for SysML supports *product breakdown structure, organic architecture,* and *allocation on components* views. As discussed in Sect. 2.2, ADF identifies SysML BDD and IBD model kinds for the product breakdown structure and organic architecture views. The *allocation* concept is used for allocating requirements and functions to components [14]. The *product breakdown structure* identifies and allocates the system functions to physical components. The *organic architecture* defines the components of the system, their interfaces, and connections, which satisfy the system's technical requirements (e.g., cost, weight, size, authorized/forbidden use of materials) and other criteria (e.g., performance, effectiveness) [14]. Architecture models for the *allocation on components* view captures the allocation and structuring of the system requirements and functions to physical components to achieve an optimal allocation. The flows between functions are associated with the interfaces/connectors (e.g., mechanic, electric, network) between components [14].

Other Viewpoints EAST-ADL extensions are considered as other viewpoints, which are orthogonal to the main architecture viewpoints:

- **Requirements** are captured in EAST-ADL following the principles of SysML [22].
- **Variability** is realized in EAST-ADL at all levels besides the feature models on vehicle level [22].
- **Timing** is supported by the TIMMO project. It defined a methodology and representation of timing aspects in automotive embedded systems [22]. TADL defines timing constraints in all levels of abstraction of EAST-ADL2 [39].
- **Dependability** extension covers several aspects, that is, availability, reliability, safety, integrity, and maintainability [22].

3.3 Discussion

Architecture description languages (ADLs) have been developed to define automotive architectures effectively to tackle the increasing complexity and development costs [24, 34]. Although the ISO 42010 international standard [18] has defined what constitutes an ADL, the automotive ADLs have been developed without specifying the architectural elements of an ADL as defined in the ISO 42010 international standard. Therefore, we have mapped the architecture viewpoints of the automotive ADLs to the viewpoints of the automotive architecture frameworks. The mapping provides an input for further aligning the architecture elements of the automotive ADLs and automotive architecture frameworks.

We have identified common architecture viewpoints and views of these frameworks. In the following section, we integrate the viewpoints and views defined in the scope of the automotive ADLs and automotive architecture frameworks and propose a conceptual model of an architecture framework for automotive systems. Other viewpoints that are briefly presented here are also discussed in the definition of the framework in the following section.

4 Architecture Framework for Automotive Systems

This section presents the architecture framework for automotive systems (AFAS), which contains architecture viewpoints consistent with the existing automotive architecture frameworks (AFs) and the automotive architecture description languages (ADLs) as discussed in Sects. 2 and 3, respectively. The architectural elements of the AFAS are shown in Fig. 6. The AFAS viewpoints are defined based on the preceding analysis of the automotive AFs and ADLs. In addition, we studied proprietary automotive architectural models and practices and aligned the AFAS with the results based on the interviews carried out with the domain experts from an OEM. The AFAS framework thus contains architectural viewpoints complementary to automotive ADLs, automotive AFs, and proprietary approaches. The simplified architectural elements are illustrated in Fig. 6. The representation

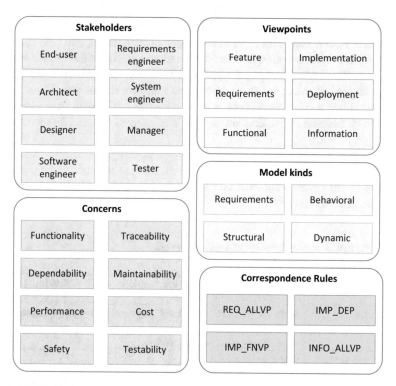

Fig. 6 AFAS overview

of the AFAS overview is in alignment with the graphical representation in the
MEGAF infrastructure [15]. In the following section, we elaborate the architecture
viewpoints and related elements of AFAS.

Feature Viewpoint Since AAF was specified as an automotive industry reference
for the vehicle line architectures [8], the feature viewpoint is not specified in
the AAF. However, we considered the feature viewpoint necessary, because even
a single feature can be configured further, such as cruise control or Bluetooth
telephone connection, which can be configured for a product or a specific vehicle.
Therefore, we revise the feature viewpoint of the EAST-ADL, which is presented
in Sect. 3. The feature viewpoint contains feature view, which specifies a vehicle
feature model. The feature model can be used as a starting point to related
requirements, use cases, and other constructs [9].

Automotive architecture frameworks and ADLs do not explicitly define the
system stakeholders for the frameworks and ADLs. General stakeholders as end
users, OEMs, suppliers, tool vendors, and research institutes are identified for
automotive architecture frameworks. End users, system architects, tier-x designers,
safety engineers, and testers or maintenance engineers are identified from the
automotive ADLs. Therefore, we have interviewed a number of domain experts
from an OEM and identified stakeholders as driver, fleet owner (fleet information
center), manager, product line manager, requirements manager, system/software
architect, designer, system integrator, developer, analyst, tester, and (external/inter-
nal) supplier. OEMs, suppliers, tool vendors, and research institutes are stakeholders
more from the organizational perspective. Therefore, we clarified more specific
roles as key stakeholders for an automotive architecture framework. We combined
the stakeholders defined for the automotive AFs and ADLs with the stakeholders
identified by the automotive domain experts. The selected stakeholders are listed in
Fig. 6.

The feature viewpoint can be used by architects, designers, requirements engineers, system engineers, managers, and testers. Feature viewpoints have a correspondence with the requirements and functional viewpoints. Table 9 summarizes
the revised feature viewpoint.

Table 9 Feature viewpoint definition in AFAS framework

Feature viewpoint	
Definition	It captures the vehicle from the perspective of the vehicle features and the world around the vehicle
Stakeholders	Architect, designer, requirements engineer, system engineer, manager, and tester
Concerns	Functionality, cost, maintainability
Architecture views	Vehicle feature view
Model kinds	Vehicle feature diagram
Correspondence rules	Correspondences to requirements and functional viewpoints

Table 10 Requirements viewpoint definition in AFAS framework

Requirements viewpoint	
Definition	It captures the vehicle from the perspective of the vehicle driver and the world around the vehicle
Stakeholders	All stakeholders (end users, architect, designer, software engineer, requirements engineer, system engineer, manager, and tester)
Concerns	Functionality, traceability, cost
Architecture views	Stakeholder requirements view, high-level requirements view, system technical requirements view, value net view
Model kinds	Requirements and use case diagram type for the stakeholder requirements, high-level requirements, and system technical requirements views
Correspondence rules	Traceability correspondences to other viewpoints

Requirements Viewpoint The requirements viewpoint is defined in the automotive AFs as presented in Sect. 2 and defined as an extension in the EAST-ADL language as discussed in Sect. 3. In the AFAS, the requirements viewpoint is included as one of the main viewpoints. We summarize below the description based on the requirements viewpoint of the automotive AFs and EAST-ADL. Table 10 presents the requirements viewpoint, which looks at the vehicle from the perspective of the vehicle stakeholders including end users (drivers and passengers) and vehicle environment.

As in ADF, the requirements viewpoint captures elicitation of stakeholder requirements and elaboration of system technical requirements. The *requirements viewpoint* supports the stakeholder requirements view, high-level requirements view, and system technical requirements view as in ADF. Since the AAF *driver/vehicle operations* view looks at the interactions, interfaces, and interdependencies between the vehicle and its end user and environment, it is considered part of the stakeholder requirements view. The *value net* view is included, as it is used to optimize the efficiency of the value creation process for an OEM, suppliers, and engineering partners [8]. We map the SysML requirements and use case diagram types for the model kinds, which can be used to model the requirements views. The use case diagram shows the interaction between users and the system. The stakeholder requirements view can also identify actors and system boundary as in ADF. The requirements viewpoint corresponds to other viewpoints to enable the requirements traceability of each viewpoint.

Functional Viewpoint The *functional viewpoint* was defined in both automotive AFs and ADLs. The functional viewpoint is considered the cornerstone of most architecture descriptions [32]. We revise the functional viewpoints of the AFs and ADLs into Table 11.

As in AAF, the *functional viewpoint* describes vehicles in terms of vehicle functions and their logical interactions. We revise the AAF architecture views of the functional viewpoint, that is, a *functional view*, which specifies a structural

Table 11 Functional viewpoint definition in AFAS framework

Functional viewpoint	
Definition	It describes the vehicle functions and their interactions
Stakeholders	Architect, designer, requirements engineer, system engineer, manager, and tester
Concerns	Functionality, dependability, cost, maintainability
Architecture views	Functional view, detailed functional view, allocation on functions view
Model kinds	BDD for the functional view, IBD for the detailed functional view, *allocation* concept for the requirements allocation on functions views
Correspondence rules	Realization and traceability correspondences to the requirements viewpoint

model that contains a number of functions or subsystems realizing features. The functional view is the first view that stakeholders try to read due to simplicity [32]. The functional architecture is described in this view, which contains a structural model kind that contains a number of functions or subsystems realizing features.

We define a *detailed functional view*, which refines the functions and their interfaces by specifying more details (similar to logical view). The ADF *allocation on functions* view is reused in this viewpoint. We revised the SysML diagram types that are defined for the functional viewpoint in ADF. In ADF, SysML activity diagram was selected for the functional breakdown structure and functional architecture views. However, it was stated that the functional requirements need SysML use case, activity, and sequence diagrams to specify the behavior of a function. It was concluded after successful application of SysML in deriving functional architectures from requirements and use cases [21].

The functional architecture represents the static view of the system; therefore behavioral diagrams are not necessary. This concurs with our view that the functional architecture needs to specify abstract functions in a static structural model independent of implementation and technological details. Therefore, we exclude the SysML activity diagram, which was part of the ADF functional viewpoint. From the ADF architecture views of the functional viewpoint, namely, *breakdown structure view*, *functional architecture view*, and *allocation on functions view*, the breakdown structure view is not selected for the functional viewpoint. The main reason is that the breakdown structure can be represented in the functional architecture without behavioral models. The allocation of requirements and features to functions is necessary for enabling traceability of the requirements and features.

Implementation Viewpoint The implementation viewpoint consists of *software*, *hardware*, and *topology* views. Therefore, the *technical/constructional viewpoint* of the AAF and ADF can be a part of the implementation viewpoint, specifically addressing the hardware view. Table 12 revises the implementation viewpoint elements discussed in Sect. 3.2.

The implementation viewpoint governs software view, hardware view, and topology view. Software view represents the software architecture, where detailed

Table 12 Implementation viewpoint definition in AFAS framework

Implementation viewpoint	
Definition	It realizes the functional architecture into software and hardware components
Stakeholders	Architect, designer, software engineer, requirements engineer, system engineer, manager, and tester
Concerns	Dependability, safety, performance, maintainability, cost
Architecture views	Software view, hardware view, topology view
Model kinds	BDD, IBD, AD, SD, and SM for a software view, BDD and IBD for the hardware and topology views
Correspondence rules	Realization correspondence to functional viewpoint (*n-to-m* mappings by realization relationships between entities in the implementation view and entities in the functional view)

descriptions and implementation of a function is realized in software components or blocks. The software components realize the functional components. Regarding the correspondence, implementation viewpoint realizes the functional view. In the hardware view, the E/E hardware architecture is represented. The hardware architecture typically consists of ECUs, sensors, actuators, and controller area network (CAN) buses. The topology view specifies the connections (buses, e.g., CAN and local interconnect network (LIN), and wires) between ECUs, sensors, and actuators.

We consider the *constructional viewpoint* discussed in the ADF in Sect. 2.2 as part of the hardware view, which is governed by the implementation viewpoint, because it decomposes a vehicle into physical components and defines their relationships and constraints. As in the functional viewpoint, all stakeholders are considered relevant to the implementation viewpoint. ADF does not specify the concerns and correspondences explicitly. However, we identified the same concerns as AAF. The correspondence is conformance according to the implicit description of architecture views of the ADF constructional viewpoints.

Deployment Viewpoint The deployment viewpoint describes the environment into which the system will be deployed, including capturing the dependencies of the system on its runtime environment [32]. Table 13 summarizes the deployment viewpoint elements extracted from the automotive ADLs. The allocation view describes the mapping between software components to ECUs. It can be in a table format.

Information Viewpoint The information viewpoint of the AAF is included in the AFAS framework, because it describes how the architecture manages and distributes information [32]. The information view is also reused for the information viewpoint. The information view describes information or data objects, their metadata, properties, relationships, configurations, and configuration constraints [8].

Table 13 Deployment viewpoint definition in AFAS framework

Deployment viewpoint	
Definition	It defines the environment into which the system will be deployed
Stakeholders	Architect, safety engineer, system engineer, tester
Concerns	Functionality, dependability, performance, safety, cost
Architecture views	Execution platform view, concurrency view, allocation view
Model kinds	Process model
Correspondence rules	Realization correspondence to implementation viewpoint

5 Conclusion

We integrated the architecture viewpoints extracted from automotive AFs and ADLs into architecture framework for automotive systems (AFAS). The main objective of the framework is to have consistent architecture description elements.

The functional viewpoint exists in both AFs and ADLs as it is a cornerstone of architecture description. However, some of the viewpoints are not directly mappable. Therefore, we analyzed the semantics of the architecture description elements and integrated feature, requirements, functional, implementation, and information viewpoints from existing AFs and ADLs. A deployment viewpoint was added to the AFAS, because an OEM plays mostly a role of an integrator or assembler by integrating software/(E/E)/hardware systems into a vehicle. Currently an architecture framework addressing continuous integration and deployment, ecosystem and transparency, and car as a constituent of a system of systems is being developed with Volvo cars as well [29]

References

1. Andrianarison E, Piques JD (2010) SysML for embedded automotive systems: a practical approach. In: Embedded real time software and systems (ERTS^2). ERTS^2 series, Toulouse, pp 1–10
2. Apvrille L, Becoulet A (2012) Prototyping an embedded automotive system from its UML/SysML models. In: Embedded real time software and systems (ERTS^2), ERTS^2 series, Toulouse, pp 1–10
3. AUTOSAR (2018) The AUTomotive Open System ARchitecture (AUTOSAR). http://autosar. org. Accessed 27 Oct 2018
4. Belategi L, Sagardui G, Etxeberria L (2010) MARTE mechanisms to model variability when analyzing embedded software product lines. In: Software product lines: going beyond. Springer, New York, pp 466–470
5. British Ministry of Defence (2012) MOD architecture framework. https://www.gov.uk/ guidance/mod-architecture-framework. Accessed 12 March 2019
6. Braun P, Rappl M (2001) A model-based approach for automotive software development. In: Workshop on object-oriented modeling of embedded real-time systems (OMER). LNI, vol 5, pp 100–105

7. Broy M (2006) Challenges in automotive software engineering. In: International conference on software engineering (ICSE). ACM, New York, pp 33–42
8. Broy M, Gleirscher M, Merenda S, Wild D, Kluge P, Krenzer W (2009) Toward a holistic and standardized automotive architecture description. Computer 42(12):98–101
9. Cuenot P, Frey P, Johansson R, Lönn H, Papadopoulos Y, Reiser M, Sandberg A, Servat D, Kolagari RT, Törngren M, Weber M (2011) The EAST-ADL architecture description language for automotive embedded software. In: Model-based engineering of embedded real-time systems. Springer, Berlin, pp 297–307
10. Eclipse (2019) PolarSys (former TOPCASED). http://polarsys.org/. Accessed 12 March 2019
11. Eclipse (2019) Sirius. http://www.eclipse.org/sirius/. Accessed 12 March 2019
12. Feiler P, Gluch D, Hudak J (2006) The architecture analysis & design language (AADL): an introduction. Tech. Rep. CMU/SEI-2006-TN-011, Software Engineering Institute, Carnegie Mellon University
13. Gerard S, Espinoza H (2006) Rationale of the UML profile for Marte. In: From MDD concepts to experiments and illustrations. ISTE, London, pp 43–52
14. Góngora H, Gaudré T, Tucci-Piergiovanni S (2013) Towards an architectural design framework for automotive systems development. In: Complex systems design and management. Springer, Cham, pp 241–258
15. Hilliard R, Malavolta I, Muccini H, Pelliccione P (2012) On the composition and reuse of viewpoints across architecture frameworks. In: The joint working IEEE/IFIP conference on software architecture (WICSA) and European conference on software architecture (ECSA). IEEE, Helsinki, pp 131–140
16. IBM (2019) Rational Rhapsody Designer for systems engineers. http://www.ibm.com/software/products/. Accessed 12 March 2019
17. ISO (1998) ISO/IEC 10746-1 Information technology – reference model of open distributed processing (RM-ODP)
18. ISO (2011) ISO/IEC/IEEE 42010 - Systems and software engineering–architecture description
19. Kang K, Cohen S, Hess J, Novak W, Peterson A (1990) Feature-oriented domain analysis (foda) feasibility study. Tech. rep., DTIC Document
20. Kruchten P (1995) The 4+1 view model of architecture. IEEE Softw 12(6):42–50. https://doi.org/10.1109/52.469759
21. Lamm J, Weilkiens T (2010) Functional architectures in SysML. Tag des Systems Engineering (TdSE)
22. MAENAD (2014) EAST-ADL 2.1.12 domain model specification. http://east-adl.info/Specification/V2.1.12/html/index.html. Accessed 12 March 2019
23. MAENAD (2014) ICT MAENAD project. http://maenad.eu/. Accessed 12 March 2019
24. Navet N, Simonot-Lion F (2009) Automotive embedded systems handbook. Industrial information technology series. CRC Press, Boca Raton
25. NoMagic (2019) MagicDraw SysML plugin. http://www.nomagic.com/products/magicdraw-addons/sysml-plugin.html. Accessed 12 March 2019
26. Object Management Group (2011) UML profile for MARTE: modeling and analysis of real-time embedded systems, version 1.1. http://www.omg.org/spec/MARTE/1.1. Accessed 12 March 2019
27. OMG (2012) Systems modeling language (SysML) specification, version 1.3. http://www.omg.org/spec/SysML/
28. OMG (2015) The Unified Modeling Language - UML 2.0. https://www.omg.org/spec/UML/2.0/. Accessed 12 March 2019
29. Pelliccione P, Knauss E, Heldal R, Ågren S. M, Mallozzi P, Alminger A, Borgentun D (2017) Automotive architecture framework: the experience of Volvo cars. J Syst Archit 77:83–100
30. Rao AC, Dhadyalla G, Jones RP, McMurran R, White D (2006) Systems modelling of a driver information system – automotive industry case study. In: System of systems engineering (SSE). IEEE, Los Angeles, pp 254–259
31. Rappl M, Braun P, Von Der Beeck M, Schröder C (2002) Automotive software development: a model based approach. Tech. rep., SAE Technical Paper

32. Rozanski N, Woods E (2005) Software systems architecture: working with stakeholders using viewpoints and perspectives. Addison-Wesley, Upper Saddle River
33. SAE International (2012) Architecture analysis and design language. http://www.aadl.info/. Accessed 12 Feb 2019
34. Schäuffele J, Zurawka T (2005) Automotive software engineering: principles, processes, methods, and tools. Society of Automotive Engineers, Warrendale
35. Sparx Systems (2019) Enterprise architect. http://www.sparxsystems.com/. Accessed 12 March 2019
36. The ATESST Consortium (2013) EAST-ADL specification. http://www.east-adl.info/ Specification.html. Accessed 12 Feb 2019
37. The Consultative Committee for Space Data Systems (2008) Reference architecture for space data systems. http://public.ccsds.org/publications/. Accessed 12 Feb 2019
38. The Open Group (2018). The open group architectural framework (TOGAF). http://www.opengroup.org/togaf/. Accessed 12 February 2019
39. The TIMMO Consortium (2009) TADL: timing augmented description language, version 2. http://adt.cs.upb.de/timmo-2-use/timmo/publications.htm. Accessed 12 March 2019
40. Vestal S (1997) MetaH support for real-time multi-processor avionics. In: Parallel and distributed real-time systems workshop, pp 11–21

Part IV
E/E Architecture and Safety

The RACE Project: An Informatics-Driven Greenfield Approach to Future E/E Architectures for Cars

Alois Knoll, Christian Buckl, Karl-Josef Kuhn, and Gernot Spiegelberg

Abstract As cars are turning more and more into "computers on wheels," the development foci for future generations of cars are shifting away from improved driving characteristics toward features and functions that are implemented in software. Classical decentralized electrical and electronic (E/E) architectures based on a large number of electronic control units (ECUs) are becoming more and more difficult to adapt to the extreme complexity that results from this trend. Moreover, the innovation speed, which will be dictated by the computer industry's dramatically short product lifecycles, requires new architectural and software engineering approaches if the car industry wants to rise to the resulting multidimensional challenges. While classical evolutionary architectures mapped the set of functions that constitute the driving behavior into a coherent set of communicating control units, RACE (Reliable Control and Automation Environment) is an attempt to redefine the architecture of future cars from an information processing point of view. It implements a straightforward perception-control/cognition-action paradigm; it is data centric, striking a balance between central and decentralized control. It implements mechanisms for fault tolerance and features plug-and-play techniques for smooth retrofitting of functions at any point in a car's lifetime.

A. Knoll (✉)
Technische Universität München (TUM) and fortiss GmbH, München, Germany
e-mail: knoll@in.tum.de

C. Buckl · K.-J. Kuhn · G. Spiegelberg
Siemens AG, CT RTC RACE, München, Germany

© Springer Nature Switzerland AG 2019
Y. Dajsuren, M. van den Brand (eds.), *Automotive Systems and Software Engineering*, https://doi.org/10.1007/978-3-030-12157-0_8

1 Introduction[1]

Over the last decades of "vehicle electronification and digitalization," it has become clear that information and communication technology (ICT) will determine the vehicle of the future. ICT is becoming the dominant factor and will drive vehicle developments by itself—people will want their cars to be equipped with ICT as powerful as it is in their offices and homes. For this reason, architectures and technologies for vehicle ICT cannot be viewed merely as a framework for gradual evolutionary innovations as they once were—they will determine the future development of cars in terms of functionality, innovation speed, and value creation. Architectures designed with these insights in mind will make new approaches and functions possible—from greater autonomy in driving to a more complete integration of the vehicle into the ICT infrastructure—and thus help significantly to achieve socio-political goals like energy efficiency or lower accident rates.

Today's automotive electronic/electric (E/E) architectures are a result of a long evolutionary process. The number of electronic control units (ECUs) has risen dramatically since their first large-series introduction into car technology in the 1970s. At present, the value added by ICT to the car is between 30% and 40%, but 80% of the innovations[2]—from entertainment systems by way of driver assistance systems to advanced engine and chassis control—are due to ICT. While the first antilock braking system (ABS) in 1978 had a small processor with a few hundred lines of code, today's luxury cars have millions of lines of code running on high- performance processors and dedicated chips. Still, however, the potential of modern ICT is far from being fully exploited by the car industry. For example, the move to full "drive-by-wire" or even "drive-by-wireless," although offering numerous advantages, has not been made because the necessary fundamental software structures are not deemed to be sufficient for this industry's standards. The development in areas like infotainment or telematics is also much faster, which has now become a real threat to traditional car manufacturers. One of the reasons is that customers want their car to keep pace with their rapidly changing ICT environment over the lifetime of the car—which will increase rather than decrease as we move to maintenance-free electric powertrains—and that static architectures will not be able to meet the dynamic, and as of now unforeseeable needs that will arise in the future.

What can be foreseen, however, is that there will be an ever-increasing speed of development in customer electronics, innovative applications (purely defined in software), new business models based on "platforms," and much more emphasis on environmental protection and resource economy. These aspects can only be taken care of if automotive original equipment manufacturers (OEMs) can keep up with the pace of the development of processing power, storage capacity, and

---

[1]Note that parts of this section, including Figs. 1 and 2, are an updated extract from [1].

[2]http://aesin.org.uk/about/about-automotive-electronics/

communication bandwidth—in other words, if they can rely on an architecture base that can make practical use of these developments.

In summary, ICT architecture is increasingly becoming a barrier—or an enabler!—to innovation.

2 A Brief History of ICT E/E Architectures for Cars

A look at the history of car electronics can be helpful to understand why the time is now ripe to develop radically new architectures (see Fig. 1). In the evolution of vehicle architectures, there is a trend for the architecture actually to accidentally become much more complex than it is essential for the achieved increase in overall functionality. Hence, new functions to be added become more and more cumbersome and hard to integrate, and the innovation activities therefore tend to lag behind. Only a substantial revision of the architecture enforced by a disruptive technology leap can bring the accidental complexity back down to its essential level.

The only way of achieving this is to raise the level of functional abstraction at which new functions can be integrated. This has already been observed in the automotive industry. To reduce emissions and improve comfort, in the 1980s it was necessary to expand the use of microcontrollers. Complexity relatively quickly

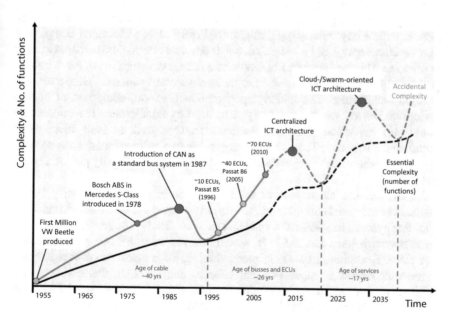

Fig. 1 Qualitative past and expected future growth of essential complexity due to increasing functionality to be supplied by electronic and software infrastructure, plotted against accidental complexity introduced by suboptimal mapping of ICT technology to vehicle technology

became a big problem because it was almost impossible to connect all these electronics modules together in a big cable harness. The solution was an abstraction of communication through shared media—buses like the CAN bus virtualized the physical connection. It thus became significantly easier to introduce new functions because integration no longer had to take place at the cable level, but at the information level.

While CAN is a networking standard originating from the automotive industry, the other important standard when it comes to architectures is Automotive Open System Architecture (AUTOSAR).[3] We are currently observing huge efforts in the industry to completely revise this standard in view of the requirements emerging from autonomous driving and an ever-increasing number of advanced driver assistance systems. This will be called the Adaptive AUTOSAR Platform, but when it will be published, it will be far from complete. Like AUTOSAR Classic, it will be subject to many revisions, and its initial version will be limited in scope.

The differences between AUTOSAR classic and RACE are manifold, and will become clear in the rest of this paper. Some differences are that the RACE RTE is dynamic to allow for plug and play. RACE provides support for fault tolerance and recovery, as well as a complete up-to-date software development environment, including separation kernels, as an underlying operating system layer. Moreover, the hardware structure based on central computers is a radical departure from the current architectures.

Notwithstanding, today's ICT architectures face problems because of the ever-increasing number of control devices. A new, centralized electric/electronics architecture, with a base middleware, might drastically reduce accidental complexity. New functions may then be integrated, not as physical electronic control devices but as software. The third step, finally, would be a further virtualization of the necessary total system of hardware and software (the hardware/software stack) into a service-oriented architecture. The underlying execution platform, composed of control devices and buses, would be entirely virtualized by middleware. The middleware would also implement non-/extrafunctional features, such as fault tolerance or communication delays. Then it would be possible to distribute functions as desired, even outside the vehicle; the car would thus become quite naturally part of a larger system.

At this point, a closer look at other sectors is in order. In the early 1980s, solutions in industrial controls and PCs proved that modular hardware and standard operating systems like MS-DOS and Unix can completely change entire industries. Open standards have resulted in increased innovations in hardware and software ever since. Economies of scale in production, with the associated cost reductions in modular hardware, made PCs attractive to end users. In the 1990s, a new architecture was introduced in aviation because of problems very similar to those in the automotive sector today. The "Integrated Modular Avionics IMA" [2] showed

---

[3]See, for example: O. Scheid, AUTOSAR Compendium, CreateSpace Independent Publishing Platform, August 2015.

that a new architecture can help reduce complexity and create a viable basis for future developments. Important concepts like centralizing and virtualizing computer architecture, local data concentrators, and "X by wire" can be adopted and adapted to the needs of automotive design—today more than ever. Robotics may also be of interest for a reorientation of ICT architecture in the automotive industry.

The logical architecture for controlling service robots, with its division into environmental perception, planning, and action, can particularly serve as a model for a logical architecture in the automotive industry. Important concepts from middleware architectures in robotics may also be of interest for the automotive sector, such as is being currently observed with the success of the Robot Operating System (ROS) [3], which in turn can only be a first step on the way to a quality-controlled industry-grade operating system for robots.

Let us now speculate briefly on how the future of ICT architectures would look if we learn from what we have observed in the past. As Fig. 2 shows, ICT architecture could thus develop in three steps. In an initial step, which is already going on today, ICT modules are integrated and encapsulated at a high level. In the second step, the ICT architecture could be reorganized with reference to all functions relevant to the vehicle. And finally, a middleware that integrates both the functions relevant to driving and the nonsafety-critical functions for comfort and entertainment would make it possible to customize vehicles for their drivers by integrating third-party software. Consequently, the automotive industry could manage the upcoming changes in two phases:

- **First Phase: Low Function/Low Cost**. This scenario is the most suitable for new market actors focusing on low-cost vehicles. The vehicle functionality and customer expectations for comfort and reliability are relatively low. The resulting requirement set is well suited for introducing a revised, simplified ICT architecture that is based on a drive-by-wire approach; actuator components are connected directly to the power electronics and the ICT. Actuators have a local energy supply and can be controlled via software protocols, reducing the number of cables and control devices.
- **Second Phase: High Function/Low Cost**. This assumes that ICT introduced in phase 1 has been optimized over the years and is now very reliable so that even customers with high expectations buy vehicles based on this architecture. This trend is reinforced by the ability to integrate new functions easily into vehicles and to customize them.

In summary, it is obvious that the necessary complexity of functions implemented by electronics and software will rise significantly:

1. Automated and autonomous driving will become state of the art very shortly. These functions are very complex and have high-performance requirements on the ECU. But even more exacting is a new requirement on E/E architectures: the system must become fail operational. While today it is state of the art to simply shut down a function as soon as an error is detected, in the future, fall-back functions must be provided to ensure that the vehicle can still be driven in a safe state even if there is a partial or even complete breakdown.

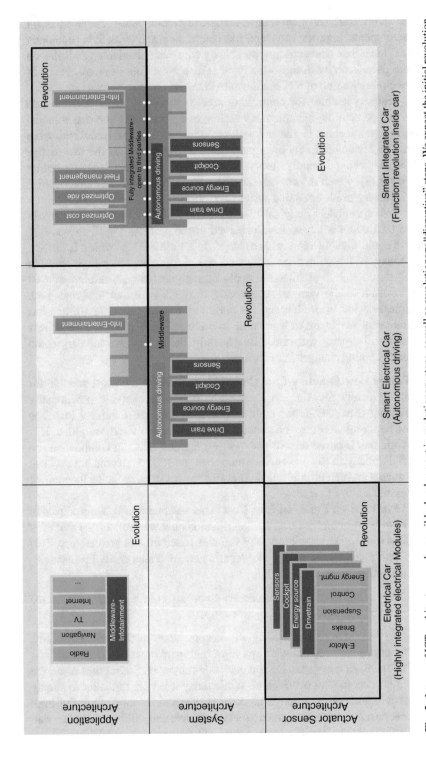

Fig. 2 Layers of ICT architectures and a possible development in evolutionary steps, as well as revolutionary "disruptive" steps. We expect the initial revolution at the level of the actuator-sensor architecture (and then evolutionary steps after that). Likewise, a disruptive development may take place at the system level and then finally at the application level

2. Connected mobility is another massive trend that will require an increasing interaction of the car with its infrastructure and with other IT-driven domains. This will not result in an increase in system complexity but generates another big challenge: security. While in the past cars were isolated systems, the growing openness of the E/E platform requires treating security as a key design criterion.

In summary, besides meeting additional functional requirements, future E/E architectures will have to provide additional extrafunctional properties, like a fail-operational system and security. Moreover, highly integrated mechatronic components are reinforcing the trend toward X-by-wire control or even X-by-wireless one. Through the introduction of middleware architectures and by encapsulating and abstracting from those mechatronic modules, integration can take place at a much higher logical level. Components to merge sensor data will become important elements of middleware architectures, along with mechanisms that ensure that safety-critical functions are separated from noncritical ones. They can be performed on a single computer without interfering with one another. This in turn will result in the centralization of all the computers in the car, similarly to server technology.

We will now look at the requirements in more detail. For an overview of the initial concepts at the time of the project's start, see [4].

3 A Set of Requirements for a New Architecture

In recognition of these trends, RACE[4] was started to establish a technology addressing the upcoming challenges. The main intention of RACE is to provide OEMs with an example of a platform technology that enables them to redesign their E/E architectures. By providing generic hardware components, a related runtime environment (RTE), and system engineering tools, OEMs can introduce new functions in most cases as software rather than the way it is today—as an ECU. This way, OEMs can benefit from faster innovation cycles of software. In the following subsections, the project goals will be listed in more detail.

3.1 Integration of New Functions in Software to Achieve Faster Development Times

The core platform will speed up the development of new automotive functions particularly related to:

- **Integration of new functions as software**: the core platform should be able to execute hardware-independent applications from all automotive domains; this includes but is not limited to body/comfort, driver assistance systems,

---

[4]For a short introduction, see http://w3.siemens.com/topics/global/en/electromobility/pages/race. aspx

power train, chassis control, and occupant and pedestrian safety. Hardware-independent applications are all those applications that do not require specific hardware components (e.g., actuators with power electronics). By integrating these functions on one single platform, RACE also addresses the trend that the functions of different domains are increasingly interacting with each other and the domain barriers vanish.

- **Software updates in the field**: users require continual updates to increase the functionality of their cars. Therefore, frequent software updates in the field of both of the platform and functions are an essential requirement. Agile development methods are to be supported.
- **Scalability and reuse**: the RACE core platform shall support the scalability to different platform variants. Reuse of applications must be supported.

This is accomplished by number of different concepts, most importantly a rigidly implemented publish/subscribe mechanism, along with a data-centric structure that not only ensures the necessary decoupling but also guarantees data consistency and sparsity—data are produced and stored at only one place across the whole system. Wherever possible, these concepts are supported by formal checks. For example, checks are made to determine whether a complete and unambiguous data flow graph can be obtained from the collection of application software components and modules. Moreover, if there is a subscriber with several potential (publishers in the computed data flow graph), then the RTE is checked to see if it has the appropriate data fusion methods available (used, for example, for redundant sensors). If this is not the case, the configuration will be aborted, and a corresponding message of the reason is generated.

However, it was beyond the scope of the RACE project to produce a complete guarantee of the overall behavior and timing of a RACE application architecture. To check whether each component is supplied with data is certainly not sufficient to ultimately ensure the functional integrity of the application architecture. This, however, is an interesting subject for follow-up research.

3.2   Enabling New Business Models by Software Updates and Opening Function Development to Third Parties

In order to meet the requirements arising from the OEMs' desire to permanently develop new business models around their cars in order to be able to react to changing market needs, the core platform shall support the integration of functions even in after-sales market. There should be no need to integrate these functions during the original design phase of the car. Moreover, the architecture should be open to the integration of third-party applications. This integration should not be restricted only to applications from today's Tier-1 suppliers but support the creation of new ecosystems (see Sect. 7).

3.3   Built-In Safety and Security

The core should be designed such that it provides the necessary safety and security for accommodating functions of the highest criticality level. It should support the development of dependable systems covering:

- Availability: readiness for correct service up to fail-operational quality
- Reliability: guaranteeing correct service
- Safety: absence of unreasonable risk up to Automotive Safety Integrity Level D (ASIL-D)
- Integrity: mechanisms to inhibit improper system alteration
- Maintainability: ability to undergo modifications and repairs
- Testability: simplifying verification strategies and analyzing misbehavior
- Security: protecting automotive software systems from unauthorized access, use, disclosure, disruption, modification, perusal, inspection, recording, or destruction

Furthermore, the core should provide the basis for functions with fail-operational requirements. The RACE core platform should support the execution of applications in fail-operational mode. Even in case of a subsystem failure, the core platform would guarantee the correct execution of the application.

The current software version of RACE runs under the PikeOS[5] hypervisor real-time operating system. This allows for third-party software to work "in isolation" and to prevent system crashes due to programming errors. While according to the above list of requirements there was some work undertaken to design a hardware security module for authentication, the implementation of security measures resulting from the integration of potentially dangerous and malignant software was not in the focus of the development work. However, it was made sure that no design decisions were taken that would prevent security measures to be implemented. By the same token, we expect this spatial and temporal separation to be a key component in meeting safety requirements: a hypervisor like PikeOS is used in many mixed-criticality environments and has stood the test of large-scale deployment and can, therefore, be expected to also meet the safety and software integrity needs in future cars.

3.4   Simplifying Migration from Other Platforms

Clearly, the core platform should support all applicable international automotive standards and state-of-the-art technologies. Furthermore, the core platform will support the collaboration between various partners by standardized data exchange

---

[5]https://www.sysgo.com/products/pikeos-hypervisor/

formats and support the integration of application software from various partners on a single ECU via a run-time environment and across the entire vehicle network.

Despite its superior functionality, production costs for a customer system platform based on the RACE core should not exceed the costs of a traditional E/E architecture. Moreover, the nonrecurring costs should be driven down to a minimum. Since these costs are hard to measure, it is important that production costs by themselves be equivalent or even lower.

4 RACE Architecture Concepts

To meet the requirements listed above, a completely new architecture has been developed, implemented as an advanced prototype, and integrated into a number of demonstrator cars (Fig. 3). The leitmotiv has been to design an architecture that fits the needs of information processing ("future cars will be computers on wheels"), capitalizes on the rich experience in Computer Science and Informatics in the design of mission-critical distributed systems, and makes it possible to keep pace with the rapid progress in methodology and tools for software design.

The central concept is that of a platform, centered about functions that are integrated very easily. The set of functions can hence change very easily and—taken as a whole—constitutes the complete functionality of the car controller across all domains. But not only that: since our focus is on communications and

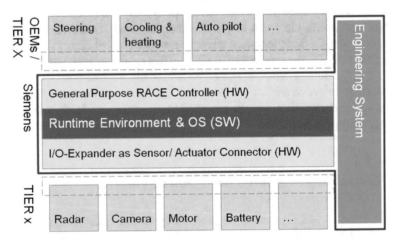

Fig. 3 Overview of the RACE architecture. The complete system consists of the controllers (GPRCs), the run-time environment (RTE) running on centralized hardware, and the IO expanders and interfaces. An integral part of the system is the engineering system, which provides all the software tools for programming the components in a certifiable way. The goal here is to abstract as much as possible from the hardware and to provide a "single system illusion" to the programmer

the communication protocols used within the car and those used outside the car are mostly identical, there is no longer a fixed barrier between the car and the infrastructure—this border becomes irrelevant.

4.1 General Structure and Communications

RACE is based on an execution platform consisting of General Purpose RACE Controllers (GPRCs), implemented as so-called Duplex Control Computers (DCCs) (see the following section), and a real-time run-time environment (RTE); see [5]. On top of this platform, OEMs can integrate their functions purely as software components. The architecture is designed to be scalable; if further processing power or resources, such as more memory, are required, additional RACE controllers can easily be added. The GPRCs communicate via a high-bandwidth, real-time Ethernet using the upcoming IEEE 802.1-TSN [6] standard.

Sensors and actuators can be integrated via common interfaces (CAN, local interconnect network (LIN), Ethernet, digital/analog input/output (IO)). If more IOs are required than what is offered by the GPRCs, IO expanders (IOXPs) can be attached to the GPRCs. An IOXP provides a number of IO interfaces, but in contrast to a GPRC it does not offer enough processing power to execute complex software functions. Its main purpose is to interface with sensors and actuators, which can provide any degree of local intelligence and/or data compression and/or preprocessing capabilities. In any case, the main benefit of using such a hierarchical system of data sources and data concentrators—as supported by our architecture— will substantially cut back on the cable harness, which may significantly reduce costs and failure probabilities.

Another concept that leads to significant cost and development time reduction is reusability. Software functions communicate via OEM-defined interfaces with each other and with the sensors and actuators. As a result, these functions can be reused across different vehicles. If an OEM so chooses, these interfaces can be shared throughout the industry (or collaboration partners), and reuse may even take place across the industry.

The system engineering tool chain includes a test system and a continuous integration solution. The tool chain is optimized for agile development. New software components can be tested seamlessly at all levels, from software in the loop, by way of hardware in the loop up to vehicle in the loop. The test system enables fault injection to test different scenarios, such as rare and typically irreproducible component errors. A configuration tool simplifies the integration and building of new vehicles, automating many steps that in the past had to be done by system integrators and were very cumbersome.

Altogether, the scalable platform, reusability of functions, and system engineering environment and tool chain offer a fast and flexible way to bring new functions or changes to future cars. We will now look at specific safety and security aspects in a little more detail.

4.2   Built-In Safety and Security

RACE implements a "Safety-Element-out-of-Context" approach as defined in ISO 26262. The run-time environment (RTE) offers a separation of all software functions to eliminate any unintended interactions between software components. As a consequence, the RACE platform can execute functions with mixed criticality on the same controller (see Fig. 4).

4.2.1   Separation Concept

To further simplify the development of safety-critical systems, there are several built-in safety mechanism patterns for different safety levels. GPRCs are currently designed with two-channel controllers, where each channel or "lane" has its own processing unit. For functions with no safety requirements (quality management), the function can be deployed on any channel. If the function has safety requirements but can be shut down in case of a failure, the function can be deployed to both channels of a GPRC. The replication can be realized as homogeneous replication (duplication the function) or diverse replication (two diverse implementations or a "productive function" and a "plausibility function/watchdog"); see Fig. 5.

The run-time environment monitors the consistency of the results on both channels and can shut down the function in case of an unwanted deviation. Functions with fail-operational requirements (meaning a function must still be operational even in case of component failures) are also available. These are implemented as a master-slave mechanism provided by the run-time environment. This mechanism allows the deployment of a function on two GPRCs to ensure the correct execution even if one GPRC fails. Depending on the required fail-over times, the system can be configured for cold or hot-standby mode.

4.2.2   Scalable Safety

All these safety patterns are based on an indication-based health-monitoring system built into the RTE. The RTE permanently monitors the data flow and execution of components. In case of a deviation from normal behavior, the RTE raises an error indication. A health monitoring component collects all these indications and determines the health status of the different fault-containment regions on application component, hardware, and network level. Several mechanisms are available to react to the failures of a component, ranging from fault masking if redundant results are available to the actual shut down and separation of the faulty component.

Besides these safety mechanisms, RACE also incorporates a configurable security mechanism, such as secure boot, authentication, authorization, and encryption. Several mechanisms of the RTE contribute to safety and security at the same time.

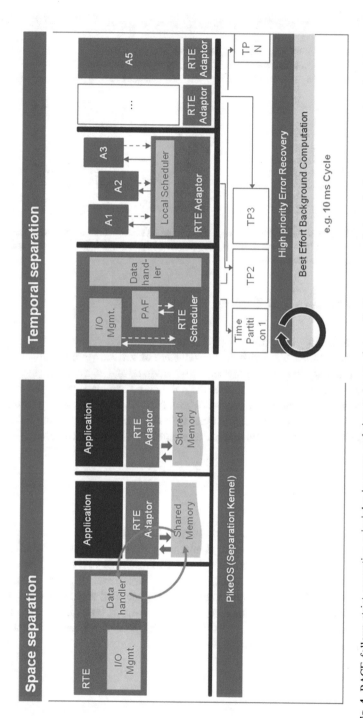

Fig. 4 RACE follows strict separation principles in space and time to ensure that applications can run independently—our approach to handling mixed criticality on one hardware platform. Currently, the underlying separation kernel is implemented in the PikeOS operating system, but other systems can be used as well

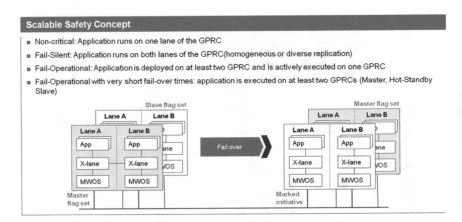

Fig. 5 Illustration of the scalable safety concept. Depending on system configuration, applications can be run that are noncritical or applications with fail-silent/operational behavior—all in mixed mode on the same physical controllers

5   Implementation and Tooling

We believe that the most important aspect when a disruptive architecture is to be introduced is the compatibility with the current state of software engineering.

This will not only result in potential cost savings and a dramatic increase in productivity, but it will also enable the OEM to keep pace with the rapid developments in the consumer electronics world. Nevertheless, it is also of utmost importance to keep abreast and keep in sync with the developments in the field of mission critical systems in general and in autonomous systems in particular.

5.1   Information Flow

Figure 6 illustrates the information flow, which implements a "perception—cognition—action" cycle. This makes it not only possible to adapt paradigms from cognitive system theory and practice, but it also results in very logical layering of the processing functions at an easily comprehensible level of abstraction. At the lowest level, the data from (smart) sensors are generated and are routed into the system through a communication layer (middleware). Likewise, through this communication layer, the signals needed for behavior generation, i.e., the data for the (smart) actuators, are also distributed.

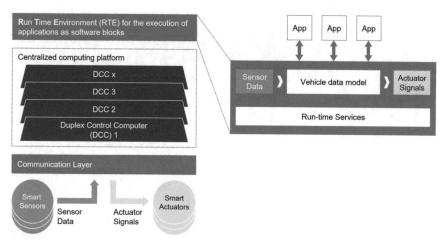

Fig. 6 Illustration of the basic implementation of a concrete car with four central Duplex Control Computers (DCCs). The general structure supports the construction of a "perception—cognition—action" architecture

   The middleware subsystem used in RACE is based on CHROMOSOME [7]. In principle, other middleware systems could also be used, such as Data Distribution Service (DDS) [8]. One layer above the communication middleware is the layer of centralized processing through the control computers; these are the dual-lane processing elements described above. Depending on the processing power that is needed, there can be an indefinite number of DCCs.

   The execution layer (run-time environment (RTE)) is responsible for the (virtual) connections between sensors and actuators. These connections can be dynamically generated, using the central data model of the vehicle, i.e., the abstraction of sensors and actuators. The RTE also provides the fail-operational services, as well as the "Plug & Play" management of all entities. The "Apps" correspond to the (retrofittable) automotive applications shown in Fig. 3.

   Figure 7 shows an example of how this general architecture may be mapped to a topology in a real car: there are two DCCs, one for the drivetrain and energy source control, the other one for braking and steering. Hardware redundancy on the communication level is achieved by double ring structure: an inner ring for the direct DCC communications and an outer ring for sensors and actuators. Both rings are doubled so that the physical failure of one ring does not lead to a complete system failure. The communication protocol is a partial implementation of [6]; the DCCs can run mixed-criticality applications. There are also provisions for ingress/egress rate limiting to isolate faulty components, and there is hardware support for the precision time protocol (PTP).

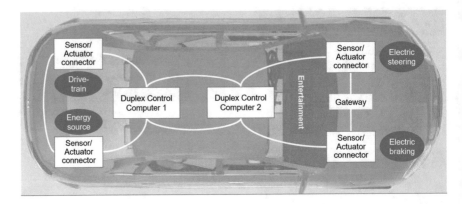

Fig. 7 Example of a topology of in a car with two central Duplex Control Computers. Inner glass-fiber ring (red) for DCC to DCC communication, outer ring (white) for sensor and actuator communications

5.2 Software Design

As mentioned above, one of the primary goals of RACE is to create a structural environment, i.e., the RTE, in which the latest developments in computer science, software engineering methods and tools, as well as insights from research into autonomous vehicles, can easily flow together. The resulting requirements on the RTE based on these project objectives are:

- The complexity of the software system should be reduced to the absolute essential minimum (see introductory section). This is realized by providing a modular development environment based on the decoupling of application modules that can establish dynamic communication links via virtual channels with guaranteed quality-of-service levels.
- Introduction of new complex functions at run time should be supported. This is realized by abstracting all automotive domain functions to the software level. New functions can hence be introduced simply by adding software modules. Clearly, this has not been done yet, but the foundations have been laid in RACE. This can be compared to the developments in the smartphone domain: they already combine a plethora of sensor modalities (vision, audition, touch, inertia . . .) in one single device, and hardly any new function needs a hardware accessory—it is all performed within the software. It is likely that at least in the sensor/communication domain, customers will expect the same from their cars.
- Plug-and-play capabilities are also mandatory. Customers will want to add new functions, or they want to attach new devices with added functionalities, such as additional entertainment equipment. If new complex functions are added (e.g., more sophisticated assistance and/or autonomy functions), this may even require the addition of one or more duplex control computers (DCCs), which means that the RTE has to be highly scalable.

- Scalability with respect to functions and computation platform is also a must. This is approached by providing automatic configuration tools and functions wherever possible and therefore provide a maximum of scalability with respect to functions and resources.
- Certifiability is of the utmost importance. We believe that in the future, formal methods will have to be applied to ensure correct system behavior in all circumstances. However, until the underlying theory is powerful enough, we must provide all the generic mechanisms for safety and security, which will have to comply with the present standards for mission-critical systems.

Finally, there should also be a clear migration path from today's architectures to an environment like RACE.

Looking at the current state of run-time environments, and starting from that point, in RACE we have added the following features: (1) data-centric design, (2) segregation in time and space for running mixed-criticality applications on cost-effective hardware, (3) configurable safety and security mechanisms that enable tailored fault detection and recovery, (4) testing environment that can inject faults and trace them, and (5) automatic configuration functions at all levels.

It is beyond the scope of this article to elaborate on all of these topics. However, we describe one aspect of the RACE implementation: the data-centric communication design based on CHROMOSOME (see Fig. 8). The state of the art is to explicitly "wire together" senders and receivers via messages over a bus system. This results in strong coupling, an inflexible topology, and very often a redundant data acquisition and processing. The basic concept for data exchange is the decoupling of physical connections and logical communication relations. Moreover, data types can be associated not only with a syntax but also with semantics (why is it there? how is it used? etc.) and attributes (e.g., how precise is it?). These properties are stored in a central "topic dictionary," which lends itself to automatic configuration because dependencies, contradictions, redundancies, etc. can be checked automatically. All of these entities are supported by powerful tools, typically adapted from open-source tools, like Eclipse, that underpin rapid access to all system variables with all associated information, stored in one central repository.

Such automatic configuration—together with the modularity that enables the "plugging together" of predefined parameterized components—can be of great help in the system integration process. At the same time, the RTE registers and analyzes different error indications, which are accessible to all the DCCs in the system (see Fig. 9). There is a very fine-grained error handling available, with several reactions possible as a result of an error/failure indication. Errors inside the RTE can be handled by error masking (if there is redundant data). The application management can decide on a graceful degradation of the application and an adjustment of performance level, can change the master/slave role assignment, and/or activate/deactivate masters and slaves. Finally, the overall platform management can deactivate DCCs, and it can determine the whole platform mode, including an emergency shutdown after driving the system into a safe state.

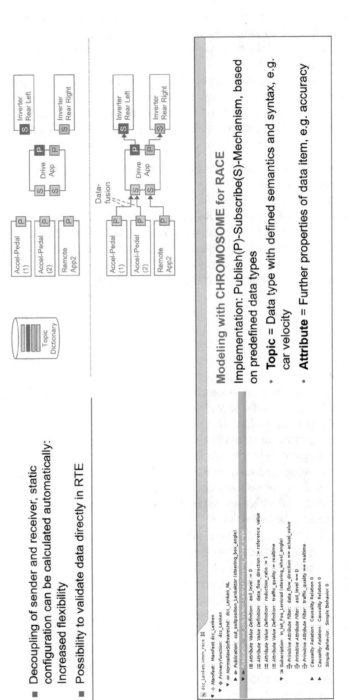

Fig. 8 Data-centric design in RACE

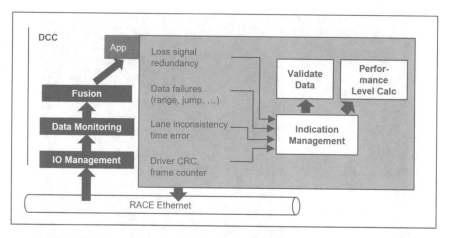

Fig. 9 The RTE registers and analyzes different error indications and makes them accessible system-wide

6 Realization on the Hardware Level

We briefly outline the design of the hardware that was developed for the prototypes. We have realized a number of successively complex testing configurations, equipped with prototypes of the hardware—starting from desktop "breadboards," by way of laboratory setups to complete cars. These cars (built by Roding GmbH, a manufacturer of small-series sports cars) were tested on a specially designed test rig (Fig. 10).

These cars have two wheel-hub motors and a steer-by-wire system without mechanical fallback, and they are a carbon/aluminum lightweight construction with a total car mass of 1250 kg. The braking system is a fully electric braking "future brake system" from TRW; the steer-by-wire subsystem was provided by Paravan GmbH. The E/E architecture is a redundant design based on RACE with an Ethernet ring structure, as depicted in Fig. 10, which uses of IOXPs for connecting sensors, actuators, and Human Machine Interface (HMI). The overall performance is 126 kW (up to 330 kW for a limited period of time); the overall torque is 1000 Nm (up to 2500 Nm for a limited period of time). Power electronics operate at a voltage level of 720 V; the battery capacity is 20 kWh.

Figure 11 shows the topology of the connections of the DCCs with all the units in the car, including the HMI and the steering wheel (an example of a redundant connection). The cars were built and programmed by a small project team. Several mission-critical functions were implemented and tested successfully. A complete set of functions, which would have enabled the cars to drive on a test site, was

Fig. 10 The cars equipped with the RACE architecture. Top: physical appearance, bottom: car on test rig with direct coupling (all four wheels) to external electric motors that can induce very realistic driving dynamics scenarios

not implemented in RACE. As a project continuation, a parcel delivery vehicle was equipped for road testing by Siemens AG after the end of the RACE project (Fig. 12).

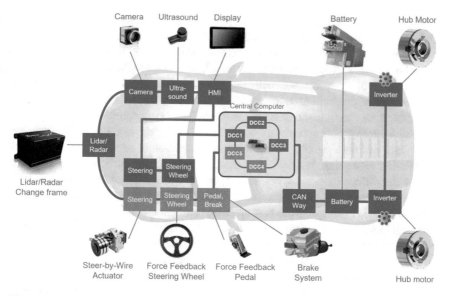

Fig. 11 The topology of the sensors, DCCs, and actuators in the prototype cars

Fig. 12 The physical realization of the DCC (left) and the IOXP (right). The DCC has two ARM A9 CPUs and four RACE Ethernet connectors, as well as two test Ethernet. The IOXP features one CPU and several I/Os: CAN, LIN, digital, and analog. It also has two RACE Ethernet and one Test-Ethernet connector

7 Deployment and Business Opportunities

There has been an ongoing discussion about the commercial viability of alternative architectures, as outlined in the introductory section of this article. Clearly there are lots of arguments in favor of evolutionary approaches and certainly just as many arguments for disruptive changes. The problem for decision makers is that the question in general is virtually undecidable because there are so many factors that influence the decision. These range from nonrecurring costs by way of integration cost for every specialty of a car series, to the cost and risks resulting from customer desires for functional update over the lifetime, to the maintenance costs of (reusable) software, and to the replacement strategies of obsolete hardware.

The major challenge of automotive OEMs is the complexity of the E/E architecture conflicting with the required faster innovation cycles in the age of digitalization. This issue can be addressed with an approach like RACE because it reduces the number of controllers, minimizes the heterogeneity of network technologies, and offers several generic services that need to be covered during application development. Another complexity issue addressed by our approach is the undesired redundancy of data signals. With each new generation of cars, the number of data signals increases significantly.[6] The reason is mainly the difficult access to data across the different domains, controllers, and network technologies. A centralized approach offers a solution to this issue, and RACE alleviates this problem further by its data-centric design and the homogeneous hardware platform. Even if RACE is not used down to the implementation and hardware level, some of its concepts may still be used as a basis to redesign the function architecture with modularity and data centricity and be reused at the application level only.

Altogether, the approach pursued here enables OEMs to build up a common architecture across the different car platforms produced by the manufacturer, including the possibility to reuse the functions across the different cars. New and innovative suppliers can enter the automotive market and readily integrate their functions. While in the past new suppliers had to provide an integrated electronic control unit (ECU) with their function (software), which made it difficult for them to enter the market, they can now deliver just the software as a safety element out of context. This makes it also possible for the OEMs to reduce their dependency on tier-one suppliers. In addition, this approach also simplifies the in-sourcing of differentiating "brand-defining" functions by the OEM.

Finally, the approach using generic hardware and a high-integration platform offers the possibility to switch from a "bill of material" business model to a "revenue over lifetime" business model. Today, the electronics are individually optimized for concrete functions, which come with their own ECU. It is hard for developers to request additional resources to prepare for later software updates—or even new functions—because these resources would need to be stretched over a very high number (typically over 100) of ECUs. With the drastically reduced number of controllers that implement application-level functions, the software deployment, support, and update mechanisms can be changed in such a way that after-sales can be significantly increased. Therefore, a modular approach like this architecture lends itself to business developments along several lines:

- The RTE is the core offering, including a configuration tool.
- The full system is offered as a modular toolbox to be used (i.e., integrated and configured) by OEMs with focus on safety up to ASIL-D and fail-operational

---

[6]Some OEMs assume that the instantaneous speed of a car (i.e., one unambiguous variable) is identified by more than 20 different functions inside the car, e.g., by direct measurement, derivation from other sensor signals, or an estimation.

mode; it includes an operating system from a preferred supplier, communication system, and hardware development support.

- The application software development is delegated to a partner network. While RACE can be used as a "neutral" and open platform, third parties can integrate their functionality easily. It offers application engineering support up to Reference implementations.
- Fully integrated RACE controllers or support in HW design and manufacturing services can be provided.
- Integration services are not within the focus today, but customers can profit from the experience and expertise/Know-How in the RACE team.

The complete substitution of a proven architecture basis in large-volume car series without changing the structures of the producers' organizations is clearly infeasible. We therefore suggest collecting experience from the producers of smaller series, e.g., of special-purpose vehicles with low production volume ($\sim$3000 cars per year), which are individualized for specific customers. An interesting example is StreetScooter,[7] a company producing specialized parcel delivery vehicles. Due to the low volume of the market, this domain is not relevant for big OEMs and smaller companies can enter the market, testing the viability of their ideas. The major challenge of these manufacturers is exactly this low volume. It is not profitable to design specialized controllers for these cars, and therefore the manufacturers depend on already existing controllers from first-tier suppliers. However, due to the low volumes, required modifications are not given very high priority.

8 Summary

RACE is designed to fully support the requirements of upcoming automotive functions. With respect to automated driving, RACE offers a fail-operational platform, allows the integration of functions with different criticality levels on one controller, and offers high performance and safety simultaneously. Especially the latter is of importance as today's high-criticality solutions are based on processor technologies with low performance. Connected mobility is addressed by a high-bandwidth network based on [6], built-in interfaces to web services (prototype), and built-in security mechanism. RACE has developed several unique properties that will be required in the future:

- *No tradeoff between performance and safety*: while today's solutions rely typically on high-performance processors for functions with lower criticality levels (up to ASIL B) and low-performance processors for functions with high criticality levels, RACE has the potential to provide full processing power up to

---

[7]http://www.streetscooter.eu/

the highest safety levels. It is also a complete solution for functions with fail-operational requirements.

- *Hardware designed for use in series production*: standard suppliers optimize the controllers based on customer requirements. This is, however, only feasible for larger volumes. RACE provides controllers ready for application in car production. The low volumes of each single application are compensated by the general applicability across different cars.
- *Solution optimized for in-house configuration and integration*: today, integration of functions is typically done by the supplier. This leads to high costs and delays in development. RACE offers the possibility to configure the platform and to integrate software components by the OEM, shortening the development times. Configuration and integration tools automate many previously tedious tasks and reduce the risk of introducing errors. With RACE, the OEM has the choice to decide whether configuration and integration are done in-house or by a service provider.
- *Qualified tool chain and infrastructure for agile development*: RACE will provide a qualified tool chain and infrastructure for agile development, reducing the development time. Similar to prototyping platforms, RACE will offer a seamless integration of development tools. The result, however, will be serial code running on serial hardware.
- *Designed for testability*: RACE RTE was designed with testability in mind. All data flows can be monitored by a nonintrusive test system guaranteeing exactly the same behavior with and without the test system. Furthermore, a fault-injection infrastructure is available to provoke specific situations simplifying the verification systems significantly. This is made possible through a dedicated and fixed scheduling time slot for testing. This approach sacrifices some time and computational power, but we consider the ease of testing that results from it a good trade-off (see [9]).
- *Designed for updates*: RACE offers direct support for integrating new functions or updating existing functions via updates over the air. Via this mechanism, automotive functions become a freestanding product. The "Plug & Play capability" allows for new business models for the aftermarket.

Acknowledgments The development of the RACE platform was supported by the German Federal Ministry for Economic Affairs and Energy (http://bmwi.de/); see http://www.projekt-race.de/en/.

The authors wish to express their gratitude to the whole team who made the development of the concepts, tools, and the cars possible in record time. This project was only possible through a real concerted team effort and a lot of passion on all sides. Clearly, an overview paper like this can only describe results at a rather high level. The authors are happy to provide additional information on request.

References

1. Bernard M et al (2010) The software car: information and communication technology (ICT) as an engine for the electromobility of the future. A study for the German Federal Ministry of Economics and Technology. Published by fortiss GmbH. http://www.fortiss.org/ikt2030/
2. Watkins CB, Walter R (2007) Transitioning from federated avionics architectures to Integrated Modular Avionics. In: 2007 IEEE/AIAA 26th digital avionics systems conference, pp 2.A.1-1–2.A.1-10. https://doi.org/10.1109/DASC.2007.4391842
3. Quigley M, Gerkey B, Conley K, Faust J, Foote T, Leibs J, Berger E, Wheeler R, Ng A (2009) ROS: an open-source Robot Operating System. In: IEEE-ICRA workshop on open source software in robotics organized by Hirohisa Hirukawa and Alois Knoll, Kobe, Japan, May 2009
4. Sommer S, Camek A, Becker K, Buckl C, Zirkler A, Fiege L, Armbruster M, Spiegelberg G, Knoll A (2013) Race: a centralized platform computer based architecture for automotive applications. In: Vehicular electronics conference (VEC) and the international electric vehicle conference (IEVC) (VEC/IEVC 2013), IEEE, October 2013
5. Becker K, Frtunikj J, Felser M, Fiege L, Buckl C, Rothbauer S, Zhang L, Klein C (2015) Race RTE: a runtime environment for robust fault-tolerant vehicle functions. In: Proceedings of the CARS workshop, 11th European dependable computing conference – dependability in practice, 2015
6. http://www.ieee802.org/1/pages/tsn.html
7. Buckl C, Geisinger M, Gulati D, Ruiz-Bertol F, Knoll A (2014) CHROMOSOME: a runtime environment for plug&play-capable embedded real-time systems. In: Sixth international workshop on adaptive and reconfigurable embedded systems (APRES 2014), ACM, April 2014
8. http://www.omg.org/spec/DDS/
9. Fröhlich J, Frtunikj J, Rothbauer S, Stückjürgen C (2016) Testing safety properties of cyber-physical systems with non-intrusive fault injection – an industrial case study. Proceedings of the workshop on dependable embedded and cyber-physical systems and systems-of-systems (DECSoS). In: Skavhaug A et al (eds) Proceedings of the workshops international conference on computer safety, reliability, and security (SAFECOMP), vol 9923, Springer, LNCS, pp 105–107

Development of ISO 11783 Compliant Agricultural Systems: Experience Report

Enkhbaatar Tumenjargal, Enkhbat Batbayar, Sodbileg Tsogt-Ochir, Munkhtamir Oyumaa, Woon Chul Ham, and Kil To Chong

Abstract The connection of different modules from different manufacturers into a single bus for the exchange of data and control is a challenge for the agricultural machinery industry using ISO 11783 standards (called ISOBUS in the market). It shows strong potential to become the de facto standard for the exchange of data between modules on the agricultural tractor. This research presents the development of an ISOBUS monitoring system and virtual terminal (VT) for agricultural vehicles. The graphical user interface (GUI) of VT is developed on the embedded system by using the Qt with cross-platform for an ARM Cortex-A9 microprocessor named by Freescale i.MX6 Quad. The GUI application programs were developed based on the Isocore-suite commercial library by the OSB AG Engineering company and certified by the Agricultural Industry Electronics Foundation. The implemented electronic control units (ECUs) and ISOBUS monitoring system were developed by the ISOAgLib open-source library, in addition to tools such as the vt-designer, the vt2iso, the CAN server, the CAN messenger, and the CAN logalizer. The implementation of ISOAgLib is fully compatible with the ISO 11783 standard. The hardware implementation is the development board for the STM32 ARM Cortex-M3 microcontroller. The implemented ECUs were experimentally tested on the ISO 11783-compliant intelligent monitoring system AFS Pro 700 for the New Holland Agriculture tractor. Also, we simulated VT-Server and implemented the sprayer, the manure spreader, the global positioning system modules with the Kvaser PCIe CAN device, and PCAN-USB device in order to analyze all CAN messages and network protocols such as the transport protocol (TP), extended transport protocol (ETP), address claiming, and request parameter group number (PGN) messages. Finally,

E. Tumenjargal (✉)
Department of Electronic Engineering, Chonbuk National University, Jeonju, Republic of Korea

Mongolian University of Science and Technology, Ulaanbaatar, Mongolia
e-mail: t.enkhbaatar@must.edu.mn

E. Batbayar · S. Tsogt-Ochir · M. Oyumaa · W. C. Ham · K. T. Chong
Department of Electronic Engineering, Chonbuk National University, Jeonju, Republic of Korea
e-mail: enkhbat@jbnu.ac.kr; wcham@jbnu.ac.kr; kitchong@jbnu.ac.kr

© Springer Nature Switzerland AG 2019
Y. Dajsuren, M. van den Brand (eds.), *Automotive Systems and Software Engineering*, https://doi.org/10.1007/978-3-030-12157-0_9

we present an ISOBUS object pool (IOP) binary file from the implemented ECU and an interpretation of IOP files shown on the CONLAB-VT.

1 Introduction

Since 2010, our team has been working on the ISO 11783 standard by implementing a virtual terminal (VT) and universal gateway for agricultural machinery. The EUREKA foundation supports that project. How can the ISOBUS network be monitored? Our development based on the parameter group number (PGN) message gives us more information on the connection between the electronic control unit (ECU) and VT implemented. An ISOBUS-PGN message is a unit number following the ISO 11783 standard and J1939. The main challenges of implementation of the VT are what does the ISOBUS really look like, what is the practical experience with Qt graphical user interface (GUI)-based application development, how compatible are the tractor systems, how can functional safety be guaranteed, what about the certification of the Agricultural Industry Electronics Foundation (AEF) and the ISOBUS plugfest? These are challenges for the entire agricultural industrial sector. Agricultural machinery control is an interdisciplinary field of study concerning the integration of mechanics, electronics, and software engineering expertise. Today, a new generation of tractors exists with capabilities so advanced that they can be entrusted with many of the roles and responsibilities that were once handled manually. This evolution in tractors is the direct result of continuing research advancements among its constituent disciplines. The ISO 11783 standard has, and continues to be, an active area of research within the agricultural engineering community. Before the advent of ISOBUS, farmers were restricted to using specific precision displays and proprietary operating systems for the control of machinery. This led to reduced efficiency owing to the lack of compatibility between proprietary solutions from various manufacturers, which often cannot communicate with each other. Farmers regularly work with tractors and other implements (components) from different manufacturers. The absence of standardization between electronic systems from these various manufacturers makes it complicated to use them, as the different proprietary solutions mean that the implements do not connect smoothly or may not connect at all. Another problem is that the tractor and each of the components require their own individual terminal for data exchange and for control of the machine. For a while, this situation was disappointing because it is not feasible to use innovative products [5, 6], hence the need for the ISOBUS standard. After the publication of the ISOBUS standard, agricultural equipment manufacturers invested a huge amount of money in the development of this standard and even more on the design of an ISOBUS-compliant product. Some tractors and many types of machineries support the ISOBUS standard nowadays, which means that this is the right time for ISOBUS to continue to grow. The ISOBUS standard provides many advantages to farm owners, namely, comprehensive tools and standardization of the control setting interface, a better overview in the cabin, simpler connection between tractor and implement via "plug in play," system reliability, in addition to cost and time-savings due to precision agriculture. ISOBUS allows the downloading of tasks of the planned field operation onto the farm management computer, along with calculations required for control. It also supports the eventual uploading to the operator's terminal (known as the monitoring system) in the cabin through an SD card, a USB, WLAN, or a wireless network. It is possible to plan and evaluate the use of resources for farmers. In addition, ISOBUS is especially suitable for cooperation between large manufacturers and small suppliers that can be eventually provided to farmers. ISOBUS will be the main connection between tractors and implements in the future; furthermore, every tractor and machinery manufacturer will fully support the ISOBUS standard. The farmer will only have to connect their ISOBUS implement, and the ISOBUS will bring up its own operator terminal. The ISOBUS group is primarily composed of eight manufacturers and three associations plus an additional 200 members. It is forecast that the number of ISOBUS device manufacturers will rise sharply and may interpret the standard in their own way, but this can lead to incompatibility problems. To address the problem, events are held where devices are tested by their manufacturers for compatibility, for example, the two annual plugfest events organized by the AEF, where manufacturers can test the plug-and-play compatibility of their devices. The AEF tests the compliance of a device with the ISO 11783 standard in their laboratories, establishes a database, and maintains transparency with regard to the functionality supported by specific products and their compatibility with others. Since the establishment of the ISOBUS standard, it has required a higher number of software systems. Software development, and especially software architecture, is the key to achieving cost savings in an ISOBUS system.

2   Background of the ISO 11783 Standard

The ISO 11783 standard establishes parameters for communication between ECUs in agricultural machinery and implements. This standard has 14 parts, with 11 published by the ISO, and was started in 1991 by an ISO committee formed by the union of two other standards, the DIN 9684 Standards Association of the German Deutsche Industrie Normen (DIN) and the SAE J1939 North American Society— Society of Automotive Engineers (SAE). The standard DIN 9684—Agricultural Tractors and Machinery or standard LBS (Landwirtschaftliches Bus-System— Mobile Agricultural Bus) was developed in Germany by groups of companies and institutions associated with the DIN. The first version of the standard was completed in 1997, with five parties, two of which had great influence on the standard ISO 11783. The SAE J1939 (2007)—Recommended Practice for Truck and Bus Control and Communication Network—was developed by the SAE J1939 Truck and Bus Control and Communications Subcommittee for applications in heavy vehicles such as buses, trucks, and construction vehicles. The working group (WG1) of the ISO 11783 first held discussions in February 1991 [4], addressing the development of a standard connector and adopting the use of the latest standard CAN 2.0B. Currently,

Table 1 Current status of the parts of the ISO 11783 standard [6]

Parts	Title	Status (international standard)
Part 1	General standard for mobile data communication	ISO 11783-1
Part 2	Physical layer	ISO 11783-2
Part 3	Data link layer	ISO 11783-3
Part 4	Network layer	ISO 11783-4
Part 5	Network management	ISO 11783-5
Part 6	Virtual terminal	ISO 11783-6
Part 7	Implement messages application layer	ISO 11783-7
Part 8	Power train messages	ISO 11783-8
Part 9	Tractor ECU	ISO 11783-9
Part 10	Task controller	ISO 11783-10
Part 11	Mobile data element dictionary	ISO 11783-11
Part 12	Diagnostic services	ISO 11783-12
Part 13	File server	ISO 11783-13
Part 14	Sequence control	ISO 11783-14

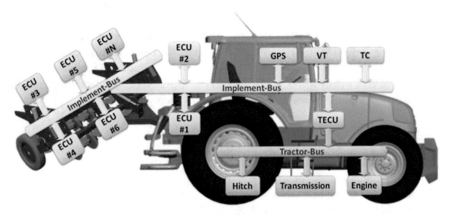

Fig. 1 Agricultural tractor network structure based on the ISO 11783 standard [1, 3, 4]

the ISO 11783 is formed of 14 documents. The current state of development of the standard is described in Table 1. Figure 1 shows a full tractor system that consists of the main network tractor bus based on SAE J1939 and an implement bus based on ISO 11783. The tractor bus is used for its own internal system ECUs such as the hitch control ECU, the transmission control ECU, and the engine ECU. It is supported by global or specific destinations in the network using its own specific source address (SA), and it can be integrated onto the monitoring system. The tractor bus may use undefined messages (data length, package size, or data description that are undefined in the standard) for the manufacturer's secret, in addition to proprietary messages specified in the standard.

Part 1 of the ISOBUS standard is the foundation of this standard. This section presents the terms and definitions used in the 13 remaining parts, the abbreviations of terms, the application of the Open Systems Interconnection (OSI) model standard, and the requirements of an ISOBUS network. Annexes of this part contain all message identifiers, address and preferred industry groups, control functions, and codes of manufacturers. Furthermore, there are also forms of codes for the requests of manufacturers, for new identifiers, or for their modification.

Part 2 of the standard ISOBUS describes the physical layer, the electrical parameters, the standard connectors, the minimum behavior of the network in cases of failure or loss of connection from the CAN bus and attached to a subdivision bit of time and examples of electrical circuits. The physical layer is based on protocol CAN 2.0 B and sets a rate of 250 kbps for serial communication. CAN 2.0 B also defines the physical environment, or the bus, consisting of four lines: two data conductors called CAN high and CAN low and two reference electrical conductors called terminating bias circuit power (TBC-PWR) and terminating bias circuit ground (TBC-GND). The standard connection between the tractor and agricultural implement is carried by the connector implement bus breakaway connector (IBBC), which is located in the tractor. The main function of this connector is to group the data pipeline with the pipeline of power (electricity). Thus, an agricultural implement tractor gets the connection to the CAN bus and 12 V power through a connector only, the implement breakaway connector (IBC), located in the agricultural implement. Both connectors are shown in Fig. 2.

This part also defines that a termination should be located at both ends of each rail, for example, in both the tractor bus and the implement bus. This termination, called terminating bias circuit, has the function of providing an electrical reference level between the pins and the CAN-H and CAN-L and promoting impedance matching at the ends of the network, through the power provided by the pins and the

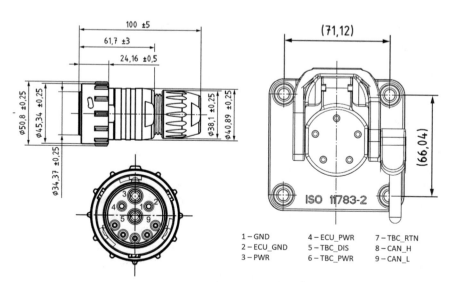

1 – GND	4 – ECU_PWR	7 – TBC_RTN
2 – ECU_GND	5 – TBC_DIS	8 – CAN_H
3 – PWR	6 – TBC_PWR	9 – CAN_L

Fig. 2 Implement breakaway connector and implement bus breakaway connector, and pin allocations

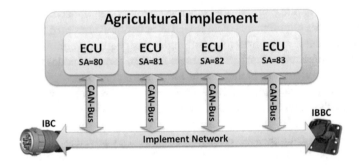

Fig. 3 Network topology in an agricultural implement

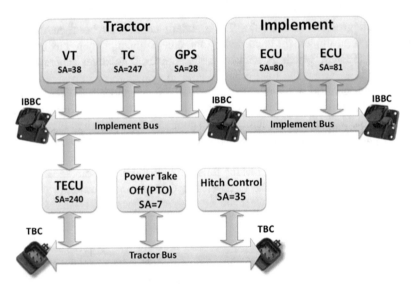

Fig. 4 Network topology on a tractor. SA, source address

TBC-PWR and TBC-GND. This standard connector allows for the communication of an agricultural implement with a tractor that has the IBBC. At the other extreme, the implement must contain a TBC as the end of the implement bus moved the tractor to the implement. As shown in Fig. 3, the tractor and the implement bus, which owns the TBC correctly in its extremes, is divided . As can be observed in the three previous configurations, the standard IBBC also has the function of performing the function of the TBC. When the IBBC implement is not connected to the connector itself at one end of the bus, it therefore provides termination of the TBC. When the IBBC is connected to the tractor implement, termination of the TBC is automatically disabled, as this site is no longer considered the end of the bus, which is located at the end of the agricultural implement bus. The agricultural implement must provide the TBC in extreme buses.

The topology of a network ISO 11783 on a tractor is shown in Fig. 4. The tractor has two buses: a tractor bus and an implement bus. The two buses are separated by the tractor ECU. Each control function communicating on the ISO 11783 data network requires an SA. There can be a one-for-one relationship of SAs with ECUs and control functions on the network. If an ECU performs more than one control function, an address is required for each control function.

Part 3 of the ISO 11783 standard describes the data link layer. We describe the format of the message board, the unit of data protocol, or protocol data unit (PDU), message types, message priority, the mechanism of access to means of communication, the arbitration process, the detection of errors, the TP, and the annexes to this party in question, a routine for message processing, the sequence of data transfer (DT) via TP, and examples of communication mode. The data link layer is also based on the CAN protocol. The identifier field is divided into smaller fields, according to the SAE J1939 standard, which adopts the extended frame format, the CAN 2.0 B. This format defines the identifier of 29 bits. The handle consists of six fields: priority (three bits), data page (one bit), reserved (one bit), PDU format (PF; eight bits), PDU specific (PS; eight bits), and SA (eight bits), as shown in Fig. 5.

The standard defines an entity called a "group parameter number," or PGN. The PGN is composed of the reserved fields, data page or PF, and PS, totaling 18 bits. Each PGN is associated with one and only one message. Thus, messages are identified by PGN, which are in the identifier field of each frame. The field data must be interpreted by the ECU according to the PGN identified. That is, for each PGN there is a protocol that defines the content and the division units from the data field. The messages are divided into two formats as the destination address. Messages can be sent to a specific address in PF 1 or to a global address (broadcast) in PF 2. The identification of the two types is made on the field PF. When the value is smaller than PF 240 (F016), the message is sent to a specific address. In this case, the PDU

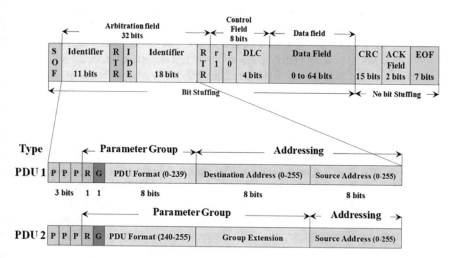

Fig. 5 The identifier field of the CAN packet

field is not considered in the PGN. In the second case, the value of the PF is greater than or equal to 240, and the PS is an extension field for the identification of the PGN. In this case, there is no specific destination, and all ECUs should receive the message. As regards the PGN field data page, the maximum number of possible PGNs to be used is doubled. Overall, $(240 + 16 * 256) * 2 = 8672$ PGNs can be transmitted. The TP is described at the data link layer. The TP is aimed at packing and unpacking messages with a size between 9 and 1785 bytes and also to manage the transport connection. There are two PGNs assigned to a TP: connection management (CM) and DT. The ISO 11783 standard currently supports five types of messages: commands, requests, broadcast/responses, acknowledgments, and group functions. The command message type categorizes parameter groups that command a specific or global destination from a source. The request message type, identified by the PGN, provides the ability to request information globally or from a specific destination. The broadcast/response message type can be either an unsolicited broadcast of information from a controller or a response to a command or request. The acknowledgment-type message is available in two different forms. The first form is provided by the CAN protocol, whereas the second form of acknowledgment is the response of a "normal broadcast," or ACKnowledgment (ACK) or negative ACKnowledgment (NACK), to a specific command or request as provided by an application layer. The group function message type is used for groups of special functions (e.g., proprietary functions, management functions, multi-packet transport functions). Transport protocol functionality is subdivided into two major functions: message "packetization" and reassembly and CN. These are described in specific subclauses, in which the term-originating controller corresponds to the controller that transmits the request to send messages and the receiving controller corresponds to the controller that transmits the clear to send messages. Messages greater than eight bytes in length are too large to fit into a single CAN data frame. Therefore, such messages are broken into several smaller packets, and those packets are transmitted in separate CAN data frames. The individual packets that consist of a large message must be identified separately so that they can be reassembled correctly; the first byte of the data field is defined as the sequence number of the packet. Individual message packets are assigned a sequence number from 1 to 255. In this part of the standard a structure called NAME is also defined, as shown in Fig. 6. This data structure is composed of 64 bits divided into various fields, which describe the function and placement of an ECU network ISOBUS. Every function NAME's describes the function that a CF performs, and its numerical value is used in the arbitration for the address. Thus, the whole ECU must have a NAME for an address on the network and be able to describe its features and functionality to all other ISOBUS ECUs on a network. The negotiation for obtaining an address is carried through four specific messages. The message "Address Claim" is used to assert, declare, or support a network address. A claim occurs when an ECU enters the network and seeks an address. The statement occurs when the address has been already used by another ECU (message "Request for Address Claimed").

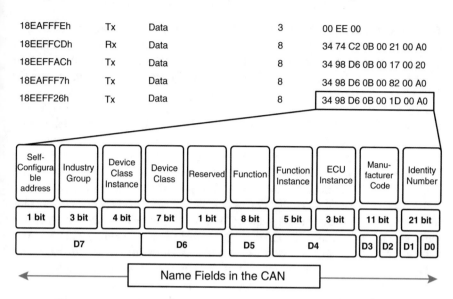

18EAFFFEh	Tx	Data	3	00 EE 00
18EEFFCDh	Rx	Data	8	34 74 C2 0B 00 21 00 A0
18EEFFACh	Tx	Data	8	34 98 D6 0B 00 17 00 20
18EAFFF7h	Tx	Data	8	34 98 D6 0B 00 82 00 A0
18EEFF26h	Tx	Data	8	34 98 D6 0B 00 1D 00 A0

Self-Configurable address	Industry Group	Device Class Instance	Device Class	Reserved	Function	Function Instance	ECU Instance	Manufacturer Code	Identity Number
1 bit	3 bit	4 bit	7 bit	1 bit	8 bit	5 bit	3 bit	11 bit	21 bit

| D7 | | D6 | D5 | D4 | | D3 | D2 | D1 | D0 |

Name Fields in the CAN

Fig. 6 NAME field of ISOBUS data

2.1 Virtual Terminal

The standard of the VT, which is used for the operation parameters set up for tractor and implements, is described in Part 6 of ISO 11783. A VT is an operator interface device that allows both information to be displayed and operators to input information. The VT is one type of CF or ECU on the implemented bus. This device is located in the tractor cab and must connect to the bus implement. The VT has a pixel-addressable (graphical) display. Information on connecting a working set (WS) is shown to the operator on the graphical display. This information is shown in display areas that are defined by data masks, alarm masks, and soft key masks. Annexes of this part of the standard contain various information about objects, events, commands, messages between the ECU and the VT, and the TP extended, or the extended transport protocol (ETP). The term WS is used for an ECU, an implement, or a group of implements either represented by a single ECU or a group of ECUs acting as a WS. Therefore, these implements can be grouped together to a WS where one ECU is the WS Master (WSM) and all others are the WS members. The WSM acts as an interface to the implement bus. A VT provides a common user interface to all WSs on the bus; it contains a graphic display with a limited set of graphical objects, a few soft keys with an icon on the display, and a means of navigating the display and manipulating values. The manner in which the display is shown in the VT is stored in the "object pool" (OP). The OP is a representation of a

WS, and it consists of objects supported in the standard. The objects may be input numbers, output numbers, bar meters, needle meters, polygon graphics, or bitmap graphics. The objects have parameters such as position, size, color, and value. The OP defines object types, the relationships among objects, and all the parameters for each object. As soon as the WS (WS) is connected to the network and powered up, the VT and WS begin to communicate. After initial handshaking and requests, the WS begins to upload the OPs to the VT screen. If the mobile system contains more than one WS, the active display can be changed on the VT [2, 5, 6]. This means that the WSM is addressed for several services (e.g., handling the communication to a VT) whereas the members are still active in performing their tasks. A WS member can receive PGNs addressed to the WSM. The view of the architecture of the WS is shown in Fig. 7. An OP is a collection of objects that completely define the operator interface for an implement and a single WS. The operator interface definition for the device of one or more implements represented by either a single ECU or a WS consists of a set of objects. Each object contains all the necessary attributes and child object references for completion of the object processing. The WS assigns a unique object ID to each object in the OP. The OP is transferred to the VT at initialization by using TP and ETP. The VT is intended to be capable of storing the OPs in a modifiable memory area. The size and number of OPs are limited only by the available memory and software design of the VT, but only one OP per WS exists, as shown in Fig. 8.

 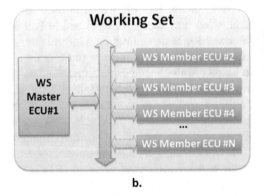

a.   b.

Fig. 7 Working set architecture. (**a**) Working set (WS) used a single ECU. (**b**) WS used a group of ECUs

Fig. 8 Object pools processing in a virtual terminal (VT). (**a**) Object pools transferred by transport protocol/extended transport protocol. (**b**) Object pools transferred to memory

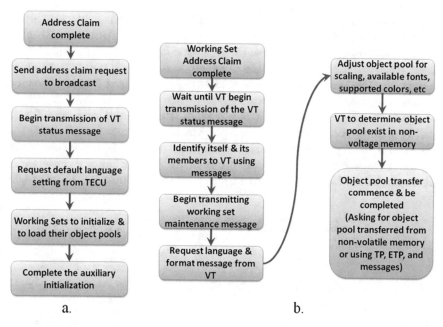

Fig. 9 Initialization process of (**a**) VT and (**b**) WS

When the OP of a WS has been deleted and a VT receives a WS maintenance message from the missing WS, it should provide a NACK message. The NACK message is sent to the SA of the WS. When a WS has been disconnected and reconnected to the VT, the WS may attempt to reload it in the OP. The status messages allow the WS to determine the health of the VT and monitor the progress of tasks in the VT. They also allow the VT to monitor the health of WSs in Fig. 9.

2.2   ISOAgLib Open-Source Library

In this research, we propose a generic ISO 11783-compatible implementation based on the ISOAgLib library [5]. The open source library written in C/C++ enables building of the ISO 11783 standard-compatible features. The ISOAgLib was initially created by Achim Spangler as LBS-Lib within the subproject of the IKB Duernast project as a doctoral student at the Technical University of Munich—crop production engineering—which started in 1999 [7]. Now he is working for the company OSB AG and leads the evolution and upgrading of the ISOAgLib according to the modifications of the ISO standards. The LBS-Lib is an open-source way of stimulating the use of open communication protocols in agriculture. With the progress in standardizing, ISO 11783 has been "upgraded" to ISOBUS as ISOAgLib and now enables users to implement ISOBUS functionality. The documentation for news, changes, main features, and the structural overview of ISOAgLib is presented on the http://ISOAgLib.org homepage. ISOAgLib is designed for easy adoption to new hardware platforms. Therefore, the greater part of the software can be used without any changes for all hardware types. The more this common software part is used, the better the quality can get. The layered structure is illustrated in Fig. 10. The ISOAgLib is designed in three layers. The top layer, which contains all algorithms to define the communication, can be used without any changes on every platform. All hardware interactions such as receiving or sending CAN messages

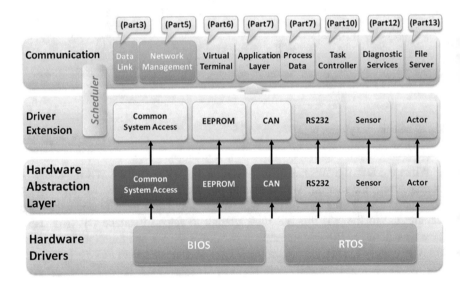

Fig. 10 System architecture of ISOAgLib

are piped through the hardware abstraction layer. This serves an abstract platform-independent interface to the communication layer and to apply itself. Each direct hardware access is handled by a hardware adoption layer, which uses mostly a simple name mapping from the BIOS or OS functions and types. By restricting any hardware adoption changes to some simple header and source files, adoption to different ECU types becomes easier.

2.3 Tool Chain

ISOAgLib is tailored to fit into the ISOBUS development tool chain of OSB. As a programming library, ISOAgLib represents the backbone of an ISO 11783 application and delivers the protocol stack function and the enhancements of vt2iso, in addition to a CAN server, as connectors to the adjacent parts of the design development process and test. Figure 11 illustrates the development structure of ISOBUS by using an OSB tool chain.

The ISOAgLib developer made a lot of useful example source codes and tutorials for the implementation of the ISO 11783 standard. In our case, we are studying version 2.2-rc5 of ISOAgLib because this version has tutorials and examples. Figure 12 illustrates the multifunction of the ISOAgLib tutorial, which is GPS ECU, display ECU, DataSource ECU, and TractorBridge ECU, on the Ethernet network using a CAN server.

3 System Architecture of the VT Server ECU

The development of the VT hardware requires a high-performance microprocessor system with an embedded operating system. We designed and implemented a development board that contains a Freescale i.MX6Q ARM Cortex-A9 (advanced RISC machine) quad-core 32-bit microprocessor, 2 GB DDR3 RAM, 16 GB eMMC flash memory, two serial ports, USB host, USB 2.0 OTG, two SD card interface, 10.4-inch TFT LCD interface with resistive touch screen, 100 Mb/s Ethernet, and two channels for analog camera interfaces. Freescale i.MX6Q is designed to provide a cost-effective, low-power capability, high-performance application processor solution for mobile devices and general applications. The features of the i.MX6Q microprocessor are ARM Cortex-A9 MPCore-based quad CPUs, Video Pro. Unit, GPU3D, GPU2D, GPUVG, image processing subsystem (IPUv3H), port USB 2.0 OTG, memory subsystem, application processor peripherals such as two CAN buses, etc., as shown in Fig. 13.

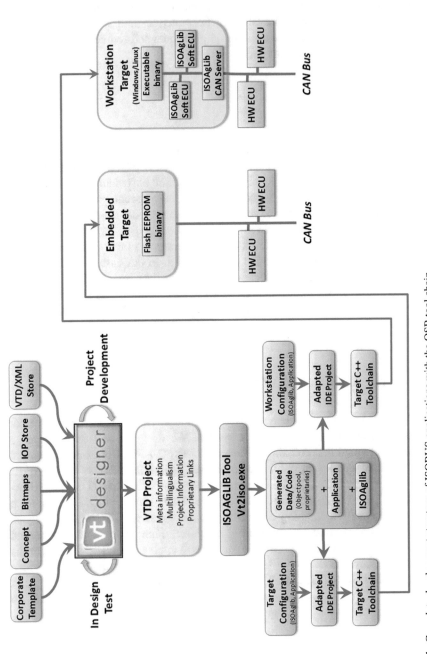

Fig. 11 Complete development structure of ISOBUS applications with the OSB tool chain

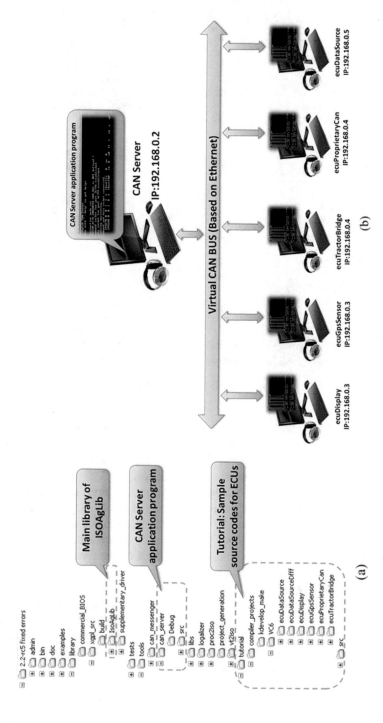

Fig. 12 (**a**) Files structures of ISOAgLib open library. (**b**) Simple working environment of ISOAgLib open library

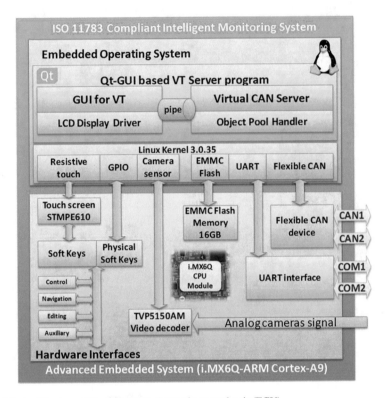

Fig. 13 Architecture of the VT server electronic control unit (ECU)

Part 6 of the ISO 11783 standard specifies a device that allows interaction between the operator and other nodes in a network based on ISO 11783 by exchanging information in a graphical display and across different input modes. Such a device is called a VT. As the user interface, the VT uses a graphical display (touch screen or not), physical keys, sound, and auxiliary inputs. The design guidelines and implementation presented in the standard seek standardization in features, but without restricting the use of different technologies and maintaining independence from manufacturers. The VT screen, like the example in Fig. 14a, is organized in a central area that can contain a data mask or alarm masks. The alarm masks are special screens that contain high-priority messages reporting anomalous or special conditions detected by the system. In total, up to five means of user interaction are provided. As Fig. 14b illustrates, there are four groups of keys on the VT: soft keys, control, navigation, and editing. Some of these are optional as they have similar functionality through the GUI. Use of ESC (escape) is provided in the standard and should be implemented to allow the user to prevent any parameter change or any editing from being done.

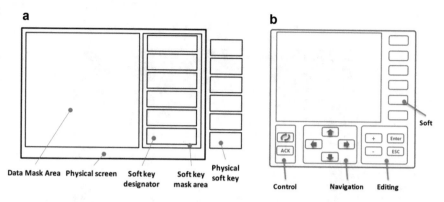

Fig. 14 (a) Example of the application of masks. (b) Keys for user interaction

Also supported is the use of simple beeps, buzzer types, or polyphonic devices. One way is through physical auxiliary inputs loosely connected to the VT and directed to independent commands on the active data mask. ISO 11783 on the network, all ECUs are categorized into different groups called WSs and each ECU contains a WSM. At the startup network, a VT server device receives a sequence of objects representing its whole segment logically, including events, benchmarks, and a complete description of the data masks and alarms. This group of objects is stored in an OP, which in turn is stored in the memory of a VT or some of the non-volatile memory of its own equipment. Being a complete logical representation of network devices, it is the access point between the application and the physical network of the VT. For example, for a given sensor update, its value on the screen of the VT sends a message by changing the corresponding property in the OP. Even if the mask contains data that are not active, the value is displayed and updated in due course, as shown in Fig. 15.

We developed an application architecture (framework) that provides the necessary support for the development of the ISO 11783 entities, providing management messages described in the standard, operable on multiple platforms, and using classes of objects that facilitate customization and expansion to meet not only the VT and task controller tasks, but also other ECUs, the tractor ECU, the network interconnection units, the WSM, and diagnostic tools. Figure 16 illustrates the sequence of procedures for configuring a WS with a VT.

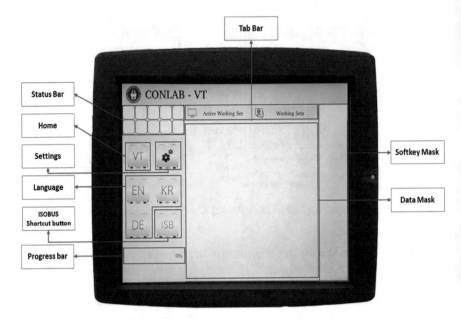

Fig. 15 ISOBUS monitoring system based on Qt. The data mask and soft-key mask comply with the standard size set in ISO 11783-6

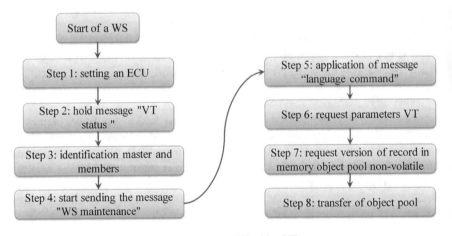

Fig. 16 Sequence of procedures for configuring a WS with a VT

The sequence of the implementation was based on the procedure of configuring a WS with a VT. The following sequence must be performed when setting an OP in a VT:

1. The WSM must complete the process of obtaining an address.
2. The WSM must wait for the transmission of the message "status of VT."

3. The WSM must identify you and the members in the group.
4. The WSM should start sending periodic "Working Set Maintenance" messages
5. The WSM must ask the message "Command Language" to VT, which contains the language information, format, and units used in the network.
6. The WSM can ask the VT sending parameters to identify their capabilities.
7. The WSM can ask the VT information about the existence of the OP in its nonvolatile memory.
8. The transfer of the OP should be initiated and completed. This can be done by asking the VT to OP load the nonvolatile memory or using a TP defined in the ISO 11783 standard.

Initialization of the VT with WSs queued as in the flowchart shown in Fig. 16. The implementation of the firmware codes came from an existing source code of the ISOAgLib library. The first step is the configuration of an ECU in the network. This is a procedure for obtaining an address. For this procedure, we use the following materials: "Part 5—Network Management—of ISO 11783 standard" and the ISOAgLib library. The library contains the implementation of an ECU configuration on the network layer in the object-oriented language C++. The second step is waiting for a message "VT status" from the VT. This message indicates that there is a VT on the network and it contains data about their occupation. In this and the second step, data contained in Part 6 of the ISO 11783 are used. The third step is the identification of the WS. This procedure is performed by two messages. One of them, the WSM, identifies the group master and contains the number of ECUs that constitutes the set. The other, called the "working set member," identifies each member of the NAME field. The PGNs of those WSs are detailed in "Part 7—Implement Message Application Layer—ISO 11783." The fourth step is the start of the cycle of sending the message "working set maintenance," as this has been shown to the VT. This message provides the recognition of the WSs of their presence at the VT. This recognition ensures the services offered by the VT. In this step, as in the second step, data contained in Part 6 of the ISO 11783 are used. The fifth step is to request the message "command language," which identifies the language, units, and format (date, time, and units) used in the network. This PGN is detailed in "Part 7—Implement Message Application Layer—ISO 11783." The sixth step is optional and refers to a request for technical information from the VT. Examples of information include the appointment of memory available for sending an OP, display size, resolution, and the quantity of soft keys. In this step, we use the data contained in "Part 6—Virtual Terminal—ISO 11783 standard." The seventh step, which is optional, also refers to the request for information on the OP version recorded in the nonvolatile memory. In this step, we used the data contained in Part 6 of the ISO 11783. The eighth and final step of this procedure is to load the OP into the VT. This can be done by asking the VT to load the existing OP into the nonvolatile memory or by sending the OP through a TP. In this step we used the following materials: "Part 3—Data link layer," "Part 6—Virtual terminal," and software tool vt-designer and vt2iso. The "Part 6—Virtual terminal—ISO 11783 standard" detail defined features of all objects and their attributes. Each attribute is defined by type (Boolean, byte, integer or floating point), size in bytes, limit values, and description. Also, this part presents the messages for managing the dispatch of the OP. The TPs used to send the OP in this development are the TP and ETP.

4   System Architecture of VT Client ECU

The implemented ECUs are called the VT client ECUs. Every client should have its own OP following the ISO 11783 standard. The ISO 11783 object pool (IOP) is a data file that contains a group of objects that graphically represent a WS in a VT. Through this resource, the WS is able to send data to the operator and also provide commands to enable features or change the settings of the same. A WS may submit only one IOP in the VT, whose objects can have their attributes modified at any time by the messages defined in Fig. 17. An IOP is a set of bytes in which the objects that define it are arranged sequentially, as shown in Fig. 18. The interpretation of the bytes in direct sequence deciphers the boundaries of each object. Figure 19 shows the structure of an IOP, a sample IOP file for the WS of a sprayer, and an example of a picture object.

Object 1	Object ID	Type	Attributes and data
Object 2	Object ID	Type	Attributes and data
Object 3	Object ID	Type	Attributes and data
... ...			
Object N	Object ID	Type	Attributes and data

Fig. 17 Structure of an ISO 11783 object pool (IOP)

Object 1=Working Set   **Object 2=Font Attributes**

```
Sprayer.iop
00000000:  01  00  00  01  01  64  00  01  00  01  b4  00  0f  00  07  00
00000010:  65  6e  05  00  17  0c  03  00  01  00  0a  00  15  b0  04  00
00000020:  00  05  00  15  3c  00  00  00  0c  00  15  3c  00  00  00  0d
00000030:  00  15  00  00  00  00  32  00  17  00  02  00  01  00  33  00
00000040:  17  00  02  00  01  00  3c  00  19  02  00  ff  ff  00  46  00
00000050:  18  00  01  aa  aa  00  64  00  01  01  96  00  1e  00  65  00
00000060:  00  00  2d  00  66  00  14  00  3a  00  69  00  0c  00  96  00
00000070:  6c  00  46  00  73  00  78  00  5a  00  50  00  6b  00  1e  00
00000080:  b4  00  b8  0b  14  00  b9  00  b9  0b  3c  00  b9  00  ba  0b
00000090:  64  00  b9  00  bb  0b  8c  00  b9  00  bc  0b  b4  00  b9  00
```

Fig. 18 Sample IOP file for a Sprayer.iop

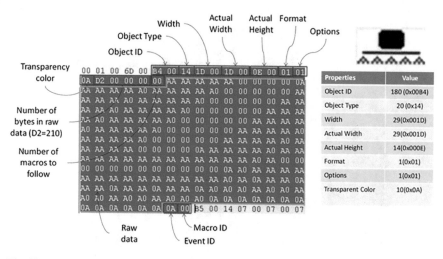

Fig. 19 Sample of a picture object in a Sprayer.iop file

The client-side ECU development is based on the ISOAgLib open-source library and some tools. Figure 20 illustrates the OP uploading state diagram. It generates hexadecimal or binary codes compatible with microprocessors and microcontrollers. In our case, we use the ARM Cortex-M3 core 32-bit microcontroller STM32F107VC.

The vt-designer [8] is a software tool whose graphical development, editing, and emulation of ISO 11783-compliant OPs are shown in Fig. 20. This tool can be imported and exported as both an IOP file and a VT project file (*.vtp). A project is created in a template folder, which contains two files and two folders. One of the file's expansions is *.vtp, which is a vt-designer project file. Another file is the XML file (e.g., Sprayer.xml), which contains OPs and their details. An XML file contains OP information such as buttons, data mask and position, name, and size, how many child and parent devices, etc. One of the folders includes picture files, and another folder contains a language file. Another tool is vt2iso, which is developed by OSB-IT and is open source. This software generates a header and includes files from vt-designer project files. Figure 21 shows a diagram of the general method of implementation of VT client ECUs.

5 Architecture of PGN Analyzer

All ISO 11783 packets, except for the request and address claim packets, contain 8 bytes of data and a standard header, which contains an index called PGN, which is embedded in the 29-bit identifier of the CAN message. A PGN identifies the function and associated data of a message. As it is possible to catch CAN messages on an ISOBUS and interpret PGN-based information, we implement a new device

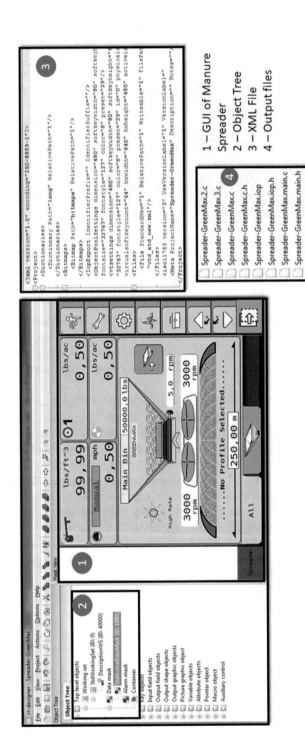

Fig. 20 Sample of an implemented ECU for Sprayer's object pool design created by a vt-designer tool

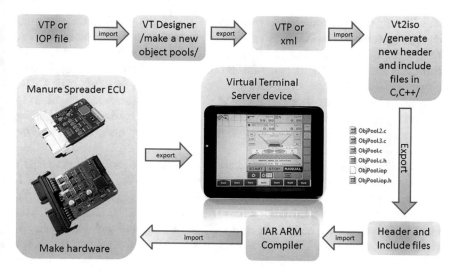

Fig. 21 Block diagram for the development of an implemented ECU with the AFS Pro 700 monitor

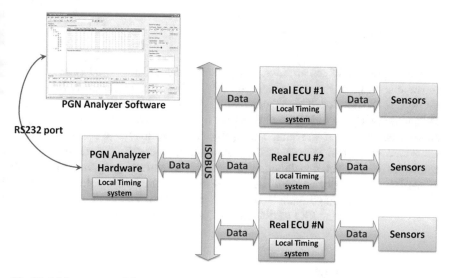

Fig. 22 Main structure of the parameter group number analyzer

ISOBUS PGN analyzer. Figure 22 illustrates the main structure of the ISOBUS PGN analyzer. The PGN analyzer, attached to the ISOBUS, reads messages, interprets them, and displays them on the screen in an easily comprehensible form. It can be used to generate messages and monitor the traffic on physical bus systems. A PGN analyzer monitors and simulates real traffic of connected ECUs and functionality of the ECUs. In fact, to support our work, we have implemented the PGN analyzer tool. The development consists of two parts: GUI of the application and firmware-level programming. The PGN analyzer is implemented on an embedded board Gold Bull where the main CPU is 32-bit, 72 MHz Cortex-M3-cored STM32F107 with two CAN interfaces. The communication between the board and the application programs in a PC is via an RS232 serial communication interface. The CAN1 channel is used to monitor the ISOBUS. The status information of the PGN analyzer is depicted on the LCD display on the board. The hardware programming is implemented at the firmware level. Figure 23 shows the main structure of the firmware-level program.

As RS232 transfers data in 8 bits, we invented a data packet to exchange a message between the PC and the PGN analyzer device. The packet structure is given in Table 2. "Command" byte defines the packet corresponding byte is "D" packet carries CAN data. In all other instances, the packet is the instruction for the PGN analyzer device. The packets start with an "@" symbol and end with a "$" symbol. The receiver via RS232 receives successful data and constructs the packet. After receiving a packet, the device recognizes its command and calls a function correspondent to the command. The device sends CAN data to the ISOBUS.

6 Experimental Results

The main purpose of research work is to implement an ISO 11783-compliant VT server, client ECUs, and a PGN analyzer device. The programming uses the library ISOAgLib. We developed global positioning system (GPS) hardware and software, the sprayer ECU, PGN analyzer device, and the VT, and the implementation has two parts of hardware and software. The hardware system for the VT is an advanced embedded board that is supported by a real-time operating system such as the embedded Linux. The development of the application-level program is based on ISOAgLib, which slows down the development of our work with the GUI of our system. The main communication parts are implemented using the ISOAgLib library according to the ISO 11783 standard, as shown in Fig. 24.

7 Conclusion and Future Work

The technical complexity of agricultural machinery will increase in the future; thus, it is becoming more important to learn and apply technologies such as the ISO 11783 standard. This standard enables the development of agricultural machines that are compatible with other ISOBUS machines. In this work, we use two kinds of embedded boards: the ARM Cortex-A9 quad-core Freescale i.MX6Q-based system for VT and the ARM Cortex-M3 core microcontroller development board for sample ECUs. Besides developing this system, we have checked the feasibility of using the ISOAgLib open-source library in a real implementation of ECUs

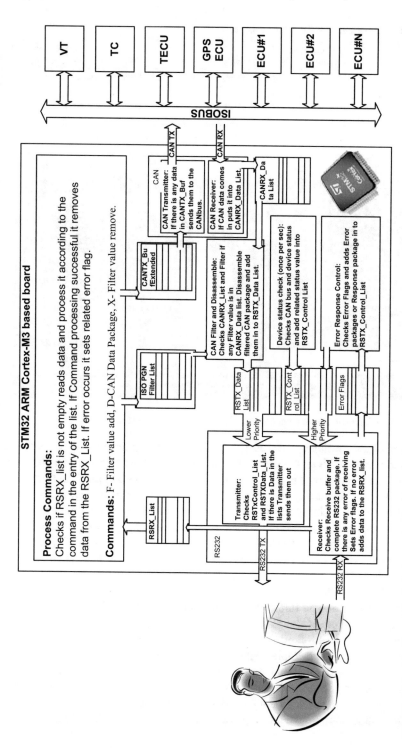

Fig. 23 Main algorithm of the buffering method between RS232 and the CAN bus

Table 2 Main structure of the RS232 data packet

	Start of serial communication	Total bytes	Command	Data	Checksum	End of serial package
RS232 data packet (from PC to device)	'@'	Total bytes	'F','X','C','D', 'R','S','P'	Data	Checksum	'$'
RS232 data packet (from device to PC)	'@'	Total bytes	'T'	Data	Checksum	'$'

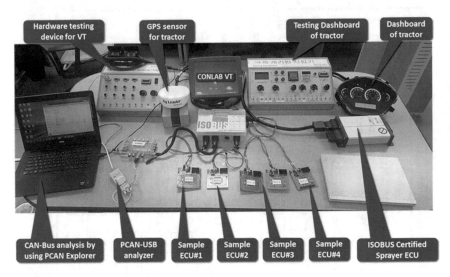

Fig. 24 Development environment for ISO 11783 CONLAB-VT and implemented ECUs

for agricultural machinery. We also tested all of the implemented ECUs such as GPS and sprayer with an AFS Pro 700–ISOBUS-compliant monitor. In addition to the successful implementation of the VT based on the ISOAgLib library and Qt GUI tool, real hardware systems consist of several development boards, which are iTOP-i.MX6q, and Huins development kits. One of the most important things is to monitor all of the devices connected to the bus. For this reason, we developed and implemented the ISOBUS PGN analyzer device firmware and GUI application program for Windows. In future work, we aim to improve further the application program for VTs according to the technical specifications of the ISO 11783. Our developed product, consistent with the standardization of the communication between a tractor and an implement, is convenient for tractor drivers and farmers. The implementation of the ISO 11783 without systematic planning is difficult and complex. There are limited amounts of ISOBUS-compatible equipment in the Republic of Korea, and implementation was difficult because of a lack of well-facilitated laboratories and real tests on a tractor and an implement. After studying source codes of the ISOAgLib library and ISO 11783 standard, we conclude that the main purpose of the development was achieved and it proved that it was possible to use our methodology for the development of ISOBUS-compatible ECUs.

References

1. Backman J, Oksanen T, Visala A (2012) Navigation system for agricultural machines: nonlinear model predictive path tracking. J Comput Electron Agric 82:32–43
2. Hyeokjae K, Enkhbaatar T, Woonchul H (2011) Implementation of virtual terminal based on CAN by using WinCE platform builder 6.0. Key Eng Mater 480:938–943
3. ISO 11783-6 (2004) Tractors and machinery for agriculture and forestry—serial control and communications data network. In: Part 6: virtual terminal. International Organization for Standardization, Geneva
4. ISO 11783-5 (2007) Tractors and machinery for agriculture and forestry—serial control and communications data network. In: Part 5: network management. International Organization for Standardization, Geneva
5. Spangler A, Auernhammer H, Demmel M (2001) Stimulating use of open communication standards in agriculture (DIN9684 and ISO11783) with capable open source program library as possible reference implementation. In: Blackmore S, Grenier G (eds) Proceedings of the 3rd European Conference on Precision Agriculture. Montpellier, pp 719–724
6. Stone ML (1999) ISO 11783 an electronic communications protocol for agricultural equipment. In: Spangler A, Wodok M (eds) Conference on agricultural equipment technology, Louisville, February 1999. IsoAgLib—development of ISO 11783 applications in an object oriented way, 2010. American Society of Engineers, Reston
7. Tumenjargal E, Badarch L, Kwon H, Ham W (2013) Embedded software and hardware implementation system for a human machine interface based on ISOAgLib. J Zheijang Univ Sci C Comput Electron 14:155–166
8. VT Designer Software. http://www.vt-designer.com

Safety-Driven Development and ISO 26262

Yaping Luo, Arash Khabbaz Saberi, and Mark van den Brand

Abstract The automotive industry has seen a rapid change in the technologies used inside the vehicles. Since the introduction of the first electronic control unit, the impact of electronics and computer science on the quality of the vehicles is increasing every year. Arguably, safety is one of the most important quality attributes of a vehicle that needs special attention during all the stages of the lifecycle of a vehicle. The overall safety of a vehicle can be seen from multiple aspects, such as passive safety, active safety, and functional safety. Functional safety addresses the hazards that are caused by malfunctioning of electrical and/or electronic (E/E) systems. There are many factors that impact functional safety such as the organization and management, the development process, the design of the systems, the system type and technologies used in it, the quality control methods, etc. The ISO 26262 standard provides the state of the art of functional safety in automotive industry. In this chapter some of the most important aspects of functional safety from ISO 26262 perspective are discussed; namely, safety management, development process, architecture design, and safety assurance are presented here.

1 Introduction

In safety-critical domains such as automotive, railway, and avionics, even a small failure of a system might cause injury to or death of people. A number of international safety standards are introduced as guidelines for system suppliers to keep the risk of systems at an acceptable level [5], such as IEC 61508

Y. Luo (✉) · M. van den Brand
Department of Mathematics and Computer Science, Eindhoven University of Technology, Eindhoven, The Netherlands
e-mail: y.luo2@tue.nl; m.g.j.v.d.brand@tue.nl

A. Khabbaz Saberi
TNO Technical Sciences/Automotive, Helmond, The Netherlands
e-mail: arash.khabbazsaberi@tno.nl

© Springer Nature Switzerland AG 2019
Y. Dajsuren, M. van den Brand (eds.), *Automotive Systems and Software Engineering*, https://doi.org/10.1007/978-3-030-12157-0_10

(multiple domains) [19], ISO 26262 [20] (automotive domain), DO 178C (avionic domain) [40], CENELEC railway standards (railway domain) [8–10], etc.

In the automotive domain, currently the ISO 26262 standard, which is a goal-oriented standard for safety-critical systems within the scope of road vehicles, is the state of the art. Since its introduction in 2011, ISO 26262 has attracted more and more attention in the automotive domain. A number of safety-driven development methods have been proposed based on this standard [24]. In this chapter we first introduce some basic concepts in the ISO 26262 standard (Sect. 1), and then we discuss safety management in Sect. 2 and the safety lifecycle (Sect. 3) in the context of the standard. Furthermore, a brief comparison of several safety architecture patterns is given in Sect. 4. Finally, as compliance with safety standards is a basis of safety assessment, a number of model-driven techniques, designed for supporting safety assessment, are presented in Sect. 5.

1.1   ISO 26262

ISO 26262 is an adaptation of the generic IEC 61508 standard, which focuses on electrical/electronic (E/E) systems but provides a general design framework for safety-critical systems [26]. Similar to IEC 61508, ISO 26262 is a risk-based safety standard. It provides a risk-driven safety lifecycle for developing safety-critical systems in the automotive domain. In the standard the risk of hazardous situations is qualitatively assessed. This is done to avoid or control systematic failures and to detect or control random hardware failures.

The ISO 26262 consists of ten parts as shown in Fig. 1. Part 3 to Part 7 correspond to the safety lifecycle, while Parts 1 and 2 and Part 8 to Part 10 provide the additional information related to interpretation of the main parts. The ISO 26262 standard is structured based upon the V model. Parts 3 to 7 construct the primary V cycle for the whole system development. The main goals of Part 3 are to identify system hazards and risks through hazard analysis and risk assessment (HARA) and define safety goals and functional safety concept (FSC). Part 4 focuses on the system-level development, integration, and validation. In this part, technical safety requirements (TSRs) are derived based on FSC. Moreover, Parts 5 and 6 have their own (smaller) V cycles for hardware and software development, respectively. In these two parts, more detailed safety requirements are derived from TSRs. These safety requirements are assigned to concrete subsystems or components for implementation. Finally, Part 7 covers the release of the system for production.

1.2   Functional Safety Definition

Functional safety is easy to understand, yet difficult to formally define. The ISO 26262 standard gives a definition of functional safety using a number of other

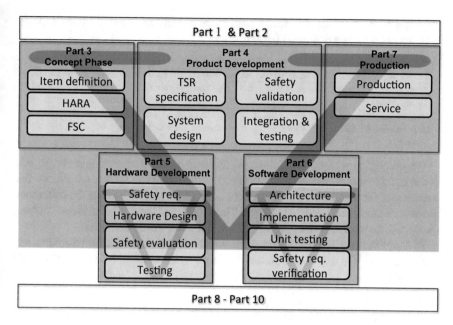

Fig. 1 An overview of the ISO 26262 V model

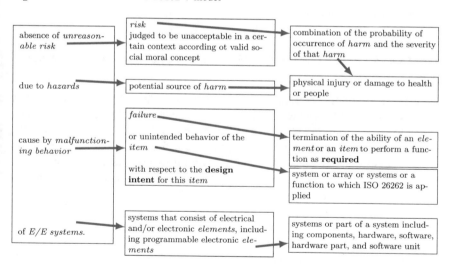

Fig. 2 The overview of functional safety definition in ISO 26262

concepts which have their own complex definition. An overview of the full definition is shown in Fig. 2. The definition of functional safety simply reads as follows:

Definition 1 "Absence of *unreasonable risk* due to *hazards* caused by *malfunctioning behavior* of E/E systems."

The phrases in *italic* are elaborated with more definitions. After compiling all the defined concepts, the definition transforms to:

> Absence of combination of the probability of occurrence of physical injury or damage to the health of people and the severity of that *harm*, judged to be unacceptable in a certain context according to valid societal moral concepts due to potential sources of *harm* caused by termination of the ability of an *element* or an *item* to perform a function as **required** or unintended behavior of the *item* with respect to the **design intent** for this *item* of systems that consist of electrical and/or electronic *elements*, including programmable electric *elements*.

We can agree that this definition is difficult to grasp at first glance! To add to the complexity, there are also some side notes attached to this definition. For instance, in the definition of *failure*, there is an important note about the difference between required and specified failures. The ISO 26262 standard considers incorrect specification also a source of failure, which is a quite strong definition.

To simplify the definition, and ease the burden of understanding functional safety, a simplified yet less accurate approach is taken to define functional safety. There are a few vital implications in ISO 26262 definition of functional safety, namely:

1. Functional safety depends on the design intent.
2. Functional safety tackles (in scope of ISO 26262) failures of E/E systems.
3. Functional safety is only applicable to hazards that cause harm to people (and not damage to property).

Considering these points, we propose a shorter definition for functional safety:

Definition 2 Operating correctly with fail-safe or fail-operational strategies to prevent hazards.

In this definition, operating correctly means doing what is intended, referring to the design intent. In other words, correct behavior reflects the design intent. Moreover, the design should be able to cope with possible failures, thus preventing hazards from happening. In this definition, two major strategies for coping with failures are identified: fail-safe that means reverting to a safe state which no longer provides the required functionality fail-operational that means reverting to a safe state while some variation (may be degraded) of the functionality is still provided.

The proposed short definition, in comparison with the longer definition in ISO 26262, lacks the notion of *risk*. Therefore, this definition does not convey the goal of reducing the risk that is addressed in the long version. However, this definition makes it easier to understand functional safety in a pragmatic manner.

1.3   Functional Safety Goals

Failures of the E/E systems are recognized (the focus of functional safety) as the primary cause for hazards. There are various ways for categorization of failures. One generic way is to classify them into two types: random (hardware) failures and

Fig. 3 The relation between failures and hazard

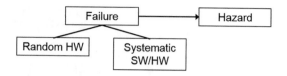

systematic failures. Random hardware failures are unpredictable failures that occur during lifetime of a hardware part [20]. These failures are only relevant to hardware parts and do not apply to software units. Systematic failures are, on the other hand, deterministic and have a certain cause (usually the design of the system). These failures can happen in both hardware and software elements. An example of these failures is a software bug/error. An overview of this categorization and the relation between failures and hazards is shown in Fig. 3.

From the definition of functional safety, it can be inferred that failures (may) cause hazards. Consequently, the goal of functional safety is to reduce the risks of hazards. This goal can be refined into two subgoals by categorizing the types of failures: (1) preventing systematic failures and (2) mitigating random failures.

Preventing systematic failures implies that the development process of safety-critical systems should be carried out in such a way that the human errors (i.e., the primary cause of systematic failures) or other contributing factors do not lead to an unresolved failure. This goal is achieved by defining a predictable process for the development of safety-critical systems. The mechanism for ensuring achievement of this goal includes reviewing work products, analysis, and testing.

We mitigate random hardware failures during design by analyzing possible failures and using detection and reaction mechanisms known as *safety mechanism*. The random hardware failures are a probabilistic phenomenon, and they are unpreventable. Hence, there should be mechanisms in the design that detect these failures and act to prevent failures from creating hazards.

Furthermore, it should be possible to systematically provide evidence about the achievement of the abovementioned subgoals.

In summary, there are three main subgoals for functional safety:

1. Preventing systematic failures
2. Mitigating random failures
3. Showing (providing evidence) that the previous goals have been achieved.

These goals are achieved by a combination of controlling the development process, design, verification and validation, and documentation.

2 Safety Management

Safety engineering is complex, especially in a multidisciplinary domain such as the automotive industry. It involves a wide variety of tasks that are typically carried out by multiple people with various skills and experiences. The safety-related tasks are

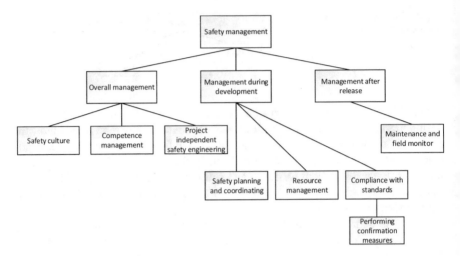

Fig. 4 The overview of safety management parts

the activities that are performed during the safety lifecycle (referred to as safety activities in ISO 26262 vocabulary). As defined by ISO 26262, the safety lifecycle is the entire duration of time that a safety-critical system exists in any phase, from the concept phase to the decomposition phase. Moreover, the safety activities are highly dependent on each other, as well as other non-safety activities related to development, testing, and production of a safety-critical system. Therefore, safety management is a necessity to ensure systematic and smooth realization of all the safety-related activities. Since management attention is required for realization of safety activities, ISO 26262 also provides some guidelines on the most critical considerations for safety management.

Safety management is divided in three main parts in ISO 26262 Part 2: overall safety management, safety management during development, safety management and after release for production. An overview of safety management is shown in Fig. 4. The overall safety management considers the project-independent aspects of safety engineering. This includes safety culture, competence management, quality management, and definition of project-independent development process. Safety culture has been the main focus of safety engineering in several industries after the Chernobyl disaster in 1986. Safety culture is described in more details in the following subsection [15].

The goal of safety management during development and after release for production is ensuring safe realization of a safety-critical system. This is (in most cases) done by ensuring compliance with a safety standard such as ISO 26262.

The ISO 26262 standard recommends assignment of a safety manager to the development of a system. The primary goal of the safety manager is to coordinate all safety activities during the safety lifecycle. Planning and coordinating of safety activities and resource management are typical responsibilities of safety managers.

The responsibilities of a safety manager overlap with those of a project manager, in tasks such as planning and resource management. The difference is that the safety manager is involved only with the safety-related planning and resources in the overlapping tasks. There are tasks with no overlap for these two roles too. For example, cost management is only a task of the project manager. Another example is risk management and project control. While both the project manager and the safety manager perform these tasks, they have different focus. The project manager performs risk management for project risks, whereas the safety manager cares for the system risks. Similarly, both roles perform project control, but the safety manager cares only about control mechanisms that impact safety. The ISO 26262 refers to these control mechanism as *confirmation measures*. These activities are performed to increase the trust in the development process with respect to safety-related issues. Confirmation measures are described in more details in the rest of this chapter.

Lastly, safety management after release for product is responsible for planning of maintenance and field monitoring for possible undiscovered failures.

2.1 Safety Culture

Safety culture is one of the key elements of overall safety management in ISO 26262 [20]. In general safety culture requires the organization to provide the proper environment for people involved in safety activities. Safety culture is defined by [39] as follows:

> The set of enduring values and attitudes regarding safety issues, shared by every member of every level of an organization. Safety Culture refers to the extent to which every individual and every group of the organization is aware of the risks and unknown hazards induced by its activities; is continuously behaving so as to preserve and enhance safety; is willing and able to adapt itself when facing safety issues; is willing to communicate safety issues; and consistently evaluates safety related behavior.

An overview of the contributing factors to safety culture is shown in Fig. 5. These key factors are the result of aggregation of the aspects considered in the literature [39, 44]. The descriptions of these factors are as follows:

Management commitment is the willingness of the organization at every level (from top to down) to invest effort in safety and their genuine positive attitude toward safety. The ISO 26262 standard emphasizes on this factor in Part 2: 5.4.2.1 and 5.4.2.2.

Justness (only considered in [39]) is the extent to which behavior according to functional safety is encouraged and rewarded by the organization. Moreover, there should be a "no blame" culture where in event of an accident, solutions are sought instead of blaming the responsible person. The ISO 26262 also mentions this matter in Part 2: 5.4.2.1.

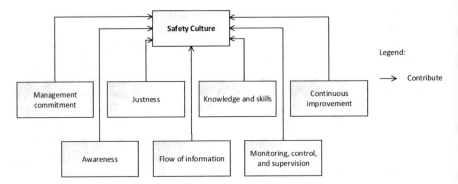

Fig. 5 Safety culture contributing factors

Awareness is the level of individuals' appreciation of their role and impact on functional safety and on safety in general. Moreover, the understanding of the risks involved in their work for themselves and others is also a part of awareness. The ISO 26262 standard addresses the issue of roles in Part 2: 5.4.2.2.

Flow of information is the accessibility of new information for the right people through transparent communication. For instance, if there is a new hazardous situation identified during a recent test, the information should be easily provided to others, to be considered if applicable in their projects. In ISO 26262-2 5.4.2.3, the flow of information is mentioned as explicit communication of functional safety anomalies. The ISO 26262 standard even takes flow of information further by stating that there should be a process for resolving functional safety anomalies in Part 2: 5.4.2.4.

Knowledge and skills (similar to "behavior" in [39]) are the extent of individuals' knowledge of safety engineering processes and activities and, in particular in this case, the ISO 26262 standard. This factor is more important in a research and development environment. General appreciation of the relevant knowledge and skills is needed in an organization to allow effective implementation of functional safety. Several clauses of ISO 26262 can be linked to this aspect of safety culture such as Part 2: 5.4.2.5, and 5.4.2.6.

Continuous improvement (the same as "adaptability" in [39]) is the willingness of an organization to learn from their experiment and improve on the way of working of the organization. Continuous improvement is also mentioned in Part 2: 5.4.2.7.

Monitoring and control (only considered in [44]) is the existence of supervision mechanisms concerned with safety and the visibility of these mechanism in the organization. Moreover, the extent of availability of the required authority to execute functional safety activities is also part of this aspect. The supervision issue can be traced in ISO 26262 Part 2: 5.4.2.8.

2.2   Safety Culture Metrics

A model for safety culture maturity is introduced by [17]. Another related work has been done by [12] on safety culture maturity model. An overview of the maturity model is shown in Fig. 6. Similar to the Capability Maturity Model (CMM), the safety culture maturity model has five levels. The general idea is that an increase in the level shows improved safety culture maturity.

The first level, indicating the worst safety culture, is when an organization considers safety as a burden. There are typically no processes in place for handling safety issues, and the members of the organization only care about not getting in trouble. The second level is applicable when there are some processes for safety but not strictly followed by the members. It could be that the management of the organization states that safety is important, but it is not believed by the members. In the next level, that is, the calculative level, the safety processes are followed and the members are more involved in the safety issues. Nevertheless, the safety processes are not believed to be critical. In the proactive level, both the management and the members believe in their safety processes, and all hazards are addressed systematically. In the last level, safety is deemed an organization value. Both members and management are constantly improving the safety. More details can be found in [16].

The ISO 26262 standard does not provide any recommendations on the safety culture maturity level. Therefore, it is the companies' ambition that drives the target with respect to their safety culture maturity level. Identifying the safety culture maturity level of a company and maintaining or improving it is therefore also the responsibility of that company.

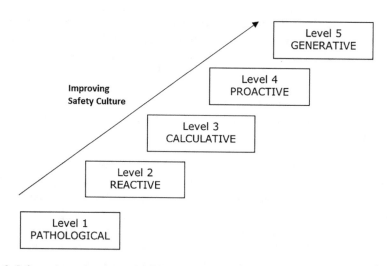

Fig. 6 Safety culture maturity model [39]

Depending on the level of safety culture maturity, the actions needed for improving or maintaining the safety culture differ. Changes in areas such as management support, processes, and training are required for improving the safety culture. More information on this topic can be found in [18].

2.3 Confirmation Measures

Ensuring compliance with safety standards is one of the important responsibilities of the safety manager during development. The ISO 26262 standard created mechanisms, referred to as confirmation measures, for ensuring compliance with this standard. These measures include confirmation reviews on selected work products indicated by ISO 26262, functional safety audit, and functional safety assessment. Depending on the Automotive Safety Integrity Level (ASIL) assigned to the system of interest, these measures shall be performed by the indicated people. This indication can be a different person (than the creator of the work product), a person from a different team within the same organization, or a person from an independent (with respect to management structure) organization. Some examples of the confirmation measures are shown in Table 1.

3   Safety Lifecycle: Integrated V Model

The ISO 26262 standard is the collection of best practices in the automotive industry; thus, it has a number of practical considerations that is specific to this domain. The best example of these considerations is the differentiation between the functional safety concept and the technical safety concept. These two phases correspond to the functional view and physical view in system engineering development. The reason for separating these two phases, which in other domains are carried out in parallel, is due to the special considerations between the original equipment manufacturers (OEMs), and their suppliers (Tiers). In case of different settings for the development chain, or developing a Safety Element out of Context (SEooC), there is a possibility of using a more effective development process by tailoring the safety lifecycle.

The ISO 26262 standard does not provide recommendations for a development process of the functionality required for the system under development. It defines a safety lifecycle that contains all the activities related to functional safety. Indeed, there is an underlying assumption about another process (seemingly going on in isolation from the safety lifecycle) for designing the system. This process is referred to as quality management (QM) in ISO 26262. The safety lifecycle of ISO 26262 requires some information about the functionality of the system from an external process. For instance, the preliminary architecture which is a prerequisite for the functional safety concept needs to be provided via an external source (ISO 26262

Table 1 Example of confirmation measures, and ASIL dependent and independent level [20]

Confirmation measures	Degree of independency applied to ASIL				Scope
	A	B	C	D	
Confirmation review of the hazard analysis and risk assessment of the item (see ISO 26262-3:2011, Clauses 5 and 7, and, if applicable, ISO 26262-8:2011, Clause 5) Independence with regard to the developers of the item, project management, and the authors of the work product	I3	I3	I3	I3	The scope of this review shall include the correctness of the determined ASILs and quality management (QM) ratings of the identified hazardous events for the item and a review of the safety goals
Confirmation review of the safety plan (see 6.5.1) Independence with regard to the developers of the item, project management, and the authors of the work product	–	I1	I2	I3	Applies to the highest ASIL among the safety goals of the item
Confirmation review of the item integration and testing plan (see ISO 26262-4) Independence with regard to the developers of the item, project management, and the authors of the work product	I0	I1	I2	I2	Applies to the highest ASIL among the safety goals of the item

The notations are defined as follows:

–: No requirement and no recommendation for or against regarding this confirmation measure

I0: The confirmation measure should be performed; however, if the confirmation measure is performed, it shall be performed by different person

I1: The confirmation measure shall be performed, by a different person

I2: The confirmation measure shall be performed, by a person from a different team, that is, not reporting to the same direct superior

I3: The confirmation measure shall be performed, by a person from a different department or organization, that is, independent from the department responsible for the considered work product(s) regarding management, resources, and release authority

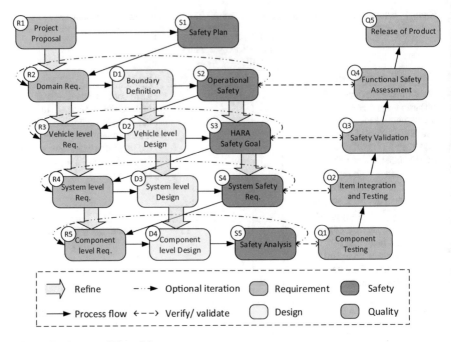

Fig. 7 The integrated V model

Part 3: 8), or (non)functional requirements should be included or referenced in the specification of technical safety requirements (ISO 26262 Part 4: 6.4).

This means that the whole development of a system (both functionality and functional safety) cannot be solely based on the safety lifecycle recommended in ISO 26262. The safety lifecycle addresses the functional safety-related developments, yet the functionality of the system is not addressed in this process. Therefore, there needs to be a development process in which the desired functionality is considered.

Considering the mentioned issues, there is a need for alignment of the safety process and the engineering process that creates the functionality of the system.

A model of the integrated V model for functional safety is shown in Fig. 7. The color codes in the model are used to differentiate between functional and safety perspectives of development: blue and yellow colors are used for functional design, orange is used for safety parts, and violet is used for verification and validation activities related to both functional and safety parts of the design.

In the integrated V model, the requirement development is modeled in a separate flow (as opposed to the traditional V model) to emphasize the hierarchical structure of requirements and to enforce gradual development and refinement of requirements based on higher-level requirements and design.

The safety lifecycle of ISO 26262 is simplified in various ways in the integrated V model. To start with, the production phase of ISO 26262 is removed completely, which, in turn, reduces some related activities too. Furthermore, the safety requirements hierarchy is slightly modified by merging functional safety requirements and technical safety requirements into system safety requirements. In addition, the dual V model (Vee of Vee) of ISO 26262 is reduced to a single V model. In other words, the development of hardware and software (which is followed in separate V models in ISO 26262 as shown in Fig. 6) is merged in the main V model; this change results in a reduction of verification and validation activities too.

This process model is especially useful for nonconventional automotive companies (i.e., companies that do not work according to the norms of the automotive industry) who require ISO 26262 compliance, simply since this process does not reflect the norms within automotive industry. Moreover, it could also be useful for conventional automotive companies that require a lightweight process for a specific project. This can be because of different reasons, for example, projects with a tight time to market requirement that does not allow full process coverage or early development projects that require tight coupling of design and safety processes. The description of the steps of the integrated V model is as follows:

Project proposal (R1): The proposed integrated V model starts with the project proposal, in which the general goals of the project, customer wishes, application, the project business plan, etc. are reflected. This step also contains the planning of the development activities for designing the system under development.

Safety plan (S1): In this step, the safety plan is made according to ISO 26262 guidelines. The safety plan contains the planning of all the safety activities related to the safety of the system.

Domain requirements (R2): Here, the domain requirements as well as customer wishes are described (refined) in the form of high-level requirements.

Boundary definition (D1): Based on the high-level requirements, the design steps are initiated by defining the system. In this step, the system is defined in interaction with its environment.

Operational safety (S2): Following the definition of the system, operational safety starts where the safety-critical behavior of the item is defined. It should be noted that the operational safety is not part of ISO 26262 safety lifecycle. The goal of operational safety is to deduce the high-level nominal behavior requirements of the item from a safety point of view. At this point iterations over high-level steps are made in order to reflect the possible changes that may be needed for satisfying operational safety requirements. After completion of this step, functional safety assessment (Q4) should be planned.

Vehicle-level requirements (R3): Here, the vehicle-level requirements are defined by *translating* customer wishes and the high-level requirement to functional and nonfunctional vehicle-level requirements. Moreover, the requirements based on the boundary of the system are also addressed at this point.

Vehicle-level design (D2): During vehicle design, the internal functions are designed to address the vehicle-level requirements and operational safety requirements. D2 is the equivalent of functional architecture design (the same as preliminary architecture in ISO 26262).

The combination of the steps R2, R3, D1, D2, and S1 composes the item definition from ISO 26262.

Hazard analysis and risk assessment and safety goals (S3): This is a step from ISO 26262, that is, hazard analysis and risk assessment (HARA). This step is performed following the guidelines of the standard and results in safety goals. The resulting safety goals may need iteration over the vehicle-level design. When the vehicle-level steps are finished, test cases should be designed based on functional requirements and safety goals to be performed for system/safety validation (Q3).

System-level requirement (R4): Afterwards, the system-level requirements are described, containing both functional and technical requirements, by refining the higher-level requirements.

System-level design (D3): Next, the design is further refined by system-level design. The architecture designed in D2 is detailed to satisfy the requirements from R4 and S3.

System safety requirements (S4): Following the system-level design, system safety requirements are described by refining S3 based on D3. Moreover, the system safety requirements are verified by doing qualitative functional safety analyses such as FMEA, FTA, etc. Similar to previous levels, iterations are made after S4 for revising the design in D3 with respect to requirements in R4.

Item integration and testing (Q2): Afterward, integration tests should be designed based on the system-level design and system-level (safety) requirements.

Component-level requirements (R5): During this step, the hardware and software requirements refine both the system-level safety and non-safety requirements (S4 and R4).

Component-level design (D4): In the component-level design, the components of the system are detailed and implemented.

Safety analysis (S5): In safety analysis step, the functional safety analyses are performed on the system. The analysis is used to verify the safety of the system in a quantitative manner.

Component testing (Q1): The components tests are designed in this step based on requirements in R5.

Quality (Q1–Q4): Finally, following the steps Q1–Q4 verifies and validates the design against the requirements, and the system is ready for delivery in Q5.

The integrated V model introduces a simplified version of the ISO 26262 standard lifecycle matched with a development process. Synchronization points between the two processes are clearly defined. Furthermore, iteration points within the same level are defined. Additionally, the defined design levels facilitate hierarchical architecture design and requirements elicitation.

4 Safety Architecture Patterns

Besides the development process (discussed in the previous section), which is the major contributor to the first goal of functional safety (preventing systematic failures), the architecture design also has an important role in achieving the first two goals. Good architecture design reduces the chances of making mistakes (primary source of systematic failure) during implementation. Moreover, it can provide proven solutions for mitigation of random failures. Therefore, in this section, this topic is discussed and some guidelines for choosing a suitable architecture pattern are proposed.

Decisions about the system architecture have a great impact on characteristics of the system under development. Furthermore, since architecture design is usually done at the early stages of a project, it is important to consider safety and specifically functional safety in the architecture design. One of the recommendations of ISO 26262 is to use well-trusted design principles for system architecture. Traditionally, architecture principles are stated using architecture patterns or styles [7]. The ISO/IEC/IEEE 42010 standard [21] recognizes architecture patterns as a fundamental mean for expressing design. Moreover, adherence to architecture patterns is considered as a form of redundancy in other domains too [43].

There are various architecture patterns for safety-critical systems in the literature [11], for example, Protected Single Channel, Homogeneous Redundancy, Heterogeneous Redundancy, Safety Executive, and 3-level Safety Monitoring (also known as E-Gas). An analysis on the impact of these safety pasterns on cost, reliability, safety, negotiability, and execution time has been provided in [4].

The Protected Single Channel Pattern improves safety by monitoring the input data and checking the data integrity and optionally monitoring the outputs. The Homogeneous Redundancy Pattern improves safety and reliability by copying the main channel and switching between them in case of failure. Duplex, Triple Modular, etc. are different variations of this pattern. The Heterogeneous Redundancy is similar to Homogeneous Redundancy except that each added channel is developed independently; therefore it is one of the most expensive patterns. The Safety Executive Pattern can switch to a secondary channel to bring the system to safe state in case of a failure in the main channel. The 3-level Safety Monitoring Pattern is widely used in the automotive industry because it provides a cost-effective safety solution. This pattern monitors the internal states of a system in the first level and monitors the inputs and outputs in the second level. The third level is dedicated to the nominal functionality of the system.

In this section, we compare some of the mentioned patterns. The comparison is done from five perspectives: reliability, safety, cost, modifiability, and impact on executive time. For each of these aspects, we define different classes to facilitate the comparison. All comparisons are carried out between a basic system and a system developed according to a pattern.

Reliability
This aspect shows the relative improvement in the system's reliability achieved by a pattern. The reliability of a pattern is assigned to one of the reliability classes R1, R2, and R3, in accordance with Table 2.

Safety
This aspect indicates the safety recommendations or improvement that a pattern could contribute to. The safety of a pattern is assigned to one of the safety classes S1, S2, and S3, in accordance with Table 3.

Cost
This aspect gives the implications on costs of a pattern, which include the recurring cost per unit and development cost of the pattern. The cost of a pattern is assigned to one of the cost classes C1, C2, and C3, in accordance with Table 4.

Modifiability
This aspect indicates the degree to which a system developed according to a pattern can be modified and changed. The modifiability of a pattern is assigned to one of the modifiability classes M1, M2, M3, and R4, in accordance with Table 5.

Impact on Execution Time
This aspect shows the effect of a pattern on the total time of execution at runtime. The impact on executive time of a pattern is assigned to one of the impact classes T1, T2, and T3, in accordance with Table 6.

Table 7 shows the results of our comparison of some of the patterns. From reliability perspective, only Triple Modular Redundancy Pattern can improve the reliability of a basic system. From safety perspective, Triple Modular Redundancy Pattern can bring most safety improvements, while Safety Executive Pattern brings lowest safety improvements. From cost perspective, we could see that among these patterns, Sanity Check Pattern, Protected Single Channel Pattern, and 3-Level Safety Monitoring Pattern are low-cost patterns, and Triple Modular Redundancy Pattern and Safety Executive Pattern are high-cost patterns. From modifiability

Table 2 Classes of reliability

Class	R1	R2	R3
Description	Lower than a basic system	The same as a basic system	Higher than a basic system

Table 3 Classes of safety

Class	S1	S2	S3
Description	Lower impact	Small improvements	Incremental improvements

Table 4 Classes of cost

Class	C1	C2	C3
Description	Low cost	Reasonable cost	High cost

Table 5 Classes of modifiability

Class	M1	M2	M3	M4
Description	The same as a basic system	Very simple to be modified	Can be modified with easy steps	Can be modified with extra cost

Table 6 Classes of impact on executive time

Class	T1	T2	T3
Description	No effect	Little influence	Increase in the execution time

Table 7 Results of comparison with other patterns

Pattern name	Reliability	Safety	Cost	Modifiability	Impact on execution time
Triple modular redundancy	R3	S3	C3	M1	T2
Sanity check	R2	S2	C1	M2	T1
Safety executive	R1	S1	C3	M3	T1
Protected single channel	R1	S2	C1	M4	T2
3-Level safety monitoring	R2	S2	C1	M3	T3

perspective, Triple Modular Redundancy Pattern and Sanity Check Pattern are easier to be modified than the other four patterns. Finally, from impact on execution time perspective, only 3-Level Safety Monitoring Pattern causes big influence. Note that the comparison results of these patterns can be different, if they are compared from other perspectives or in different contexts. Therefore, when engineers choose which patterns to be used or applied, they could validate our comparison in the context of their use scenarios. Besides, sometimes engineering can also build their own patterns based on the existing ones.

5 Model-Driven Design for Safety Assessment

In the previous sections, we discussed the main factors for achieving the first two goals of functional safety. In this chapter we explore the third goal. We specifically discuss how to provide evidence for achieving the first two goals.

The safety standards describe generalized approaches to identifying hazards and risks, design lifecycle, and analyses and design techniques. Therefore, when applying such standards for a specific application, a significant degree of interpretation of those standards may be necessary.

The process for developing safety-critical systems in these safety domains is manually checked for compliance with the standards. This checking process is referred as safety assurance and certification. Due to the amount of manual work involved, safety assurance is usually costly and time-consuming. Moreover, when a system evolves, some of the existing safety assurance data needs to be regathered or revalidated. To address this, model-driven techniques have been applied to facilitate safety assurance. We divide these techniques into three categories: modeling safety standards, modeling safety argumentation, and modeling support for safety case assessment.

5.1 Modeling Safety Standards

Models of safety standards are widely used for understanding and communicating among engineers and software developers. However, there are a number of significant challenges to deal with. Firstly, the modeling process suffers from subjectiveness issues. In some domains (such as the automotive domain), there is no authority providing an interpretation of the safety standard, and the modeling process is mainly performed by experts based on manufacturer requirements to ensure sufficient quality. Thus, the whole process of extracting information from the safety standards becomes subjective. Furthermore, when a new version of the standard is released, the models need to be updated or modified by persons who may have not created these models. Due to the invisible modeling process, most of the previous work needs to be redone. Secondly, standards are represented in natural language, with the resulting inevitable manual work of interpretation becoming more costly and less reliable. It also increases the difficulty of identifying the reusable information from the safety-related artifacts developed during the safety lifecycle. Thirdly, standards themselves contain inconsistencies. There are a number of synonyms used in the standard, which makes it impossible to generate the models from the standards automatically. Sometimes, standards are even in contradiction with themselves [42]. For example, in ISO 26262, formal methods are merely recommended, while the use of semiformal methods is always highly recommended. However, the standard does mention formal methods and formal notations at a number of places. Finally, any formal model should support the demonstration of compliance with the safety standard, both for the development process and for the diverse artifacts created during product development. We advocate that standards need to be universally understandable and expressed in a language that is simple, well structured, but strict. For this goal, we believe that in the future it should be possible to transform standards into models automatically and vice versa.

Work to date has generally involved conceptual modeling of standards for understanding. A conceptual model for the aeronautic standard DO 178B is created to improve communication and collaboration among safety engineers and software engineers [50]. A conceptual model of the generic standard IEC 61508 for electrical and electronic equipment is proposed for the development of compliant embedded software [38]. Also, a study on process modeling has been done in the context of ISO 26262 [25]. All of these studies refer to compliance with the standards from

Table 8 Methods and tools used for each model

Model	Purpose	Extraction method	Possible description methods
Structure model	Showing the structure of the standard	Manual modeling of the table of content	UML, Ecore, ontology
Conceptual model	Capturing the main concepts or terms used in the standard and their relations	Snowball approach	UML, Ecore, ontology
Process model	Demonstrating the required process described in the standard	Mapping between standard concepts and SPEM elements	SPEM, BPMN

a specific point of view. However, the modeling process is still subjective, which may lead to inconsistencies of the models after future modifications. Furthermore, the *traceability* of the source of the models is not covered: no one knows where the concepts and relations in the models come from, except the expert who has identified or defined them.

To address this, three kinds of models are proposed for the safety standards. The structure model and the conceptual model are introduced to support unambiguous understanding of the standard; the process model supports the demonstration of compliance of the process of the project with the process described in the standard. Due to the different characteristics and aims of ISO 26262 models (structure model, conceptual model, and process model), different methods are chosen to extract and describe these models. Most of the selected description methods in Table 8 are widely used in the industry.

The structure model of the standard can be obtained by modeling the table of content. For the conceptual model, we defined the Snowball approach for extraction [28]. The results of the structure model and conceptual model can be represented as an Ecore model, an UML model [34], or an ontology. For the process model, we have used Software and Systems Process Engineering Meta model (SPEM) [35] as the description language and the SPEM supporting tool Eclipse Process Framework (EPF) [3] for visualization. Besides, other formal process languages can also be used for constructing process model, such as BPMN [45].

5.2 Modeling Safety Argumentation

A safety case is a well-structured argument for justifying that a system is safe. In [6], a safety case is defined as:

Definition 3 "A documented body of evidence that provides a convincing and valid argument that a system is adequately safe for a given application in a given environment."

In some international safety standards, explicit safety cases are required for safety-critical systems. For example, ISO 26262 stimulates the use of safety cases to demonstrate the product safety [41]. Besides, MOD Def Stan 00-55 [33] for safety-critical software in defense equipment requires producing safety cases with explicit safety requirements.

Typically, safety cases are represented in free text, but in this way, the structure of the safety cases might be unclear, which allows for inconsistencies and confusion [31, 32]. To address this, modeling techniques are introduced to facilitate safety case construction and to increase the understandability and confidence in the claimed safety assurance [41]. For instance, techniques originally from model-driven development are used for representing concepts in safety cases, such as ontologies and SBVR models. Goal Structuring Notation (GSN) is introduced as a graphical modeling approach for safety case construction [23]. With the increase of safety-critical software and systems, such as cars, more and more GSN-based safety cases are developed. The reusability of GSN-based safety cases becomes another challenge. People want to reuse safety cases whenever it is possible. Informal reuse of safety case elements occurs, like "Copy and Paste" of the textual safety case documents between projects. A number of problems with informal reuse are listed in [22]. For example, it may cause inappropriate reuse, lack of traceability, or lack of consistency. To prevent these problems, safety case patterns are introduced as an approach to reuse of common structures of safety cases.

5.2.1   Safety Case Construction with Controlled Language

As more and more users (argument readers and writers, such as safety engineers, or safety assessors) are involved in safety case development, common understanding of the meaning of a safety case element is important. If it is not the case, the confidence of a safety case can be misplaced. To address this, some research has been done on the understandability of safety arguments. In [14], assured safety arguments are proposed as a clear argument structure to demonstrate how to create clear safety arguments. Besides, in [13], a precise definition of context in GSN arguments is proposed to achieve a better understanding. However, the content of a safety case element is still documented by natural language. The ambiguities caused by using a natural language are still unsolved.

We have proposed a methodology to use an SBVR-based controlled language [36] to support the development of clear safety arguments [30]. By using a controlled language, all the concepts (noun concepts and verb concepts) in a safety case are well-defined in a SBVR vocabulary. Argument readers can check the definitions or examples of those concepts to get a common understanding of them. In this way, the understandability of safety arguments can be improved. Besides, a model transformation has been implemented to generate SBVR vocabularies from EMF conceptual models, which can be obtained via modeling safety standards (Sect. 5.1). Then the manual work involved in vocabulary development can be reduced.

An overview of the proposed approach is shown in Fig. 8. There are three phases: conceptual phase (P1), vocabulary phase (P2), and modeling phase (P3). In the conceptual phase (P1), a conceptual model of the target domain will be manually built from scratch using the Snowball approach [28] or semiautomatically refined from other conceptual models [29]. The conceptual model will be used as an input of the vocabulary development. The meta model that we use for describing conceptual models is the Ecore meta model. After this, a model transformation will be carried out to transform the conceptual model from an EMF format to an SBVR specification. Then in the vocabulary phase (P2), users (argument writers) can build their own vocabulary based on the generated SBVR model. Note that users can also skip the previous phase and start by creating a new SBVR vocabulary. Finally, in the modeling phase (P3), the vocabulary will be used to facilitate the safety case construction.

5.2.2 A GSN Editor with SBVR Functionality

To construct safety cases in GSN with vocabulary support, we have integrated the SBVR functionality into the GSN editor. As the result, the noun and verb concepts defined in a vocabulary will be highlighted while safety engineers edit a GSN element. Figure 9 shows a screenshot of the GSN editor. When a GSN element is edited, a list of suggested concepts is given via content assistant. For example, after typing "p," a list of concepts in the vocabulary that start with "p" is provided. In this way, the number of errors, such as ambiguities of a safety case, can be reduced. Users can always look into the vocabulary to check the definitions of nouns and verbs used in their safety cases to avoid misunderstanding.

5.3 Safety Case Assessment

Currently, different industries have different processes for assessing safety cases. To the best of our knowledge, there does not exist a general and formal manual which describes how a safety case is assessed. After a restrict literature study, we have found four sources that have mentioned safety case assessment from different angles. These descriptions are not only specified for GSN-based safety cases, but also applicable for textual safety cases.

5.3.1 Overview of Safety Assessment Approaches

The first source is a safety case assessment manual for Gas Conveyors' Safety Cases provided by HSE [1]. In this document, they described a framework for assessing GSMR (Gas Safety Management Regulations) safety cases. In the Gas and Pipelines Unit, a safety case is assessed in two stages. The first stage is the registration stage.

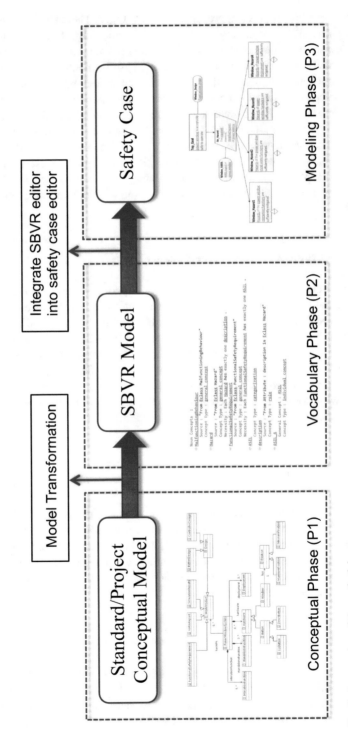

Fig. 8 An overview of the methodology

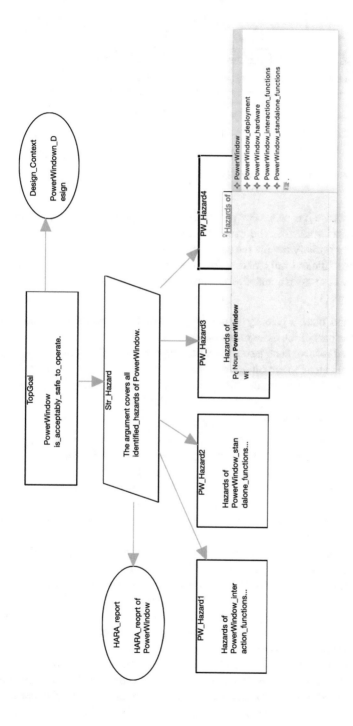

Fig. 9 Illustration of the GSN editor with SBVR functionality

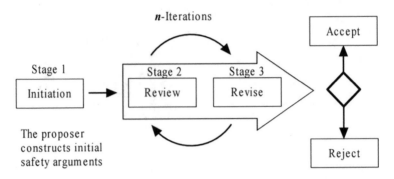

Fig. 10 Safety argument review process [49]

After a safety case has been received by the administration team (AT), the member of the AT checks whether the case is complete as described by its own content list. Then they initially review the safety case to determine whether it is reasonable and contains sufficient information for assessment. The second stage is the main part of safety case assessment. During this stage, the assessor should complete the following steps [1]:

- Identify and clarify priority issues which should be examined further and/or resolved as part of the assessment process.
- Discuss and resolve such issues with the proposer of a safety case.
- Reach formal agreement on improvements required.
- Reach a decision, where possible, to accept the safety case and record why.
- Provide reasons, in writing, for rejecting a safety case.
- Identify inspection topics.

The second source is a safety case review process introduced in [48, 49] (Fig. 10). This process includes three stages: Initiation, Review, and Revise. Stage 1 is the initial development of a safety case which is done by safety case developers. When the safety case is completed, developers submit it to safety case assessors. Assessors need to review the safety case and give feedback for revising it if necessary. This review and revise stage may be repeated several times until a judgment is proposed. The judgment can be either "Accept" or "Reject."

The third source is the description about the goal and needs of safety assessor in high-level requirements [2] of OPENCOSS. The goal of safety assessors is *to assess whether a safety demonstration of a product or assurance demonstration of a system or component is acceptable.* The principal needs of a safety assessor are *to view the baseline artifacts of safety case, to improve locating deficiencies and inconsistencies in the safety-critical system, and to cooperate with safety managers or other safety case assessors.*

The fourth resource is a presentation by Ola Örsmark on the third Scandinavian Conference on System and Safety [37]. It describes three topics around functional safety: the objectives and outcomes of safety case assessment, the benefits of delivering a good safety case, and the tips for developing effective argumentation. They stated, *The objectives of safety case assessment are to evaluate whether the reasoning about the functional safety of the product is valid and to get an independent statement that the claim about the functional safety is reasonable.* This is consistent with the objectives in the third source. Simply, the assessor is required to evaluate a safety case and then to provide a recommendation which gives judgments on the claims. The outcomes of a safety case assessment could be identified strengths and weaknesses of a safety case, a recommendation of the judgments, and required corrective actions.

5.3.2 An Alternative Safety Assessment Process

From the aforementioned four resources, we obtained an insight of safety case assessment process. This helped us to understand the responsibilities of safety assessors and to identify the user actions during the assessment process. However, the activities in these processes are not specified, especially in the review stage. To make the steps in the review stage explicit, we propose a detailed process flow for the review stage. It is designed for general safety case assessment which is independent on safety case formats. In other words, this process can be applied to both textual and graphical safety cases.

Figure 11 shows the detailed steps of our proposed process. There are four key steps in the safety case assessment: (1) prepare for review; (2) validate logic and structure; (3) evaluate quality; (4) and record and give feedback. In the first step, *Preparation*, the assessor receives a safety case which is developed by safety case developers. Hereby, we assume that the safety case is submitted with additional information wherein the purpose of the safety case and its background are introduced. Before starting the assessment, a number of preparations should be done. The completeness and consistency of the safety case are checked by the assessor, for example, whether there are undeveloped elements. Besides, they also need to check the format of the safety case to select corresponding tool used for the review process.

In the second step, *Logic and structure validation*, the assessor should initially review the safety case and determine whether its logic and structure is reasonable

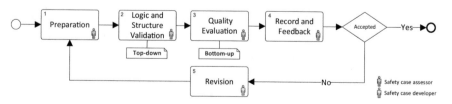

Fig. 11 The proposed safety case assessment process

and valid. We propose to start reviewing the safety case top down, because it is a natural and logic order of reading GSN-based safety cases.

After each element is reviewed from a logic and inference aspect, the next step, *Quality evaluation*, is to evaluate the elements from a quality aspect. We suggest to qualify a safety case bottom up, because the most important information is stored in the evidence. In this step, the assessor reviews the content of each evidence document and provides an evaluation for it. Depending on the quality level assigned to the evidence, the assessor can determinate whether the goal supported by the evidence is qualified. In this way, every element in a safety case can be evaluated via its children. Eventually, the top goal can be evaluated based on the quality level of all its branches.

Finally, in the fourth step, *Record and feedback*, the assessor needs to record and summarize evaluation results, then gives feedback to developers. Besides, a final judgment can be provided to safety case developers. If the judgment is accepted, then the assessment process finishes. If the judgment is not accepted, the safety case needs to be revised by developers and reviewed again by assessors.

5.3.3 The AGSN Editor

To support the proposed process, we have developed an editor (AGSN editor). In the AGSN editor, all assessment steps are supported. Figure 12 shows a screenshot of the AGSN editor. For more detailed information of the AGSN editor, we refer to [27].

Firstly, the assessment status is provided to facilitate the validation of logic and structure of a safety case. In Fig. 12, we could see that in the properties view, the assessment status of the selected element is shown. Secondly, the quality level is designed to facilitate the quality evaluation of safety case elements. Similar to the assessment status, the quality level of an element can also be added or modified via the properties view. To support recording and giving feedback, recommendations, statistical reports, and evidential reasoning (ER) score calculations are provided. A recommendation can be directly used to give feedback. The statistical report helps the assessors to analyze the assessment and evaluation results and to provide an overview of the assessment. Finally, the ER algorithm [46, 47] is applied to calculate an overall quality evaluation of a safety case. In Fig. 12, the degree of belief of the selected element is also shown, which is calculated based on the evaluation results of its evidence. Besides, the ER score of a safety case can also address the uncertainties in evidence. Uncertainties in evidence can affect assessors' confidences in the evaluation process. Showing the confidence degree of an evaluation makes the evaluation more credible and objective.

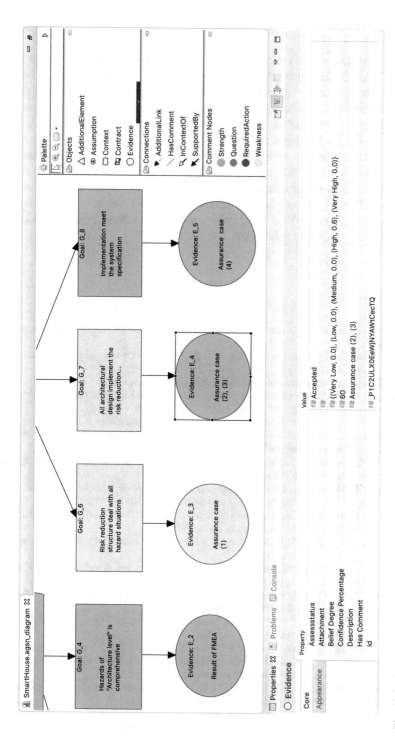

Fig. 12 A screenshot of the AGSN editor

6   Conclusions

As more and more manufacturers in the automotive domain start to comply with the ISO 26262 in their projects, in this chapter we have discussed a number of recent research directions regarding this standard. We provided a brief introduction of several basic concepts in the standard. Then we discussed safety management parts in details. After that we presented an integrated V model based on ISO 26262 to emphasize the hierarchal structure of requirements and to enforce gradual development or refinement of requirements based on higher-level requirements and design. Then, we briefly showed a comparison of a number of safety architecture patterns. The results of this comparison can be used as implementation or extension suggestions. Finally, as model-based techniques have been used to support safety assessment, we also described our current research on modeling safety standards, modeling safety cases, and safety case assessment.

References

1. (2011) HSE: safety case assessment manual. http://www.hse.gov.uk/gas/supply/gasscham/gsmrscham.pdf
2. (2013) OPENCOSS: Deliverable D2.2 – high-level requirements (report). http://www.opencoss-project.eu/node/7
3. (2015) Eclipse process framework project. http://www.eclipse.org/epf/
4. Amroush A (2010) Design patterns for safety-critical embedded systems. PhD thesis, Aachen University, https://doi.org/10.1016/B978-1-85617-707-8.00006-6
5. Armengaud E, Bourrouilh Q, Griessnig G, Martin H, Reichenpfader P (2012) Using the CESAR safety framework for functional safety management in the context of ISO 26262, Embedded real time software and systems
6. Bishop P, Bloomfield R (1998) A methodology for safety case development. Industrial Perspectives of Safety-critical Systems, P194-203 Cited by 201
7. Bushmann F, Meunier R, Rohnert H, Architecture SW (1996) Pattern-oriented software architecture, vol 1. Wiley, Chichester. https://doi.org/10.1192/bjp.108.452.101
8. CENELEC (1999) EN50126: railway applications – the specification and demonstration of reliability. Availability, maintainability and safety (RAMS)
9. CENELEC (2000) EN50129: railway application – safety related electronic systems for signaling
10. CENELEC (2011) EN50128: railway applications-communication, signaling and processing systems-software for railway control and protection systems
11. Douglass BP (2002) Real-time design patterns: robust scalable architecture for real-time systems. Addison-Wesley Professional, Boston
12. Fleming M (2001) Safety culture maturity model. Technical report, The Keil Centre. ISBN 0 7176 1919 2. www.hse.gov.uk/research/
13. Graydon P (2014) Towards a clearer understanding of context and its role in assurance argument confidence. In: Computer safety, reliability, and security. Lecture notes in computer science, vol 8666. Springer International Publishing, Cham, pp 139–154
14. Hawkins R, Kelly T, Knight J, Graydon P (2011) A new approach to creating clear safety arguments. In: Advances in systems safety. Springer, London, pp 3–23

15. Hecker S, Goldenhar L (2014) Understanding safety culture and safety climate in construction: existing evidence and a path forward. In: Safety culture/climate workshop. The Center for Construction Research and Training. CPWR, Washington, DC

16. Hudson P (1999) Safety culture – theory and practice. The human factor in system reliability – is human performance predictable

17. Hudson P (2001) Safety culture: the ultimate goal. Flight Safety Australia (October):29–31. http://82.94.179.196/bookshelf/books/1091.pdf

18. Hudson P (2001) Safety management and safety culture the long, hard and winding road. In: Occupational health & safety management systems: proceedings of the first national conference, p 3. http://www.ohs.com.au/ohsms-publication.pdf#page=11

19. International Electrotechnical Commission (2010) IEC 61508 functional safety of electrical/-electronic/programmable electronic safety-related systems, Geneva, Switzerland. https://www.iec.ch/functionalsafety/

20. ISO (2011) ISO 26262 road vehicles – functional safety. ISO, Geneva, Switzerland

21. ISO (2011) ISO/IEC/IEEE 42010: systems and software engineering – architecture description, pp 1–46. https://doi.org/10.1109/IEEESTD.2011.6129467

22. Kelly T, McDermid J (1998) Safety case patterns – reusing successful arguments. In: IEEE colloquium on understanding patterns and their application to systems engineering (Digest No. 1998/308), pp 3/1–3/9, cited by 41

23. Kelly T, Weaver R (2004) The goal structuring notation – a safety argument notation. In: Proceedings of Dependable Systems and Networks 2004 Workshop on Assurance Cases Cited by 257

24. Khabbaz Saberi A, Luo Y, Cichosz FP, van den Brand M, Jansen S (2015) An approach for functional safety improvement of an existing automotive system. In: Systems Conference (SysCon), 9th Annual IEEE International, pp 277–282

25. Krammer M, Armengaud E, Bourrouilh Q (2011) Method library framework for safety standard compliant process tailoring. In: 37th EUROMICRO Conference on Software Engineering and Advanced Applications. IEEE, Piscataway, pp 302–305

26. Langheim J, Guegan B, Maillet-Contoz L, Maaziz K, Zeppa G, Phillipot F, Boutin S, Aboutaleb I, David P (2010) System architecture, tools and modelling for safety critical automotive applications – the R&D project SASHA. In: ERTS2 2010, embedded real time software & systems, Toulouse, France, pp 1–8

27. Li Z (2016) A systematic approach and tool support for assessing GSN-based safety case. Master's thesis, Eindhoven University of Technology

28. Luo Y, van den Brand M, Engelen L, Favaro J, Klabbers M, Sartori G (2013) Extracting models from ISO 26262 for reusable safety assurance. In: Safe and secure software reuse – 13th international conference on software reuse, vol 7925. Springer, Berlin, pp 192–207

29. Luo Y, van den Brand M, Engelen L, Klabbers M (2014) From conceptual models to safety assurance. In: Conceptual modeling. Lecture notes in computer science, vol 8824. Springer International Publishing, Cham, pp 195–208

30. Luo Y, van den Brand M, Engelen L, Klabbers M (2015) A modeling approach to support safety assurance in the automotive domain. In: Progress in systems engineering. Advances in intelligent systems and computing, vol 1089. Springer International Publishing, Cham, pp 339–345

31. Luo Y, van den Brand MGJ, Engelen L, Klabbers M (2015) A modeling approach to support safety assurance in the automotive domain. In: Progress in systems engineering, vol 1089. Springer International Publishing, Cham, pp 339–345

32. Luo Y, van den Brand MGJ, Kiburse A (2015) Safety case development with SBVR-based controlled language. In: Proceedings of third international conference on model-driven engineering and software development

33. MOD (1997) Defence standard 00–55 part 1. http://www.software-supportability.org/Docs/00-55_Part_1.pdf

34. OMG (2005) Unified Modeling Language 2.0: superstructure specification

35. OMG (2008) Software and systems process engineering metamodel specification. http://www.omg.org/spec/SPEM/2.0/
36. OMG (2013) SBVR: semantics of business vocabulary and rules (version 1.2)
37. Örsmark O (2015) Will your safety case pass an ISO 26262 assessment? http://safety.addalot.se/2015/programme
38. Panesar-Walawege R, Sabetzadeh M, Briand L (2011) Using UML profiles for sector-specific tailoring of safety evidence information. In: Jeusfeld M, Delcambre L, Ling TW (eds) 30th ACM international conference on conceptual modeling (ER). Lecture notes in computer science, vol 6998. Springer, Heidelberg, pp 362–378
39. Piers M, Montijn C, Balk A (2009) Safety culture framework for the ECAST SMS-WG. European Commercial Aviation Safety Team (ECAST). https://www.easa.europa.eu/sites/default/files/dfu/WP1-ECASTSMSWG-SafetyCultureframework1.pdf
40. RTCA (2011) RTCA DO-178C: software consideration in airborne systems and equipment certification
41. Safety Case Repository (2013) Safety case repository. http://dependability.cs.virginia.edu/info/Safety_Cases:Repository
42. Sternudd P (2011) Unambiguous requirements in functional safety and ISO 26262: dream or reality? Master's thesis, Uppsala University
43. van den Brand M, Groote JF (2013) Software engineering: Redundancy is key. Sci Comput Program 97:75–81. https://doi.org/10.1016/j.scico.2013.11.020
44. Warszawska K, Kraslawski A (2015) Method for quantitative assessment of safety culture. J Loss Prev Process Ind:323–330. https://doi.org/10.1016/j.jlp.2015.09.005, http://www.sciencedirect.com/science/article/pii/S0950423015300309
45. White SA (2004) Introduction to BPMN. IBM Cooperation 2(0):0
46. Yang JB, Singh MG (1994) An evidential reasoning approach for multiple-attribute decision making with uncertainty. IEEE Trans Syst Man Cybern 24(1):1–18
47. Yang JB, Xu DL (2002) On the evidential reasoning algorithm for multiple attribute decision analysis under uncertainty. IEEE Trans Syst Man Cybern A Syst Hum 32(3):289–304
48. Yuan T, Kelly T (2012) Argument-based approach to computer system safety engineering. Int J Crit Comput Based Syst 3(3):151–167
49. Yuan T, Kelly T, Xu T, Wang H, Zhao L (2013) A dialogue based safety argument review tool. In: Proceedings of the 1st international workshop on argument for agreement and assurance (AAA-2013), Kanagawa, Japan
50. Zoughbi G, Briand L, Labiche Y (2011) Modeling safety and airworthiness (RTCA DO-178B) information: conceptual model and UML profile. Softw Syst Model 10(3):337–367. https://doi.org/10.1007/s10270-010-0164-x

Part V
C-ITS and Security

Introduction to Cooperative Intelligent Transportation Systems

Johan Lukkien

Abstract In this chapter we compactly introduce the overall system architecture and standards of countrywide cooperative intelligent transportation systems. This is an introduction to the next three chapters that take three different application perspectives on C-ITS, namely, intra-vehicle, inter-vehicle and countrywide. The focus lies on architecture and on security and privacy: protecting assets, safety, and functionality.

1 Introduction

The concept of connected and intelligent vehicles has been a promise for long, but it is moving toward realization rapidly. Cooperative intelligent transportation systems (C-ITS) comprise a range of technologies designed to improve safety and comfort on the road by collaborative applications built on top of interconnected vehicles. Within the vehicle itself, collaborative networking has been around for long; however, networks between vehicles have reached maturity and initial deployment only recently. Global data collection about vehicles to enable services like traffic management has been increasing as well. However, data collection is currently limited to several companies that collect this data through a special application or to agencies that manage road systems and can observe vehicles from outside.

The biggest step in recent years is the automation of the execution of functions in response to an observed (upcoming) situation. Between no automation and fully automated driving a series of five steps and later six steps were defined by the

J. Lukkien (✉)
Department of Mathematics and Computer Science, Eindhoven University of Technology, Eindhoven, The Netherlands
e-mail: j.j.lukkien@tue.nl

© Springer Nature Switzerland AG 2019
Y. Dajsuren, M. van den Brand (eds.), *Automotive Systems and Software Engineering*, https://doi.org/10.1007/978-3-030-12157-0_11

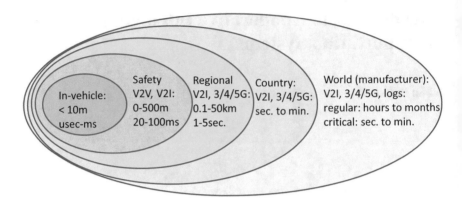

Fig. 1 Network scopes in vehicle networking and correspondent requirements on timing. From regional onward, services are mostly information oriented and less control oriented. Manufacturers have access to data in several ways: by building in a telecom connection, by reading out log files, and by dedicated bandwidth in V2I (vehicle-to-infrastructure) communication

NHTSA[1] and the SAE [3, 4]. For the coming years, automation systems until levels 3 and 4 are foreseen. In level 3, the driver hands over safety-critical functions to the vehicle under certain traffic and environment conditions. The driver is still there to take over control. In level 4, vehicles are designed to perform all functions for an entire trip, but within the "operational design domain" of the vehicle (i.e., restricted to certain scenarios). Systems supporting this are named *advanced driver assistance systems*. The special category of fully automated driving under all circumstances is studied intensely but is expected to be delayed for a number of years, not the least because of societal acceptance and legal issues. Current automated driving relies on intense sensing and analysis; when networking is added to this, it will rapidly improve further.

2   Vehicle Networking

Vehicle networking includes both real-time control and information-oriented services. Figure 1 gives an impression of several scopes in terms of distance and timing requirements. Going to the right in the figure, the focus moves from (safety-critical) real-time control to information services.

Within the vehicle, networking standards depend on vendor and application domains. Figure 2 gives an overview of in-vehicle domains and their requirements from the perspective of a series of metrics. In-vehicle networks include variants of

---

[1]NHTSA: National Highway Traffic Safety Administration. SAE: Society of Automotive Engineers.

	Flexibility	Predictability	Dependability	Bandwidth	Confidentiality
Powertrain	low	high	high	high	N/A
Chassis	some	high	high	high	N/A
Body/Comfort	some	some	some	low	N/A
Telematics	high	some	low	high	high
Passive Safety	low	high	high	high	N/A

Fig. 2 Vehicle electronics is commonly described using five separate in-vehicle domains. Vehicle networks take care of connecting electronic components within these domains. Picture taken from [1]

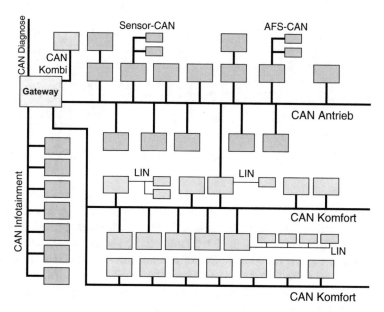

Fig. 3 The network topology of a VW Passat using several different standards: variants of CAN as well as LIN, for different application areas in the vehicle. Other network technologies include MOST, Byteflight, Flexray, and Ethernet. Picture taken from [1]

CAN, MOST, Byteflight, LIN, and FlexRay; more recently, real-time Ethernet has been adopted, first mainly for multimedia. Figure 3 gives the layout of a VW Passat network (some years ago) showing the different domains as described in Fig. 2. Both figures are taken from [1]. The different domain put different performance requirements and also must not be interfered in an uncontrolled manner.

Communication *between* vehicles is based on the IEEE 802.11p[2] version of wireless LAN. On top of that, the European standards organization ETSI has defined a series of standards describing usage of frequency bands, message routing,

---

[2]https://standards.ieee.org/standard/802.11p-2010.html.

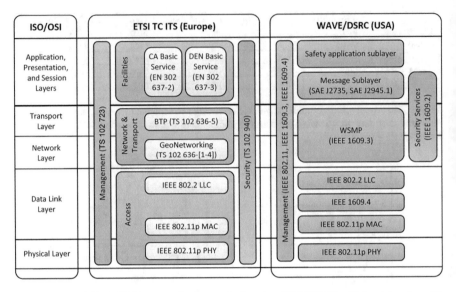

Fig. 4 EU and US V2V/V2I communication stacks based on IEEE 802.11p. Picture taken from [5]

and semantics named *ETSI TC ITS*.[3] Standardization in the USA has resulted in the WAVE standard (Wireless Access in Vehicular Environment, IEEE 1609).[4] In contrast to the WAVE standard, the ETSI family of standards supports geonetworking (routing messages between vehicles), several different channels (to increase flexibility and reliability), and several different transport mechanisms. With respect to safety applications, they both support a concept of periodic multicasting in which vehicles transmit messages in their one-hop neighborhood.

The current main focus lies with a series of selected safety applications out of some 80 possible ones that concern particular safety-critical conditions on the road, for example, *lane change warning, collision warning*, and *emergency brake lights*. The mode of operation is that vehicles emit broadcast messages[5] describing their relevant state and that they react to similar messages from other vehicles and from infrastructure. This is called vehicle-to-vehicle (V2V) and vehicle-to-infrastructure (V2I) communication, respectively. Figure 4 taken from [5] gives an overview.

Besides IEEE 802.11p, upcoming 5G telecom networks are a candidate for V2V and V2I because, in comparison with 4G, 5G technology aims to provide predictable high performance in terms of latency and signal reliability which are vital for safety-critical applications. It will take a couple of years though before that

---

[3]ETSI Technology Cluster ITS, https://www.etsi.org/technologies-clusters/technologies/automotive-intelligent-transport.

[4]https://standards.ieee.org/standard/1609-12-2016.html.

[5]WAVE: BSM defined by SAE J2735 on top of WSMP, and CAM messages in ETSI TC ITS.

standard is deployed, and there is a debate as to whether it will be reliable enough for safety-critical applications. An interesting comment by Nokia CEO Rajeev Suri at the 2015 Mobile World Congress[6] was that network neutrality does not allow to handle safety-critical traffic differently from other traffic. This would mean that no guarantees can be given to safety-critical applications.

Regional and country-level vehicle applications are much less defined and are currently experimented with. A basis for regional networked services is formed by V2I. Applications include, for example, warnings for an upcoming situation on the road or finding a free parking spot or a charging pole. The idea behind this approach is that local network traffic can remain local.

Country-level and manufacturer-defined applications have a "car as a sensor" purpose. Typically, telecom networks are being used. Some manufacturers equip their vehicles with telecommunication transceivers and also fleet owners do this. There are debates in this context about ownership (control) of such data. One might think further of people voluntarily providing data through their phones in exchange for useful services. Interesting examples include car sharing and reservation of a charging pole. There is little standardized, neither about the data collection (except for the limited OBD2 port) nor about storage and application development. Figure 5, taken from [2] gives a conceptual overview.

3   View on C-ITS

It can be helpful to look at C-ITS as being a *system of systems*. At its essence this term refers to the composition of systems into a larger whole while the composed systems retain a private and independent goal. This results among others in managerial independence (with respect to the composition) and competition of control [2] (between the subsystems and the composition). Vehicles represent independent systems with embedded ICT serving the purposes of driving, comfort, and safety. Cooperative applications like accident prevention or cooperative cruise control run on top of the electronic systems in vehicles while the original purpose of these vehicle systems remain. As argued in [2], special care should be taken at the place where systems are connected, managing access to internal resources and protection of functions.

Besides the technical issues of standardization and implementation, there are two aspects that require attention. First, the entire system has a principle setup of *collaboration*. Within the vehicle, devices connected to the networks are trusted by virtue of them being there. The point of view is that if an attacker has access to the vehicle then all bets are off anyway. However, when a vehicle is equipped with a network connection, this is no longer true and some famous breaks-in have

---

[6]https://www.cnet.com/news/nokia-knocks-net-neutrality-self-driving-cars-wont-get-the-service-you-need/.

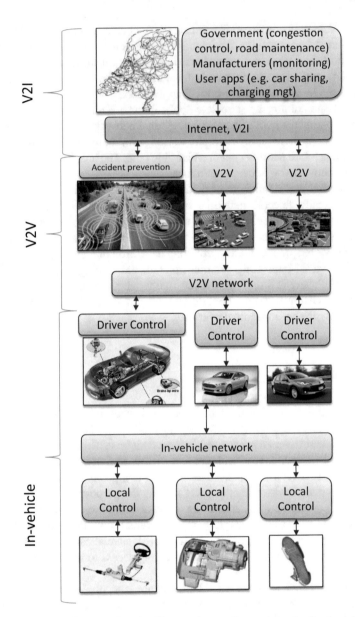

Fig. 5 A conceptual diagram of an intelligent transportation system. Applications like traffic management operate on collected data from vehicles. Besides the managerial applications, more consumer-like applications are possible as well, for example, car sharing, ride sharing, energy optimization, and so on

been reported allowing to remotely take over control of the vehicle. A similar consideration pertains to V2V and V2I communication: the basis on which a decision is taken must be authentic, and no attacker must be enabled to convince a vehicle to act upon false information. In addition, the new electronic functions should not make the driver or owner of the vehicle more traceable. Security mechanisms are required to protect safety and privacy. The new European General Data Protection Regulation (EU 2016/679) takes steps in this regard.

Second, stakeholder concerns must be much better addressed than currently is the case. For example, application development is commonly done by vehicle manufacturers, while this is neither their core interest nor their primary business. Similarly, roadsides are owned and managed by governmental departments who are not really interested in profitable service provisioning. These stakeholder concerns must be better analyzed in order to understand stakeholder position but also to guarantee progress.

4 Overview

Following are several contributions that each address different aspects of networks and vehicles. One examines security concerns within the vehicle considering that intruders may add equipment to the internal networks and considering that vehicles are connected to the Internet in some way. The other contribution addresses security and privacy concerns within connected vehicles as well as services provided within the regional domain. They identify architectural choices and elements vital to the quality and success of future C-ITS systems.

References

1. Keskin U (2009) In-vehicle communication networks: a literature survey. Tech. rep., Eindhoven University of Technology
2. Lukkien J (2016) A systems of systems perspective on the internet of things. ACM SIGBED Rev 13(3):56–62
3. NHTSA (2018) Policy on automated vehicle development. https://www.transportation.gov/briefing-room/us-department-transportation-releases-policy-automated-vehicle-development. Accessed 14 Nov 2018
4. SAE (2018) Taxonomy and definitions for terms related to driving automation systems for on-road motor vehicles (j3016). https://www.sae.org/standards/content/j3016_201806/. Accessed 14 Nov 2018
5. Tielert T (2014) Rate-adaptation based congestion control for vehicle safety communications. PhD thesis, Karlsruhe Institute of Technology

In-Vehicle Networks and Security

Timo van Roermund

Abstract Vehicles are experiencing a rapid evolution: mechanical systems are rapidly extended, or even replaced, with electrical systems, leading to highly computerized vehicles. Wireless connectivity, such as telematics and vehicle to everything (V2X), is being introduced to help connect vehicles with the world around them. The information exchanged via these interfaces is used to improve, among others, safety, convenience, comfort, and efficiency.

However, adding connectivity is at the same time opening the Connected Car to a multitude of security problems. Modern vehicles are highly complex cyber-physical systems with a high degree of automation and loads of (valuable) data. As such, they are an attractive target for hackers. But until recently, it was not possible for hackers to attack vehicles remotely, at large scale. The wireless connections are changing the game—as they form new entry points for hackers into the vehicle networks and systems. Most vehicles that are currently on the road were not designed with security in mind. And consequently, there were a few big-impact vehicle hacks last year that made headlines in the mainstream media.

To properly prepare vehicles for their connected future, concepts such as security by design, privacy by design, defense in depth, and life-cycle management must be applied. In this chapter, we will present a structural approach to applying these principles to in-vehicle networks.

1 Introduction

The automotive industry is rapidly evolving, and the car is transformed from a simple mode of transport to a personalized mobile information hub. New technologies like V2X communications, telematics based on cellular connectivity, Near Field

T. van Roermund (✉)
NXP Semiconductors Germany, GmbH, Hamburg, Germany
e-mail: timo.van.roermund@nxp.com

© Springer Nature Switzerland AG 2019
Y. Dajsuren, M. van den Brand (eds.), *Automotive Systems and Software Engineering*, https://doi.org/10.1007/978-3-030-12157-0_12

Communication (NFC), radar, and multistandard digital broadcast reception are rapidly adopted by vehicle manufacturers.

All these electronic functions bring great benefits to the driver, increasing comfort, convenience, safety, and efficiency. But these features come with new risks, too. Modern vehicles are gradually turning into "smartphones-on-wheels," which continuously generate, process, exchange, and store large amounts of data. Their wireless interfaces connect the in-vehicle systems of these "Connected Cars" to external networks such as the Internet, enhancing consumer experience by enabling new features and services. But this also makes the Connected Car vulnerable to hackers who attack the vehicle by seeking and exploiting weaknesses in its computer systems or networks. In fact, several studies, for example [1], have already warned some years ago that hacking into a car is possible, and more recently hackers indeed demonstrated that they could gain remote control over vehicles [2, 3]. The same day, U.S. Senators Markey and Blumenthal introduced an automotive security bill [4] that would establish federal standards to both secure vehicles and protect user privacy. This bill followed after Senator Markey's earlier report [5] that stated that the technology systems and data in today's cars and trucks are vulnerable to theft, hacking, and the same invasions faced by any technical system today.

Such hacks pose a threat to the reliability and safety of the car—the hacker can potentially take control over the car—as well as to the privacy of the driver—the vehicle data can be used to build a profile of its user(s). Therefore, steps need to be taken now: the Connected Car must be secured, to ensure the correct functioning of all in-vehicle systems, as well as user privacy. This implies a paradigm shift in the design of in-vehicle electronics. Traditionally, there has been a strong focus on (functional) *safety*, meaning that, for example, the brakes should function correctly under all circumstances. Safety will remain equally important in the future, but the increasing amount of electronics and software in vehicles will additionally require *security*, to protect the vehicle against hackers.

2   Connectivity: Driving the Need for Security

Until recently, cars have been isolated from their environment and from the Internet. The only exception was the (mandatory) interface for vehicle diagnostics, but because this on-board diagnostics II (OBD-II) port is a wired interface, installed in the vehicle's interior, it could (at least somewhat) rely on the physical protection offered by the vehicle cabin.

But things are changing rapidly: most modern cars allow smartphones to be paired via Bluetooth with the car radio for hand-free phone calls or to play music. And it does not stop there: many modern cars are wirelessly connected to the Internet, for example, to enable additional services in the car and, to a certain extent, provide for remote control over the car such as remote unlocking and starting. To improve safety, these cars will furthermore be equipped with eCall and V2X communication technologies, complemented by Advanced Driver Assistance

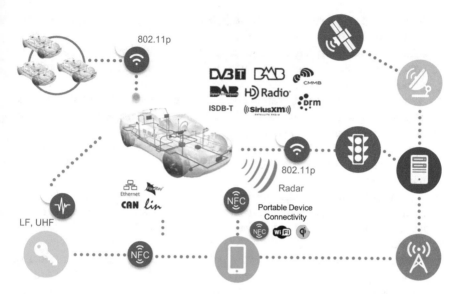

Fig. 1 Various interfaces of the Connected Car

Systems (ADAS) that offer advanced driving assistance features, and ultimately these cars will be self-driving (Fig. 1).

These technologies bring great benefits to the driver: for example, *comfort* is increased because you can remotely enable the air conditioning systems to cool the cabin shortly before driving home, in summer time. *Convenience* can be increased because, for example, your in-car entertainment system is seamlessly synchronized with your phone and via your phone to your media collection. Last but certainly not least, the introduction of ADAS helps to increase *safety and efficiency*, for example, by using information from nearby vehicles to prevent collisions or by using information provided by road infrastructure or the cloud to reduce travel time.

2.1 Potential Risks

Despite all these advantages, these new features come with new risks, too. For example:

- Traffic information provided by infrastructure and the cloud should be trustworthy. You do not want someone to broadcast traffic jam warnings for his own route, thereby "clearing his route" and gaining time, at the cost of others.
- Autonomous Emergency Braking systems can prevent a crash or reduce the impact speed of a crash by applying the brakes independently of driver input. But it should not be possible for hackers to activate this system by sending fake

V2X messages to a vehicle or by manipulating the safety-critical communication inside the vehicle.

- eCall systems bring rapid assistance to motorists involved in a collision by automatically sending post-crash information to the emergency call center. You do not want third parties to get access to this personal data.
- Automated Vehicle Identification allows the car to identify itself for seamless access to a parking or a toll road. When not protected, hackers could steal, for example, personal data, including payment details.
- Car-sharing systems allow access to a vehicle via a smart card or mobile device. If not protected well, a thief might be able to abuse this system to gain access to the vehicle.
- The OBD-II port offers diagnostic and reporting capabilities, allowing one to rapidly identify and remedy malfunctions within the vehicle. But attackers may use it to gain access to the in-vehicle network, potentially even remotely (via Bluetooth or cellular dongles).
- Pay-as-you-drive insurance schemes may allow you to reduce your monthly insurance premium because you only pay for the miles you actually drive and based on your driving style. However, you don't want unauthorized third parties to get access to the same data and get insights into your driving habits.
- Car owners may want to chip-tune their engine, to increase performance of the engine. Manufacturers (OEMs) may want to protect against such manipulation by the vehicle owner because it may have negative consequences for the reliability and the emission levels.

2.2 The Connected Vehicle: An Attractive Target for Hackers

It is important to understand *why* hackers are now aiming their effort at the Connected Vehicle. We will explain this using Fig. 2.

Hackers generally need an *incentive* to hack a system. For one group, the researchers, it is clearly academic: they aim at warning the industry and the public—like in the case of the Jeep hack—that vulnerabilities exist that allow them to take control over (parts of) the vehicle. But probably the bigger incentive is a financial one—modern vehicles contain a lot of valuable data, which forms an attractive target for criminals who may want to monetize it.

Modern vehicles are inherently *vulnerable* due to their complexity. A good indicator for the vulnerability of systems is the complexity of their software—modern vehicles can contain up to 200M lines of code. If we (conservatively) assume 1 bug per 1000 lines of code and at least 1 security vulnerability per 1000 bugs, we can conclude that a modern vehicle may have (at least) hundreds of security vulnerabilities. By comparison, modern computer and smartphone OSes are less complex and security vulnerabilities are found almost every day.

Lastly, the vehicle's (Internet) connections form an easy *entry point* for hackers as it removes the need to be in physical proximity of a vehicle to be able to attack

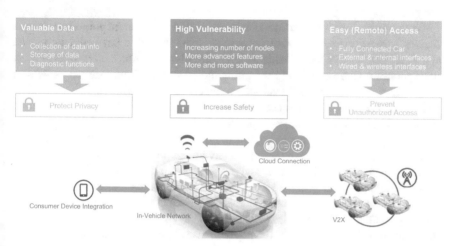

Fig. 2 Security challenges for the Connected Car

it. It may even enable hackers to remotely exploit systematic vulnerabilities at a large scale (e.g., complete fleets). Hence, the cost (effort, time) for hackers decrease, while the benefits (gain/impact) increase.

The main reason why vehicles, until recently, were not actively targeted by hackers was the lack of connectivity. But that is changing rapidly, as explained before. Due to the introduction of wireless interfaces, the Connected Car is exposed to new risks as they open the door for *remote* attacks. Consequently, manufacturers cannot solely rely on the physical protection offered by the vehicle's physical protection (steel and glass). The vehicle's electric and electronic systems themselves must (also) be protected against cyberattacks.

2.3 The Challenge

The range of attacks that a Connected Car faces is extensive and diverse: it varies from relatively simple attacks, in which, for example, malicious messages are sent to a vehicle, to more sophisticated attacks in which hackers may open electronic control units (ECUs) and reverse engineer their microcontrollers and software.

A first reason for that is that there is not a single, well-defined hacker. In fact, there are various attackers, with different motivations, skill levels, and resources. For example, there may be (academic) researchers who try to take (partial) control over the vehicle for scientific reasons. Or there may be (organized) criminals with large budgets who want to steal valuable data from a vehicle for financial gain or take control over vehicles to cause damage to society. But the threats do not only come from third parties: the car owner himself may be the "attacker" who wants to unlock extra features or gain (engine) performance, through "chip tuning."

Furthermore, the attack surface, that is, the sum of the different points (the "attack vectors") where an unauthorized user can attack the system, is large: attacks may be mounted directly from the in-vehicle electronics network, from user devices such as smartphones that are coupled to the infotainment system, from external devices in proximity, such as other V2X-equipped vehicles, or from the cloud.

Finally, the impact of a successful hacking attack may also widely differ. In certain cases, a hacker may target a specific vehicle, causing (limited) damage to that vehicle only. But a hacker may also find an exploit that can be abused over complete series of cars. When such an attack can easily be reproduced by others, for example, because the hacker publishes tools and instructions on the Internet, the impact, and likely also the (financial) damage, is obviously much larger. For example, a large-scale attack at random vehicles could easily have an economic impact because it has the potential to severely disturb traffic in a large geographical region. Also, the costs for car manufacturers could be high because of potential recalls and associated brand damage.

The big challenge for vehicle manufacturers and their suppliers is therefore to jointly implement solutions that block this wide variety of hackers with different motivations, resources, and skill levels and using many different attack vectors, in a cost-effective way.

3 No Safety Without Security

System performance and reliability has had (and will always have) high attention from vehicle manufacturers, with a strong focus on safety hazards. By adding wireless interfaces to their cars and connecting their vehicles to external networks and devices (such as smartphones), manufacturers are, however, suddenly confronted with new threats that stem from an uncontrolled and evolving environment. They are faced with intentional hazards caused by hackers who do not obey to any rule. On the contrary, they will do whatever it takes to achieve their goal. Also, their attacks will only get better over time: their knowledge level continuously increases, and their (hardware and software) tools get more and more sophisticated.

Like safety, security is a quality aspect—threats of either type can have a negative impact on the reliability and safety of the Connected Car. But there are also important differences.

The ISO 26262 "functional safety" standard addresses systematic failures and random hardware failures. Such safety threats are quite predictable—systematic failures are deterministic, and random hardware failure rates can be predicted with reasonable accuracy—and the nature of the hazards will not change over time. Furthermore, it is considered unlikely that multiple failures occur simultaneously in safety engineering.

Security threats, on the other hand, are generally less predictable, and they will also change over time. Furthermore, hackers do not hesitate to manipulate various parts of a system simultaneously if that increases the chance of a successful attack.

Consequently, security threats are not necessarily covered within a safety framework such as ISO 26262 but require dedicated frameworks and standards (more on that later in this chapter).

Overall, security threats still form a large challenge to the automotive industry. Outside the automotive industry, standardized frameworks such as Common Criteria are used to provide customers assurance that a product's security attributes can be trusted and that the customer's security needs are protected. Such frameworks are, however, new to the automotive industry, and it will likely take some time, as well as adaptations, before they are widely embraced.

4 Applying Best Practices

To properly prepare vehicles for their connected future, concepts such as security by design, privacy by design, defense in depth, and life-cycle management must be applied. In this chapter, we will elaborate on these concepts.

4.1 Defense in Depth

A typical vehicle hack consists of several smaller steps. It usually starts with finding a vulnerability (a "bug") in a system that is remotely accessible. But once you get, for example, into the telematics unit of a typical car that is on the streets today, you have a good chance of getting into just about any other part of the car, such as the ECUs that control engine speed, braking, cruise control, etc.

It is good practice to use multiple security techniques to mitigate the risk of one component of the defense being compromised or circumvented. In the example above, the first line of defense is to protect the telematics unit itself. But the security architecture should furthermore be designed in a way that an attack on an individual ECU does not scale to other ECUs in the vehicle, for example, by isolating critical ECUs and their networks from noncritical ECUs and their networks using firewalls.

4.2 From Afterthought to Integral Approach

To successfully protect the Connected Car from attacks, a paradigm shift is needed in automotive vehicle design: security must become part of the entire life cycle of the vehicle. It needs to become an *integral* part of the design process, as opposed to an afterthought, because security is only as strong as the weakest link. This calls for security by design and privacy by design, which may also have a significant impact on the architecture and on the in-vehicle electronics. The design of in-vehicle network architectures was, until recently, primarily driven by safety, bandwidth, and cost optimizations—and not by security. However, this is rapidly changing, and in-vehicle networks are now adapted such that systems with similar criticality are clustered in separate (physical or virtual) networks to better isolate highly critical safety systems from, for example, in-car entertainment systems.

Furthermore, the security architecture requires regular maintenance. This requires active life-cycle management for the security architecture, and its components, over a long time (the average lifetime of vehicles is well over 10 years).

4.3 Adoption of Existing Technologies

Traditionally, the automotive industry has been conservative in adopting features offered by consumer electronics. But the fully Connected Car is finally becoming a reality, and it will likely redefine the entire automotive industry. Vehicle manufacturers must find ways to deliver the advanced features their customers demand into their "smartphones-on-wheels." Similarly, they will also need to embrace security solutions that are widely used in smartphones and IT infrastructure but that are relatively new to the automotive world. Examples of such technologies are firewalls, intrusion detection and prevention systems, virtualization technologies, and secure firmware updates.

However, existing technologies and solutions cannot always be applied one on one—vehicles are, unlike IT systems, cyber-*physical* systems, and safety therefore is, and will remain, a key concern in automotive. The existing technologies and solutions must be tailored to automotive needs to make sure that they do not violate safety requirements.

4.4 Risk Analysis

Security always comes at a cost. Basically, it is an insurance premium you pay upfront, to prevent potential damage later, if the vehicle would be attacked.

To ensure that costs and benefits are well balanced, one needs to perform a risk assessment upfront when designing an actual vehicle electronics architecture. Since security risks are a function of threats, vulnerabilities, and potential impact, one needs to identify all possible threats and assess, for each threat, the potential impact of a successful attack, as well as the vulnerability of the device to such attack. Finally, one must set priorities for the different risks to determine which risks to address first and then define countermeasures to mitigate those risks.

One popular method for reasoning about computer security threats is the STRIDE [6] system, developed by Microsoft. It provides a mnemonic for security threats in six categories: Spoofing of user identity, Tampering, Repudiation, Information disclosure (privacy breach or data leak), Denial of service, and Elevation of privilege. The same classification can also be applied to connected vehicles, to help structure the threat analysis for these vehicles.

5 How to Secure a Vehicle

The Connected Car and the presence of hackers are now parts of life—hence, security must be an integral part of the design of the Connected Car as security is as weak as the weakest link. Vehicle security is a big topic, but it can be broken down into manageable chunks.

We can break it down on two different axes—a vehicle architecture axis and a time axis (see Fig. 3).

5.1 The Vehicle Architecture Axis

We can look at the IT industry for guidance to solve the problem of vehicle security. The key point is defense in depth—never rely on just one line of defense, but assume that has been breached, to reveal another layer of defense. Then assume that has been breached to reveal another layer, etc.

There is, for example, a common myth that adopting a system of individual unique secret keys for every vehicle is sufficient, but that assumes the perfect impenetrable system, and one of those has not been designed yet. Unique keys alone are not sufficient to protect the vehicle. At best, they prevent scaling of the attack to other vehicles.

5.2 The Time Axis

Security needs to be designed into the vehicle architecture from the very start, and it must furthermore be maintained throughout the vehicle's entire life cycle. Contrary to common belief, security is much more than prevention only. To secure a vehicle, one must:

- **Prevent unauthorized access**, for example, using machine-to-machine authentication and gateway firewalls, to ensure that hackers cannot access and tamper with the (safety critical) nodes in the vehicle
- **Detect intruders**, for example, secure boot of the controller, to validate that the software is (and remains) genuine and trusted
- **Reduce impact** of any determined intruders who did manage to gain access, for example, by isolating the network domains, to prevent that a compromised

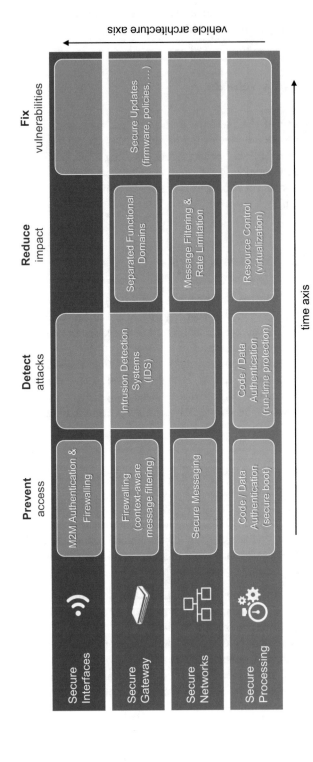

Fig. 3 Breaking down the topic of vehicle network security

infotainment unit in one domain can be used to control, for example, the brakes in another domain

- **Fix vulnerabilities**, for example, enable full vehicle Over-the-Air (OTA) update capability through the secure gateway, to fix vulnerabilities before they can be exploited (at a large scale) by hackers

6   A Multilayer Security Framework

To properly prepare vehicles for their connected future, secure vehicle architectures should implement the following four layers of security that together provide the right level of protection:

1. **Secure interfaces** that securely connect the vehicle to the external world
2. **Secure gateway(s)**, providing domain isolation within the vehicle network
3. **Secure networks**, providing secure communication between control units (ECUs)
4. **Secure processing** units that securely implement the features of the connected car

Each layer provides a unique form of protection while adding to the defenses of the overall system. Together, these layers provide a strong defense against hackers throughout the complete car electronics architecture. Figure 4 provides a brief overview of some of the possible countermeasures per layer.

Let us now take a closer look at these four layers.

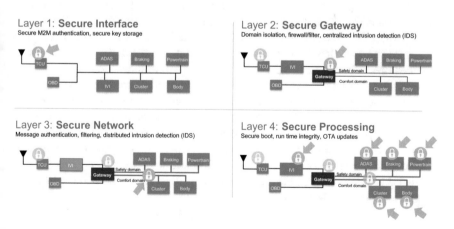

Fig. 4 The 4 layers of security within a vehicle network

6.1 Layer 1: Secure Interface

The common networks in vehicles today are largely unprotected. If a hacker gains access to the telematics control unit (TCU), he or she can potentially send spoofed CAN messages and gain control of safety critical items, like brakes.

To secure the Connected Car, these external communication channels therefore must be protected against unauthorized access, for example, by applying strong machine-to-machine authentication, against data theft, for example, by encrypting the data, and against manipulation, for example, by authenticating the messages that are exchanged to protect their authenticity and integrity.

6.2 Layer 2: Secure Gateway

As we saw with the Jeep hack [2] in 2015, once hackers obtain access to an insecure network, they can send messages anywhere. This is where layer 2, the secure gateway, plays its part. A central gateway ECU separates the external interfaces (e.g., TCU and OBD) from the in-vehicle network and furthermore divides the network into separate functional domains, with the gateway firewall deciding what nodes can legitimately communicate with other nodes.

In the Tesla Model S hack [7] of 2015, the protection offered by the gateway was highlighted as a key security feature for modern vehicles. In the Jeep hack, hackers could switch off brakes remotely due to the lack of a gateway. In the Tesla hack, the worst they could do was sound the horn.

The first true gateway was introduced into some high-end vehicles 8 years ago. Since then, as the amount of data being transferred between ECUs in the vehicle has significantly increased, the gateway functionality has become more complex, and more commonplace in our vehicles. In its current form, the central gateway provides many functions, linking data and signals from the various nodes around the vehicle, converting the various automotive communication protocols.

One of the most important functions of the secure gateway is the firewall that separates the external interfaces from the safety-critical inner vehicle network and furthermore provides a means to control the communication inside the vehicle network, between different domains and subnetworks. The gateway engine is a contextually aware routing function that determines, by several increasingly sophisticated checks, which messages are currently legitimate and will pass through the gateway onto the destination.

6.3 Layer 3: Secure Network

With the network now split into domains, the attack surface of the architecture is significantly reduced, but the subdomains are still vulnerable to attack. Layer 3 protects the subdomains through various network-level safeguards, such as:

1. Device authentication—the authenticity and integrity of ECUs is checked before they can communicate over the internal network.
2. Message authentication—messages are extended with a cryptographic code to guarantee an authentic sender and that it was received unaltered.
3. Message encryption—the risk of data and identity theft can be reduced by encrypting the messages that are exchanged between different ECUs inside the vehicle.
4. Intrusion detection—anomaly detection in the network traffic allows early detection of attacks, thereby enabling further countermeasures, for example, to prevent such attacks from spreading to other vehicles.
5. Rate limiting and traffic shaping—the number of messages that an ECU can send in a certain time frame is restricted, to prevent Denial-of-Service attacks.

6.4 Layer 4: Secure Processing

Last, but certainly not least, the "brains" of the Connected Car must be protected. These brains are formed by up to (and in some cases, over) a hundred individual computers (ECUs) that together implement the control functions in the car, including many advanced (automated) driving functions. These ECUs continuously generate, process, exchange, and store large amounts of valuable (sensitive) data.

To protect these functions and the data, modern microcontrollers feature secure boot and real-time integrity checking schemes to guarantee that the code image is authentic, trusted, and unaltered. Furthermore, mechanisms for controlled lockdown of the MCU and ECU through manufacturing are employed to lock out, for example, debug access, which would be invaluable to hackers.

As a logical consequence of the fast growth of software in vehicles, there is also a trend to reuse hardware [8], by integrating multiple software stacks, sometimes with different criticalities and often originating from different vendors, on the same microcontroller or central processing unit (CPU). Therefore, modern (high-end) automotive microprocessors also offer hardware-enforced process and resource isolation, for example, in the form of virtualization techniques, to isolate the various software stacks from each other and thereby to prevent that one vulnerable software stack can be misused to attack the other software stacks.

6.5 Which Layers to Apply and in Which Order?

The four layers are presented here as logical sequential four steps. However, depending on the OEM architecture, it may be that layer 4 is instigated prior to layer 3 or that layer 1, including strict machine-to-machine authentication, is initially only implemented for certain interfaces (e.g., V2X), while the vehicle network relies on the firewall of the gateway first for other interfaces (such as Bluetooth connections between smartphones and the infotainment system). Decisions like that would need to be driven by individual vehicle threat analysis.

7 Hardware Trust Anchors

System integration and bring-up will be the major time-to-market challenge for many OEMs. It is also likely that improvements are needed during the lifetime of the car, not only because its complex systems will need to be fine-tuned over time but also because the Connected Car is actively interacting with its environment, which itself will evolve during the lifetime of the vehicle. Therefore, many of the features of modern cars will need to be implemented in software, rather than hardware, because it not only reduces the development time (and cost) but also significantly reduces the need for expensive recalls as firmware updates can be provided over the air. The same holds for the security architecture and its implementation: for some parts, software is the better choice, for example, because of the need for updatability in the field.

But software is not enough—and hardware support is needed: this may be because of performance reasons but more often also for security reasons: updatability of software is, on the one hand, a powerful feature that allows the manufacturer to manage the product during its entire life cycle. But on the other hand, the same updatability also provides hackers with a means to manipulate the product. Therefore, there must be some immutable (hardware) root of trust to safeguard most critical parts of the system from being manipulated.

And there are more reasons why hardware is needed. For example, (software) security implementation should be isolated from other, less-trusted code, for example, by executing the security software at a higher privilege level (system vs. user mode), enforced by hardware. Other examples include side-channel resistant crypto implementations, which usually need specific hardware, and on-chip security implementations such as HSMs that protect cryptographic keys from software attacks by moving the control over those keys to the hardware domain. These examples illustrate that dedicated security hardware is needed as a basis, or as "trust anchor," for protecting the system.

We can distinguish at least three different categories of hardware security solutions that play a role in securing in-vehicle networks:

- *Secure micros*; to secure the "brains" of the car, the newest general-purpose microcontrollers and processors in the market are equipped with implementations of the Secure Hardware Extension (SHE) or Hardware Security Module (HSM) specifications. An SHE or HSM implementation is an on-chip extension to any given microcontroller, which can be used to protect software (secure boot, secure update) and data (secure storage, secure communication). It moves the control over cryptographic keys from the software domain into the hardware domain to protect those keys from software attacks. Depending on the implementation, such micros may also provide certain resistance against side-channel analysis and fault injection attacks.
- *Security companions*, such as secure elements, which are dedicated security microcontrollers offering state-of-the-art security to protect against advanced attacks, including physical attacks and reverse engineering: the security of these products is usually also proven via third-party security assessments, for example, via Common Criteria certification.
- *Function-specific secure ICs*: this comprises various solutions. For example, security features can also be implemented inside a Controller Area Network (CAN) transceiver or inside an Ethernet switch. Such security solutions at network level can provide for example, authentication and encryption of messages, detection of anomalies in the network traffic, (context-aware) message filtering, and rate-limiting mechanisms.

8   Life-Cycle Management

It is not enough to only design and implement a secure vehicle electronics architecture; it must also be maintained over the life cycle of the vehicle to keep up with the vehicle's evolving environment. After all, security is a moving target: the knowledge, skills, and tools of hackers improve over time. The Connected Car thus needs active maintenance, including key management and secure firmware updates.

8.1   Key Management and Crypto Agility

To prevent attacks from being scaled from one device (or vehicle) to a complete network of devices (or vehicle fleets), ECUs will need unique cryptographic device keys. These keys need to be managed during their life cycle. For example, existing keys may need to be replaced with newly created keys every now and then, and the existing keys need to be destroyed. In some cases, the new keys are created outside the vehicle, for example, in a cloud server, and need to be securely distributed from these cloud servers into the vehicles.

And also, cryptographic algorithms eventually may become obsolete and be replaced with new ones. Vehicles shall be sufficiently protected during their entire lifetime, which is approximately 15 years. The use of open and preferably standardized crypto algorithms and security protocols, as well as the key sizes that are generally expected to be sufficiently secure during the vehicle's entire lifetime, is a good starting point. But sooner or later, a move to larger key sizes, or maybe even different algorithms, will be needed.

Key management and cryptographic algorithm agility are thus important aspects of the life-cycle management of a Connected Car.

8.2   Secure Firmware Upgrades

Another clear reason why active maintenance is needed is that security vulnerabilities may, and will, be found after initial deployment. Modern vehicles average around 40 microcontrollers and millions of lines of code, creating massive software complexity. Such complex systems cannot be perfect and will contain vulnerabilities. Furthermore, flaws might be discovered in widely used security building blocks that were deemed "secure" when the vehicle was designed. For example, serious flaws have been discovered in the Transport Layer Security (TLS) specification, as well as some of its implementations [9–11]. This affected many Internet-connected devices, which all became vulnerable from one day to another and needed to be patched urgently.

Security flaws can affect individual models, or even worse, complete series of cars or even all cars and OEMs must have the ability to quickly, seamlessly, and securely mitigate the risk through updating the vehicle software, preferably without the need to visit the garage. The ability to perform OTA (Over-The-Air) software updates for every ECU in the vehicle is now demanded, and justified, by the number and cost of vehicle recalls in the last few years.

9   Standardization

There is also a need for standardization, to create a common (baseline) set of security requirements for products, as well as for the underlying engineering processes.

On the process side, one can think of security processes, policies, and requirements that apply to the product life-cycle stages—from development, via deployment, to maintenance. Early 2016, SAE's Vehicle Electrical System Security Committee [12] published a cybersecurity guidebook (J3061) to address this. More recently, ISO and SAE joined forces and created a new work item to create a standard on "Cybersecurity engineering" for "Road Vehicles." This standard, ISO/SAE 21434, is targeted for publication in 2020 and will likely replace SAE's J3061. It will address security in various engineering processes, including product development (concept, design, implementation), production, operation, maintenance, and decommissioning—similarly to how ISO 26262 addresses functional safety in the same processes.

Solid security engineering processes are necessary but not enough. One also needs solid technical security specifications. SAE is currently working on a specification for hardware-protected security (J3101), which may partially close this gap. Furthermore, Automotive Open System Architecture (AUTOSAR) introduced the "Crypto Stack" in version 4.3 of its software architecture specification. This stack offers logical crypto services such as "Symmetric Encryption" or "Signature Verification." It furthermore provides requirements for and a specification of security mechanisms that can be used to protect communication on in-vehicle networks— nicknamed SecOC (Secure Onboard Communication)—and it adopts the ETSI ITS security standards for V2X communication as part of its stack for secure Off-Board Communication (V2X). As it is not uncommon for straightforward mistakes to be made in security architectures and implementations, a seamless integration of such security features into a well-reviewed specification like the AUTOSAR software stack is highly beneficial. But further standardization efforts will be needed, for example, to come to interoperable key management and software update solutions.

Lastly, product security evaluation, and in the future possibly also certification, requires attention. Something based on Common Criteria could potentially form the basis for such framework; however, adaptations may be needed to make it fit the automotive requirements and constraints.

10   Conclusions

We are in a new era of vehicle complexity and connectivity, and with that a new era of ingenuity and resourcefulness for car hackers. The Connected Car, as part of a smarter world, is highly connected to and constantly interacting with its environment, which brings enormous promises for increased comfort, safety, and efficiency.

But it also raises questions regarding security and privacy: like all connected devices, it also becomes a target for attackers. Therefore, the security of the vehicle electrical architecture is vital to ensure the safety of the vehicle occupants, and the industry must respond to this threat. The four-layer security framework, introduced in this chapter, provides a holistic approach for securing the complete vehicle architecture. This approach ultimately protects the well-being and privacy of the users of the vehicle.

References

1. Comprehensive experimental analyses of automotive attack surfaces. *CAESS*, August 2011. http://www.autosec.org/publications.html

2. Hackers remotely kill a jeep on the highway – with me in it. *WIRED*. http://www.wired.com/2015/07/hackers-remotely-kill-jeep-highway/

3. Remote exploitation of an unaltered passenger vehicle. Dr. Charlie Miller and Chris Valasek. http://illmatics.com/Remote%20Car%20Hacking.pdf

4. Sens. Markey, blumenthal introduce legislation to protect drivers from auto security, privacy risks with standards & "cyber dashboard" rating system. http://www.markey.senate.gov/news/press-releases/sens-markey-blumenthal-introduce-legislation-to-protect-drivers-from-auto-security-privacy-risks-with-standards-and-cyber-dashboard-rating-system

5. Tracking & hacking: security & privacy gaps put American drivers at risk. Ed Markey, United States Senator for Massachusetts; February 2015. http://www.markey.senate.gov/imo/media/doc/2015-02-06_MarkeyReport-Tracking_Hacking_CarSecurity%202.pdf

6. The STRIDE threat model. https://msdn.microsoft.com/en-us/library/ee823878(v=cs.20).aspx

7. Hacking a Tesla model S: what we found and what we learned. *Lookout Blog*. https://blog.lookout.com/blog/2015/08/07/hacking-a-tesla/

8. Consolidation in vehicle electronic architectures. Roland Berger Strategy Consultants. http://www.rolandberger.com/media/publications/2015-07-23-rbsc-pub-consolidation_in_vehicle_electronics_architecture.html

9. The 'Heartbleed' security flaw that affects most of the Internet. *CNN.com*. http://edition.cnn.com/2014/04/08/tech/web/heartbleed-openssl/

10. How the poodle computer bug impacts business. *Fortune.com*. http://fortune.com/2014/11/12/poodle-bug/

11. FREAKing hell: all windows versions vulnerable to SSL snoop. *The Register*. https://cve.mitre.org/cgi-bin/cvename.cgi?name=CVE-2015-0204

12. Vehicle electrical system security committee. http://www.sae.org/works/committeeHome.do?comtID=TEVEES18

Security for V2X

Marc Klaassen and Tomasz Szuprycinski

Abstract In this chapter, a high-level description of security is presented for car-to-car communication. Future cars will become safer and more robust using different sensors and technologies. Radar, Lidar, cameras, and connected cars will create a bubble around the car that a smart CPU box can manage, with or without the presence of the driver.

Security is an integral part of a robust communication system. Safety is not only enforced through packet error rates or transmit performance. In the first section of this chapter, the significance of security is discussed. Why do we need it, and what impact does it have? Subsequently, the requirements are described in more detail. Privacy, for example, is key for becoming a success in this connected world.

Optimal utilization of the channel throughput is considered beneficial in any communication system. At the same time, messages must be authentic. In this chapter, the security scheme, based on elliptic curves, is discussed, as well as public key infrastructure. The last sections are referring to standardization work done so far and what the near future has to bring.

1 Introduction

Cooperative Intelligent Transport Systems (C-ITS) is an umbrella name for different ITS technologies implemented using active cooperation based on communication between ITS actors. These actors can both be vehicles, as, for example, in a collision avoidance application. One actor can also be infrastructure based, for example, a traffic light disseminating SPAT (Signal Phase and Timing) information to vehicles. Having an agreed, standardized way to communicate information offers a wide range of potential applications, especially in the areas of safety, road and fuel economy, and driver comfort. The impact on safety alone makes C-ITS worth

M. Klaassen (✉) · T. Szuprycinski
NXP Semiconductors, Eindhoven, The Netherlands
e-mail: marc.klaassen@nxp.com; tomasz.szuprycinski@nxp.com

© Springer Nature Switzerland AG 2019
Y. Dajsuren, M. van den Brand (eds.), *Automotive Systems and Software Engineering*, https://doi.org/10.1007/978-3-030-12157-0_13

considering as according to the World Health Organization (WHO), road injuries have an associated governmental cost of about 3% of gross domestic product (GDP), with roughly 1.25 million people killed yearly due to traffic accidents.

The international IEEE 802.11p communication standard is intended to be one of the standards for C-ITS, having special focus on the short-range links, for example, directly between vehicles. C-ITS is also referred to as V2X: vehicle-vehicle (V2V)/vehicle-infrastructure (V2I) communication. IEEE 802.11p creates a highly dynamic, ad hoc network using a reserved 5.9 GHz band. IEEE 802.11p is often referred to as DSRC (Dedicated Short Range Communication), although also other communication technologies fall under the DSRC umbrella, such as the European 5.8 GHz standard used for road tolling gates.

From a security point of view, the V2X interface is a big risk to the car. Every wireless interface opens the door for remote attacks. One does not need to be in direct neighborhood of the vehicle to get access to its internal systems. In other words, manufacturers can no longer solely rely on the physical protection offered by the vehicle's chassis. There is a risk that these systems get hacked during driving, which creates a serious threat to the reliability and safety of the car. There are even examples where crucial controls of the car can be taken over. Furthermore, there is of course the privacy of the car. Data stored inside the navigation system, for example, should not be accessible to hackers.

2 Use Cases and Requirements for C-ITS

Different applications and use cases will have different requirements. Key topics to consider are latency, geographical coverage, and required bandwidth. Before discussing the main security requirements, first the use cases are presented. These can be roughly divided into three different categories based on different latency requirements:

1. **Relaxed latency** (>1 s): dissemination of pseudo-static information used mainly for informing the driver. This is typically (although not necessarily) information known by a traffic management center and distributed by infrastructure, referred to as "V2I" or "I2V": vehicle—infrastructure communication.

   Example: a Road Works Warning application, intended to inform the driver for an upcoming road works (e.g., "right lane blocked"). This also can be used as input in, for example, navigation systems.

2. **Real-time noncritical latency** (100 ms–1 s): dissemination of dynamic information. This is typically (although not per se) information generated either locally or by a traffic management center and distributed by infrastructure. Most use cases also fall under the V2I category but deal with real-time data.

   Example: the Green Light Optimal Speed Advice (GLOSA) application assisting the driver in optimizing speed toward an intersection. It is based on traffic light Signal Phase and Timing (SPAT) messages. Data are generated either at a traffic control center or locally at the intersection as traffic light timing can be autonomous based on traffic dynamics and, for example, pedestrian crossing requests.

3. **Real-time critical (safety) latency** (≤ 100 ms): event-based information that requires low latency to be able to realize adequate response behavior, either by the driver or by the autonomous car. This will often be information communicated directly between cars and hence is grouped under the vehicle-to-vehicle (V2V) category.

   Example: collision avoidance applications, based on the neighboring vehicle dynamics as received by U.S.A. BSM (Basic Safety Message) or European CAM (Cooperative Awareness message) and DENM (e.g., electronic emergency brake light). Initially only used to inform the driver, on the longer term these will also be used for direct control of a vehicle's behavior.

In a nutshell, V2X-enabled vehicles and road-side units broadcast messages that can be received by any device with a V2X receiver that is within range. The receiver can use these messages to identify hazardous situations and to alert the driver or to prevent accidents autonomously. The car can create a local dynamic map (LDM) out of all that information to visualize the surrounding environment on the screen.

The first main requirement from a security point of view is that the trustworthiness of these messages must be ensured. The receiving end must rely on the fact that the origin of the message is truly from the driver in the other car and not from a hacker sitting alongside the road. Emergency vehicles might get special privileges using V2X, for example, right of way or traffic light control. It can be imagined that the keys associated with these emergency vehicles are very desirable for the people with wrong intentions. Therefore, messages are authenticated using digital signatures, proving their origin and integrity to the receiver.

The second requirement is that the privacy of the driver must be protected. The car is continuously transmitting its location, speed, and direction (CAM message). Without measures, it would be easy to track this car. A solution is that each vehicle regularly changes its identifier ("pseudonym") to make it harder to identify or follow.

V2X is by nature a system where equality must be ensured. This means that every car must be given the opportunity to broadcast in all directions. The V2X system used is by specification ad hoc. Cars assess the channel if it is free and subsequently transmit their CAM or BSM messages. To allow as much transmitting cars as possible the data payload needs to be small. The main content of the channel is safety information. The security information must be appended in an efficient manner. The use of public-key crypto simplifies that, because only the nonsecret public key is exchanged. Certificate authorities, as part of a larger public key infrastructure (PKI), authorize vehicles and road-side units to send messages by issuing certificates describing their digital identity and their permissions.

The privacy requirement comes also with extra challenging requirements. The implementation in a car must ensure that the keys that it uses to create the digital signatures cannot leak. Otherwise, they could be misused to send fake messages.

Such attacks could easily be scaled: a single compromised key could be cloned and as such impact an important infrastructure area. A tamper-resistant Secure Element therefore provides the right level of protection for such keys.

It is possible that within a safety channel, 500 (even a 1000) cars are visible on the Local Dynamic Map (LDM). The main challenge is performance. Each individual vehicle sends messages at a low rate, in normal situations, no more than ten per second. In case of a traffic jam, this can be reduced to, for example, two per second. Nevertheless, the receiver can receive hundreds of messages per second in crowded traffic situations with many nearby vehicles and Road Side Units (RSUs). The signature verification for these messages imposes a big computational load and requires high-speed crypto accelerators to be integrated into the receiver.

3 V2X Communication

3.1 Ensuring Trust Using ECDSA

As stated before, trust and privacy are the main requirements for a successful secure car-to-car communication system. A scheme is needed that provides a high level of trust. The security architecture must remain trustworthy for the lifetime of the car and as such a system that can be upgraded in future. The authentication scheme selected is based on the Elliptic Curve Digital Signature Algorithm (ECDSA). Elliptic curve handling contains very strong cryptographic strength with the main advantage that the key is rather short in contrast to the Rivest–Shamir–Adleman (RSA) handling, for example. In computer science, RSA is widely deployed, leaving key lengths of more than 2048–4096 bits. In the ITS communication system, the throughput of the channel is limited. Order of magnitude is around 6–12 Mbps. This yields around 500–1000 of messages per second, where each message is around 250 bytes (considering the efficiency of the network). To keep the payload for safety messages optimized, an authentication scheme must be selected containing low overhead. Signature overhead for ECDSA is 64 bytes, while for RSA with equivalent strength it would be 256 bytes.

The message in the transmit path is accessible to all but needs to be tamper proof. A secure element is drawn in Fig. 1, where the keys are stored (secure zone). There are multiple implementations possible for the definition of a secure zone. Some companies, like ARM, propose Trustzone; for example, there are microprocessors with a dedicated security architecture and hardware accelerators and, finally, separate ICs shielded from the application software. The choice is determined by the attacks allowed.

The public key exchange mechanism is based on an asymmetric scheme Fig. 2. An Elliptic Curve Cryptography (ECC) key pair is generated in the secure zone (e.g., a secure element), subsequently the public key is included in a certificate and stored outside of the secure zone. With the private key, the message is being signed

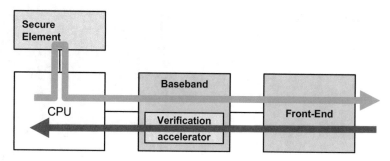

Fig. 1 Basic security dataflow

Fig. 2 Message authentication using digital signatures

and a signature is appended. The certificate, including the public key (paired with the secret key), is transmitted along with the message. At the receiver, the message is verified using this key. The receiving party is talking to someone completely unknown; therefore, to manage trustworthiness a public key infrastructure must be specified, including Certification Authorities.

3.2 Privacy of Sender

The V2X system needs to preserve the privacy of the user. It is achieved through so-called pseudonym identities. The pseudonym identity is used to sign messages being transmitted without revealing the real identity of the sender and yet providing sufficient amount of information to verify trustfulness of the sender and authenticity of the messages. Pseudonym identities are provided by Pseudonym Certification Authority, upon verification that the user is registered in V2X system.

Usage of pseudonym identity is ensuring there is no direct connection between the information transmitted and the actual user, although if the user keeps on sending messages using the same pseudonym identity, it will allow other users to track his movements. For example, when knowing the start and finish location, we can deduct who was driving based on knowledge of home and work addresses. To prevent such leakage, each user gets multiple pseudonym identities, and he needs to change them frequently to minimize the risk of getting tracked and identified.

Fig. 3 Identities across V2X
stack

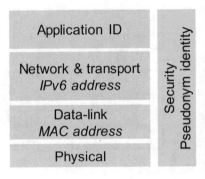

The V2X stack comprises of multiple layers, many of them with their own identity. The stack is presented in Fig. 3. It is important that when switching identity, all identities on the complete stack are changed simultaneously. During the change, message transmission must be prevented as partial change would allow for the linking of different identities. The identifiers for different layers are derived from the currently chosen pseudonym identity.

The pseudonym identities are provided in batches covering a few months. Each batch contains a number of identities valid simultaneously within a specific week. At each start of a vehicle, the software stack chooses an unused identity and keeps on using it for the few first kilometers. Subsequently, another identity is chosen, and that continues until the end of the journey. There is no specific mechanism for selecting an identity; some regions have decided for random selection, while others are putting more specific requirements on usage and reuse strategies.

Due to security concerns such as Sybil attacks, the amount of simultaneously valid identities must be limited. A Sybil attack consists of the simultaneous use of multiple pseudonym identities by the attacker. For example, transmitting a lot of fake messages can make the recipient car believe that there is a traffic jam ahead. Other limiting factors are cost of creation of the identities and transport and storage of them. Depending on the driving profile, different amount of identities is needed to preserve privacy. Analysis performed states that up to 100 identities per week are needed to protect standard driving profile defined based on statistical data, not covering taxis, trucks, etc.

4   Public Key Infrastructure

A public key infrastructure (PKI) is a set of hardware, software, people, policies, and procedures needed to create, manage, distribute, use, store, and revoke digital certificates. In cryptography, a PKI is an arrangement that binds public keys with respective user identities by means of certificates issued by a certificate authority (CA).

Fig. 4 Simplified V2X PKI
system [4]

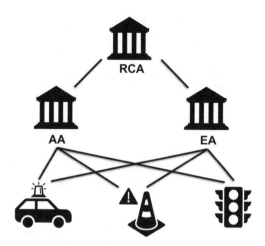

Certificates bind a subject's identity with its public key(s) and (optional) properties. The subject is, for example, another CA or a car's ITS station. Binding is done by means of a signature. A certificate is basically a signed message, which contains all the data explained below.

Typically, there are two public keys in a certificate: one for signature validation (of messages sent by the subject) and one for message encryption (of messages sent to the subject). Additional properties inside a certificate can be, for example, the lifetime of the certificate ("valid from," "valid until") or the permissions (e.g., "may use light bar" for an ambulance).

The certification is usually done in a chained approach; there is a root certificate and subordinate certificate handling.

A proposal for the PKI in ITS is shown in Fig. 4. A layered scheme is presented that enables the possibility to differentiate between regions and countries. In addition, operational aspects are considered in the model.

There are three layers defined:

1. Root CA

    The Root certificate is the trust anchor. It assures the trustworthiness of the Enrollment Authority (EA) and Authorization Authority (AA). There will be a few of these Root CA institutions available: one for the USA, one for the EU, for example.

    Manipulation of Root certificates has major consequences. It impacts the complete security level of the ITS system.

2. Enrollment Authority (EA) and Authorization Authority (AA)

    Based on the enrolment process, different parties can operate the Enrollment Certificate handling. Parties such as car manufacturers or governments. Every EA has a certificate that is signed with the private key of the RCA. The EA authenticates an ITS-S and grants it access to ITS communications. The AA provides an ITS-S with authoritative proof that it may use specific ITS services [4].

The EA is introduced to distribute enrollment certificates and spread the trust anchors over the regions/manufacturers. It is expected that there will be maximum hundreds of EAs present. The enrollment certificates will not be broadcasted together with safety messages.

The AA, also known as Pseudonym Certification Authority (PCA), issues Authorization Tickets (also known as Pseudonym Certificates) used for safety message broadcasting, keeping the privacy requirements in mind. As explained before, each ITS-S will consist of multiple simultaneously valid Authorization Tickets.

3. ITS Stations

The ITS stations (cars, road infrastructure) sign and authenticate the messages using the appropriate keys and certificates. Each ITS station will contain a Canonical Identifier (like a vehicle identification number), an Enrollment credential, and a lot of Authorization Tickets (Fig. 5).

What does a transaction look like? The transmission of messages does not contain the same security every time. As discussed before, bandwidth is scarce and there must remain a level of trade-off between throughput and convenience. The receiver is not always receiving the full certification path. Sometimes it must wait for a retransmit; sometimes it has to request.

If the receiver does not have all missing certificates in its cache, it cannot verify the authenticity of the received message. It has to wait until it receives the missing certificates or actively requests them. Table 1 shows information provided with messages providing authentication of the sender.

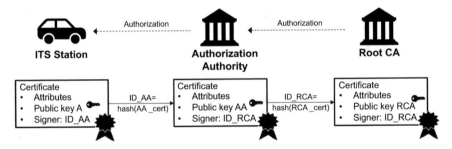

Fig. 5 Authorization and certificate chain

Table 1 Security information appended

What is included by the sender (vehicle A)?	Size	Missing certificates?	When is this done?
The digest of its own certificate (ID_AT = hash (AT))	8 bytes	AT, AA, RCA	For most messages
Its own certificate (A) only	~200 bytes	AA, RCA	Periodically (e.g., once per second)
Its own certificate (A), plus the CA certificate (AA)	~400 bytes	RCA	Only on request

Table 2 Life-cycle management

Key pair	Defined where	Task	Private	Public
Canonical	In factory	Authorization toward EA (possibly together with VIN number)	Never leaves the module	Needs to be registered by OEM
Enrollment	First-time dealership	Authorization toward AA	Never leaves the module	Needs to be registered by EA
Authorization Ticket	At service interval/on the road	Authorization to other ITS stations	Never leaves the module	Needs to be registered by AA

4.1 Life-Cycle Management

If is not enough to only *design* a secure vehicle electronics architecture, it must also be maintained during the deployment of the vehicle. A car using V2X is a highly complex IT system on wheels, which needs active maintenance, including key management and secure firmware updates. Vehicles should be sufficiently protected during their entire lifetime, which is approximately 15 years.

When considering key management, existing keys may need to be replaced with newly created keys every now and then, and the existing keys need to be destroyed. In some cases, the new keys are created outside the vehicle and need to be securely distributed into the vehicles.

Life-cycle management starts from manufacturing to sale to dealer to purchase by the consumer and ends at the car graveyard. In Table 2 the life-cycle management related to cryptographic keys is described in more detail.

4.1.1 At Production

At the OEM, a key pair C is generated. This key pair is stored locally in secure storage. The public key of this key pair is subsequently registered as a Canonical Identifier at the OEM. This is done during production in a secure environment.

4.1.2 Before or At Sales

At the dealer, the Enrollment process must be performed:

1. A pair E is created for Enrollment credentials. The private key pair never leaves the secure storage.
2. A signed certificate request is created. The signature is created with the private key of the key pair C.
3. The EA verifies with the OEM that the request is valid. The public key of key pair C is used.
4. If valid, the EA generates an Enrollment Certificate over the public key of the identity E.

4.1.3 After Sales

After sales, the pseudo certificate process is started:

1. The car creates several pseudonym key pairs P.
2. A signed certificate request is created by ITS-S, and it is signed by the private key of key pair E. Along with the certificate request, the Enrollment Certificate is sent to the AA. Part of the request proving identity is encrypted for EA.
3. The AA receives the certificate request and forwards it to EA, which in turn verifies that it is valid.

   (a) Part of request is sent to EA for verification if an ITS-S is enrolled to the system. The signature over the certificate request can be verified using the Enrollment Certificate.

4. If valid, the AA generates Authorization Tickets with provided pseudonym public keys.

4.1.4 In Operation (While Driving)

As discussed before, during operation:

1. The car creates a signed safety message. The signature is created with the private key of pseudonym key pair P. The message is broadcasted, including reference to respective Authorization Ticket (certificate with public key of key pair P). The Authorization Ticket is also broadcasted and is sent at least once per second and when requested.
2. Another car receives the safety message and verifies that it is valid. The message can be verified using an already known Authorization Ticket. If the Authorization Ticket is unknown but received, it can be verified using a public key included in the AA certificate. If the AA certificate is unknown, an ITS-S can request the certificate from the sender.

4.1.5   End of Life

At the end of the car's life, it is rather "needed" that keys and certificates are removed from the nonvolatile components within the car, simply to make sure that they cannot be stolen and reused. A kind of full-clean erase button is needed that wipes the complete car clean.

5   Standardization

The standardization for security in V2X communication is described in multiple documents created by well-known organizations: IEEE, ETSI, SAE. Input to these standardization bodies was provided by consortiums established by OEM-ers: in Europe Car-2-Car Communication Consortium and in the United States Crash Avoidance Metrics Partnership (CAMP) and Vehicle Safety Communications 3 (VSC3) Consortium.

The protocols for secure communication between vehicles are harmonized between these two main regions and described in IEEE 1609.2 Security Services for Applications and Management Messages. The SAE J2945/1 specifies On-board Minimum Performance Requirements for V2V Safety Systems, such as profiles to be used for messages and certificates in the USA, while ETSI TS 103 097 standard defines the same for Europe. ETSI TS 102 941 specifies protocols to be used for communication with PKI.

6   Conclusion

The security architecture selected for the C-ITS communication system is very effective from a channel utilization point of view. The overhead is relatively low to reach enough car-to-car communication with maximum authentication level. Next to that, the PKI architecture is tuned to the application, guaranteeing privacy and secure key management. However, not only in car-to-car but also in any other connected world application, tracking and privacy are heavily under scrutiny.

Governmental bodies are required to play a role here. The certification bodies must be instantiated. Standardization shall finalize the compliance assessment criteria. Enforcement is needed (car checks) to verify that the keys are in right order and that the system is meeting the compliance criteria.

The mentioned consortiums and standardization bodies are still busy with topics related to security in V2X communication. The work in progress covers, among others, protocols for communications with PKI (CAMP and ETSI); misbehavior detection (CAMP and ETSI); certificate revocation list (CRL) distribution (IEEE); threat, vulnerability, risk Analysis (TVRA) (ETSI, C2C-CC); and security certification (C2C-CC, OmniAir).

Bibliography

1. TS 102 731 ITS security; security services and architecture
2. TR 102 893 ITS security; threat, vulnerability and risk analysis
3. TS 103 097 ITS security; security header and certificate formats
4. TS 102 940 ITS security; ITS communications security architecture and security management
5. TS 102 941 ITS security; trust and privacy management
6. TS 102 942 ITS security; access control
7. TS 102 943 ITS security; confidentiality services

Intelligent Transportation System Infrastructure and Software Challenges

Horst Wieker, Jonas Vogt, and Manuel Fuenfrocken

Abstract Intelligent transportation systems architectures are changing. While traditional architectures have been hardwired, long-living, and regional in scope, modern-day architectures are different. In particular, cooperative intelligent transportation architectures need to be very flexible, change with speed unknown to the industry before, and have interregional or even global scope. To provide optimized benefit to travelers and other transportation stakeholders, new architectures need to be built with future use in mind. This chapter throws light as to why modern-day architectures differ from traditional ones. This knowledge is then used to identify basic goals of every future intelligent transportation systems architecture and give an outlook on how key elements of such an architecture could look like. Finally, it shows what steps are needed to introduce such an architecture.

1 Motivation

Software architectures for intelligent transportation systems (ITS) are part of a very diverse and changing environment. In this environment, a multitude of stakeholders as well as an increasing clock speed uncommon to the automotive industry is changing how architectures are planned and implemented. More traditional ITS architectures, like those for traffic management with dynamic signs on gentries or traffic monitoring over video cameras, are operated by single institutions or a joint venture of a few related institutions. For those architectures, requirements, stakeholders, and capabilities could be defined in advance, as could cost-benefit estimations. They solved defined, current problems in their areas, following design principles developed over years in various industries. In context, the trend for open interfaces has started only recently, and in many currently deployed systems, proprietary interfaces are used.

H. Wieker · J. Vogt (✉) · M. Fuenfrocken
htw saar, Saarbrücken, Germany
e-mail: wieker@htwsaar.de; jonas.vogt@htwsaar.de; manuel.fuenfrocken@htwsaar.de

© Springer Nature Switzerland AG 2019
Y. Dajsuren, M. van den Brand (eds.), *Automotive Systems and Software Engineering*, https://doi.org/10.1007/978-3-030-12157-0_14

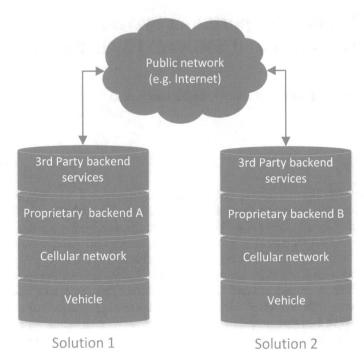

Fig. 1 Applications in a "pillar architecture"

However, current ITS architectures are changing, especially with regard to cooperative systems. Modern applications include more and more information delivered by a growing number of stakeholders. This increases the complexity of the architectures. But as more and more parts of modern ITS architectures adapt to the faster world of consumer IT devices, ITS stakeholders need also to adapt to an environment, in which technologies change more rapidly than they are accustomed to. Current applications in ITS environments are specific solutions in narrow domains. Usually, they are created by a couple of stakeholders in the context of more or less specific environments, for example, for specific regions, specific car brands, or customers of a service. These so-called pillar-based solutions as described in Fig. 1 represent closed, parallel systems in which the possible usage of synergy effects is limited.

For example, if a car manufacturer wants to provide a new advanced driver-assistance system (ADAS), based on information obtained via backend networks, to its customers, different stakeholders work together to create this system. A contractor is hired to create the vehicle side of the application, integrating it into the vehicle software environment of this specific brand. To obtain the data, a supplier creates a backend service, which receives mostly proprietary information from the manufacturer's backend as well as additionally needed data from third parties. That information may either be proprietary or standardized, mostly a combination of both. To communicate those data to the vehicle, the backend of the car manufacturer is used. This backend consists of proprietary networks, which use a cellular network to reach the vehicles.

On the infrastructure side, the process is quite similar. If a road operator wants to provide additional data to traffic participants, for example, to allow them to make better decisions, he needs to utilize different stakeholders. The road operator himself may provide data of the traffic condition. These data could be obtained from sensors, inserted by people, or extracted from forecasts made out of historical data. In this area, there exists a variety of closed, open, and semi-open data formats and interfaces, which complicate the process of data gathering. These data will either be refined by the road operator himself or by a service provider (SP); hence, the information relevant to induce the desired decisions by the traffic participants can be bundled and communicated. To communicate this information, the road operator may use broadly available and standardized mechanisms, like TMC, variable traffic signs, or radio broadcast. These mechanisms can reach a very broad audience of traffic participants, but these are limited in the amount and kind of information they can communicate. Alternatively, a more flexible distribution mechanism could be used, which allows communication of specialized data in arbitrary periods. Such mechanisms could communicate that information to an application—for example, on consumer smartphones—or provide it on a website. However, even if this approach allows for a more flexible definition of data, it is limited in the audience it can reach, as it is assumed that traffic participants are in possession of a dedicated application or actively look up the information themselves.

In both and similar scenarios, the stakeholders will of course try to reduce development costs and harness benefits of already existing solutions. However, due to the necessary combination of systems from different stakeholders, nonstandardized data formats, and a focus on narrow problem domains, reuse of those solutions, or parts of them, is limited. This results in a multitude of different solutions to the same problems, some of which are solved over and over again. These are, for example, addressing of communication sinks; structuring of data, billing, quality of service considerations; and assessing privacy and security impacts. With respect to the rising need for reliable data and communication in transportation, for example, for automated road transport or higher traffic efficiency, this will become an even greater issue in the future.

In the Internet, leading new economy companies like Apple or Google have developed their own architectures based on Internet principles. Those solutions are closed systems based on mostly proprietary interfaces. They are suitable for the special need of a developing company, but they are neither designed nor intended to interact with outside systems or integrate external services, which are not part of their own ecosystem.

To overcome those limitations, ITS architectures need to be planned with a strong focus on well-defined interfaces, standardized data formats, and well-planned modularization. This approach is already well understood and extensively used in software engineering. However, on systems level, especially in the field of cooperative ITS, it is more seldom used, as described earlier.

Separating ITS architectures in specific, clearly divided modules serves additional purposes. From an economic and organizational point of view, an additional step is needed. While a division in modules might be enough for the technical implementation of ITS architectures, a separation from the technical role and performing institution is necessary. In the example of an abovementioned automotive application, the car manufacturer is dependent on the presence of a supplier, who provides information to him. If this supplier would vanish, for example, due to a bankruptcy, the manufacturer will need to replace the old supplier with a new one. For those reasons, detailed specification, standardized production mechanisms, and standardized technology are used, so that a manufacturer can contract a new supplier to replace the old one. By separating role and role-fulfilling institution, the so-called actor, the same mechanism can be applied to all stakeholders in ITS architectures. If the actor of a role decides to stop fulfilling the role, a replacement can be found, as it is well defined what the duties and responsibilities of an actor will be for this role. Additionally, a role could be fulfilled by multiple actors, introducing redundancy as well as competition. Higher redundancy will result in a more robust architecture, as single points of failures are removed. Competition between the actors will lead to more efficient solutions as well as higher degree of innovation.

Furthermore, research has shown that, especially in the field of ITS, the investment of different stakeholders varies over time. In the C-ITS sector, the government and public agencies are seen as a main actor during the introduction of C-ITS architecture. Eventually, their investment is expected to decrease, when more and more private companies join, and a reliable market emerges. This could be compared to space exploration, where public-held institutions were the only stakeholders over half a century, until more and more private companies started to join and make space exploration more and more economically viable [16]. Therefore, ITS architectures need to foresee the flexibility to exchange actors with minimal effort and minimal-to-no disruptions to those actors related to leaving ones.

2 Goal

As already stated in Sect. 1, no single stakeholder can introduce a C-ITS architecture alone. C-ITS depends highly on the interaction and exchange of information between the different actors comprising the system. As there is a great need for cooperation, future ITS architectures need to possess characteristics, which facilitate cooperation and enhance flexibility. Those are also some of the building blocks of the Internet. The Internet has other beneficial characteristics that lead to success. Examples of these characteristics are modularity, simplicity, multi-operator structure, resilience, or the support for new services. But the Internet has also some unwanted and potentially dangerous aspects. Examples for these are unspecified quality of data, unidentifiable misuse, high degree of spam, subject of unwanted surveillance, and authorized data capturing and privacy issues. Here the idea of the research project CONVERGE step in and tries to utilize the ups and prevent the downs of the Internet architecture. The best technical solution cannot be successful in the market, if the underlying business model is not valid. Hence, an economic solution is as important as the technical solution. To enhance the chances of success, both, technical and economic solutions, should be comprised in one single architecture model.

2.1 Key Characteristics

The four key characteristics are openness of interfaces, operator independence, security, and economic feasibility.

2.1.1 Openness of Interfaces

One of the key characteristics is open interfaces. If a wide range of stakeholders should interact with ITS architecture, it is obvious that they need to have access to the specification of its interfaces. If interfaces are closed and proprietary, which means that they are not standardized at all and specified by a single stakeholder, it is almost impossible for others to join. To do so, every interested stakeholder would need to enter into a business relation with the owner of this interface beforehand. A similar problem arises, if interfaces are specified via "closed standardization." In this case, an interested party needs to become a member of a consortium or similar organization before getting access to interface specifications. This process is time-consuming and may be very expensive. Both situations, proprietary and closed standardized interfaces, provide essential barriers especially for small organizations and emerging start-ups.

2.1.2 Operator Independence

Even if the abovementioned open interfaces are a necessary step toward reliable architectures, they are not sufficient alone. If an architecture possesses open interfaces but relies on a single organization to maintain a vital part of it, it is in danger, if the organization decides to no longer support the architecture. In this case, such an organization is a "single point of failure" for the whole architecture. Similar to its central role, such an organization could lead to a monopole, which may hamper innovation.

To prevent this from happening, architectures should be planned in a way so that their implementation does not rely on the presence and willingness of a single institution to overtake a specific part. As mentioned in the motivation, it should be possible to replace organizations easily during the lifetime of an architecture. This will most probably be necessary due to different investment grades of stakeholders during various market phases.

2.1.3   Security and Privacy

Security and privacy are of course of great concern to every ITS architecture. In more traditional architectures, data are examined and prepared before it is communicated between stakeholders, for example, traffic situation based on traffic monitoring cameras. In modern, cooperative architectures, every information exchanged may be put in relation with other data. This opens the possibility to create traveling profiles for individual travelers, thus unintentionally giving away their privacy.

In unstable countries, the government might try to get a hold on the data communicated by travelers and use it to oppress opposition and restrict the freedom of its citizens. As transportation is, especially in Europe, a multinational affair, efforts should be made to create strong and reliable privacy-enhancing mechanism in each part of an ITS architecture.

To guarantee the effectiveness of privacy-enhancing mechanisms, equally strong security procedures need to be in place. Furthermore, those are even more of a concern, as they are also needed in the context of traffic safety. A modern ITS architecture, especially a cooperative one, provides a wealth of possibilities to influence traffic. If those possibilities are not sufficiently secured against misuse, an attacker could bring great harm to the transportation infrastructure as well as put the lives of travelers at risk. To prevent this, ITS architectures need state-of-the-art security mechanisms, like end-to-end security, individual link security, as well as in-station security for all parts of the architecture.

2.1.4   Economical Feasibility

When communication architectures are specified, technical feasibility is almost automatically considered. Economic feasibility is often only considered in terms of costs for an implementation of the architecture later on. This concept breaks for cooperative ITS architectures, as those rely on the presence of multiple stakeholders. To have them partake in the architecture, each stakeholder needs to be able to implement his own business models and business cases. Furthermore, architecture designers need to consider that stakeholder will need to foresee that roles will only be implemented, when the actors can associate a valuable business model with it. At a more general level, an institution will only be willing to partake in the architecture if this generates an economic benefit for them. If this is not taken into account, an architecture might be created, in which crucial roles cannot be fulfilled.

2.2   Reuse of Existing Architectures

As mentioned earlier, some of the described concepts may sound familiar, especially the grouping of technical necessary duties to technical roles as well as the prevention of single points of failure. Those concepts are already present in the modern

Internet. The Internet itself was planned decentralized and robust to failure from its early years on. Given the unregulated nature of the Internet, it also grew in such a decentralized manner. However, for the sake of safety and privacy, ITS architectures need more and better security mechanisms than the Internet currently provides. Also, such mechanisms, as well as privacy mechanisms, should be planned in the architecture itself, not be added afterward. Figure 2 shows how a base architecture uses the idea of multiple interconnected roles to combine the different parts.

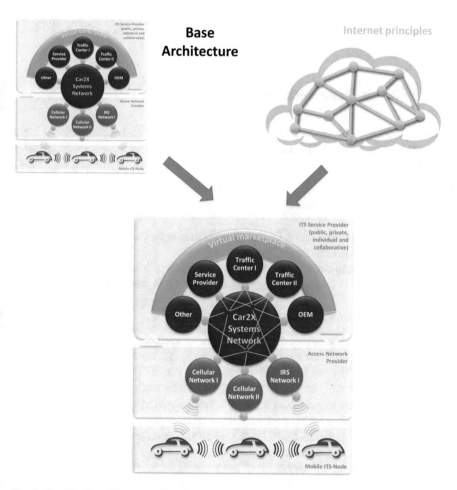

Fig. 2 Combination of Internet principles to architecture

3   Architecture

As described earlier, modern ITS architectures and software systems face multiple challenges. To overcome these challenges, various research projects have created different possible solutions. Even if those solutions combined do still not provide a comprehensive architecture, they should be considered as a baseline. First, they provide building blocks, which can be reused by an ITS architecture. Second, by providing solutions, they highlight which challenges and problems need to be addressed by a modern-day ITS software architecture. Therefore, each of the following subsections highlights a specific challenge by providing an exemplary solution to it.

3.1   *Hybrid Communication*

Unsurprisingly, communication is the most fundamental building block for a communication system. In the domain of ITS, different types of communication networks are used. The following list shows an extract of current and future technologies:

- Traffic Message Channel (TMC) for traffic messages in the very high-frequency (VHF) band
- Transport Protocol Experts Group (TPEG) messages via digital audio broadcast (DAB) or Internet Protocol (IP)
- Cellular communication (2G, 3G, 4G, and future versions)
- Dedicated Short-Range Communication (DSRC) in vehicular ad hoc network (VANET)
- Wireless Local Area Network (WLAN)

These technologies can be used in different scenarios. TMC and TPEG are classic media for wide-area broadcast in which no return channel is implemented. With the possibility to send TPEG messages via HTTP/IP, this situation has changed and return messages are possible. Cellular communication can be used for bidirectional communication between service users (SUs) and service providers. Also a (cell) broadcast was specified, but as of now, it is not widely implemented in cellular communication. DSRC was specifically developed for direct vehicle-to-vehicle/infrastructure communication.

With hybrid communication, all of those technologies should be used. This can happen in parallel, meaning that information is sent via more than one technology at the same time, or only one channel is used. In the latter case, one communication channel is selected and the information is sent over this channel, whereas a second information is sent over another communication technology. A definition of hybrid communication could be in this context: the combination of two or more communication channels for parallel or serial transmission.

For a communication architecture, it is important that different possibilities and characteristics are known inside the system, so that the best channel(s) can be selected for a certain service. In the same way, this should not be something a service provider or a service should concern itself about. A service provider should only send a message to the network, and the network itself should select the appropriate channel(s). The selection process should be done based on meta information sent with the message. Selection parameters can be time constraints, reliability, message type, urgency, bandwidth, etc. Security and privacy should not be selection parameters but remain the same, independent of communication channel.

3.2   GeoMessaging and Bridge

In the context of ITS, most of the information has geographical relevance. A warning of an end of a traffic jam behind the curve, a warning of hazardous weather condition, the information about available parking spaces in a city, or the information about the current traffic light state—all these information are relevant for traffic participants (vehicles, powered two wheelers, bicycles, …) in a certain area. The distribution of messages correlated with those information is called GeoMessaging. The idea is to send information about events or value-added services to service users in a geographical area. To enhance the probability of message delivery even in an area with difficult communication conditions, the hybrid communication approach described in the previous chapter should be utilized. Different communication networks can be used for different conditions and different information.

The distribution should reach all *relevant* participants in an area, as not every information is of relevance to every participant. For that reason, a GeoMessaging-bridge-server (GBS) should be introduced. With this system, information can not only be distributed in geographical areas but can also be filtered for topics and target groups. The principle is best illustrated with an example: imagine a road bridge on motorway that is some years old. A structural analysis showed that the road bridge cannot handle the weight of the vehicles and trucks anymore. So the road bridge is closed for truck traffic, and the allowed vehicle velocity is decreased. This information is now distributed via the GBS in the vicinity of the affected road bridge. The trucks get the information that the road bridge is closed—maybe an alternative route is suggested, too. The vehicles get information about the decreased speed limit on the road bridge. So each party just gets the information needed by them. This reduces the number of messages sent and the utilization of the communication channels. Also the number of messages a service user has to process is reduced, because only relevant information is received. In this example, the geomessaging part is responsible for the geographical distribution, whereas the bridge part is responsible for the topic (here, traffic information) and the kind of service user (here, vehicle or trucks).

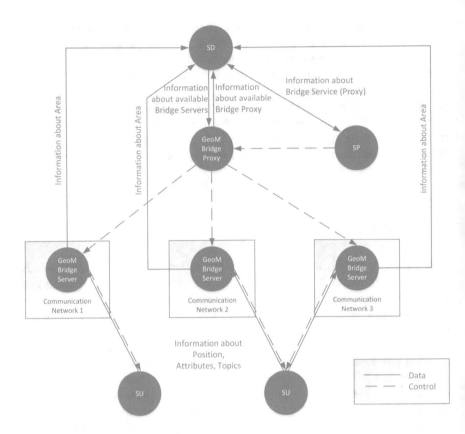

Fig. 3 GeoMessaging and bridge basic concept

Figure 3 shows the basic concept of the GeoMessaging-bridge principles. The functionality is divided into two parts. The idea is for the service user and especially for the service provider to make the usage as simple as possible. The service directory (SD) is used to publish and to search for the proxy and the server component of the GeoMessaging-bridge concept. The proxy is the interface for the service provider (SP). It can be placed at the SP itself or can be provided by a third party as a service on its own. A SP just sends a message to the GeoMessaging-bridge-proxy (GBP) within an area and the message-related topics and recipient classes. Hence, the SP does not need to have any knowledge about communication networks or different GeoMessaging-bridge-server (GBS). The GBP looks up suitable GMS for the area indicated in the message in the SD. The GBP then forwards the message and the information about the area and the user to the suitable GBS found in the SD. Every GBS and GBP should publish their information in the SD, so that they can be found.

The GBS can be responsible for one communication network or for several communication networks. It can be included in the network, or it can be placed

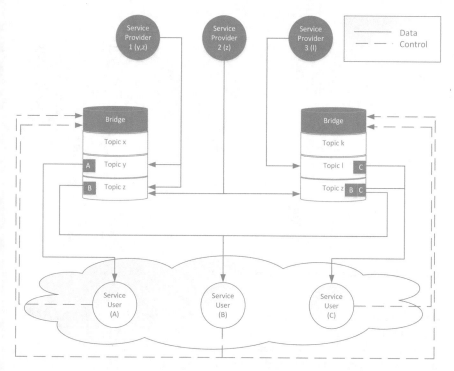

Fig. 4 GeoMessaging and bridge usage

"over the top," meaning above the network. The service users in the networks send their location and their attributes to the GBS. This could be done in the manner specified in [3]. With the information from the service user and the data attached to the messages from the service providers, the GBS can forward the messages to the SU. This can either be done via unicast, multicast, or broadcast mechanisms depending on the number of clients in the area and the underlying communication network.

The bridge mechanism is shown in Fig. 4. The service users send the information about topics (e.g., weather, roadworks, and parking) they want to receive and the user class (attributes, e.g., truck, vehicle, powered two wheelers, and human) they belong to the corresponding bridge component of the GBS. This component saves the information for comparison with incoming messages. The service users can register with different bridges with different or same topics. Hence, it could be possible to receive the same message more than once. But usually one bridge should be sufficient. The service provider sends their messages to the GMP (in the figure omitted for simplicity reasons) which forwards the message to the GMS for the area. The bridge part of the GMS saves the client information for the topics as a list. If a message with a specific topic is now received from the GBP, the GBS looks up the

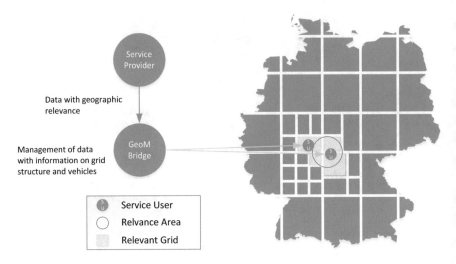

Fig. 5 GeoMessaging and bridge broadcast

list of clients for a topic and forwards the message to the clients in the list, if they are also in the defined communication area.

For example, with a cellular network, a big area can be reached within that communication network. To structure the area and to reduce traffic between the SU and the GBS, the concept of [3] can be used. In this approach (as shown in Fig. 5), a grid is laid above the geographical map. The grid structure could consist out of rectangular shapes. The size of the shapes should depend on the number of users inside an area. In high-density areas, like cities, the size should be much smaller than in rural areas. Every time a user starts its system, it once sends its position to the server. The server responds with the corresponding area dimensions. Also the identifier of the areas adjacent to the current area can be transmitted. When the SU leaves the area, it informs the server and gets information about the new area it now entered. With that concept, the client does not have to send every position change, which can be quite often considering fast velocity on motorways. This reduces the number of signaling messages between the user and GBS. To indicate the principle availability, a SU should send a keep-alive indication if no message exchange occurred in a certain amount of time.

If a message will now be sent from the GBS to all users, for example, because it is a warning message, no topic registrations should be considered. The server examines the dissemination area of the message and marks all rectangles completely inside or partly inside the area. All SU in those areas should receive the message. In this case also SU not inside the original dissemination may get the message, as they are in a rectangle which overlaps the dissemination area. This is acceptable due to the fact that the SU can again filter the messages. With the right size of the rectangles, the number of SU receiving nonrelevant messages should be acceptably small. At the same time, it accelerates the processes at the server because it has not to check for every vehicle, if the vehicle is in the relevant area. The grid size has also a huge impact on the signaling traffic. The bigger the rectangle the less signaling traffic is necessary, resulting in substantially reduced costs.

Additionally, the geographical distribution of message on the wireless link in an DSRC environment is specified by ETSI [2]. The techniques used are at least partly similar to the ones used in geomessaging.

3.3   Security

In an increasingly connected world, security plays a vital role. Commonly, the goal of security is to provide a separation of assets from threats. In C-ITS architectures, network security and information security have to be taken into account. Of course, other forms of security are of equal importance, like physical security on the premises of partaking organizations. Those security variants can at least indirectly be affected by the architecture. For example, certain architecture decisions can make hardening of a system more, or rather less, difficult. Also the possibility to distribute or centralize system functionalists can have effects outside the architecture. But those effects are out of scope for these considerations.

Due to the possible life-threatening consequences through malicious behavior in traffic-related situations, security is vital to provide a positive safety influence. In the context of C-ITS architectures, as well as the important role of transportation in our society, privacy considerations are also of great concern.

As different stakeholders have different privacy needs, three levels of identifiability are defined: anonymous, pseudonymous, and identifiable. On communication with an anonymous participant, it is impossible to determine which participant one is communicating with, whereas for a pseudonymous participant, the pseudonym in use can be determined. With the pseudonym, certain properties (e.g., authorization to use a specific service, legal capacity) can be validated. This is not possible in fully anonymous communication. However, the pseudonym shall not be traceable to a specific, identifiable participant. In the last case, identifiably, the partaking communication partners can be clearly identified and related to a real-world entity.

Additionally, the security of an architecture needs to provide an end-to-end security mechanism, to provide protection from malicious nodes in the architecture, as well as to prevent man-in-the-middle or similar attacks. Those mechanisms need to provide means to ensure nonreputable, authorized, and verifiable communication, as those are necessary to establish trust between participants. For certain communications, it is also necessary to have the possibility to perform confidential communication, which is additionally protected against eavesdropping.

A possible approach to create such a security environment, shown in Fig. 6, can be achieved, by combining Transport Layer Security (TLS) [10] for TCP/IP-based backend connections and ETSI TS 103 097 [4] secure messaging for wireless communication. In both architectures, certificates used to secure the communication are issued by a central public key infrastructure (PKI). To support pseudonymity,

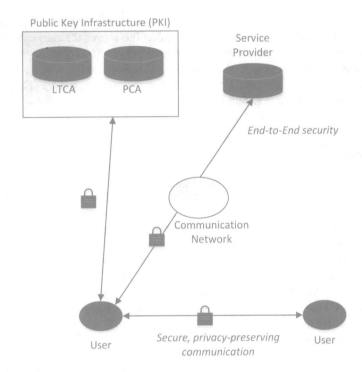

Fig. 6 Example of a security architecture

this infrastructure is separated in long-term certificate authority (LTCA) and pseudonym certificate authority (PCA). After he has been verified, a user can request for a long-term certificate within the LTCA upon joining the network. With this certificate, which does not contain any elements that could be used by a third party to identify the user, the user can request a pseudonymous authorization ticket from the PCA. The user then utilizes those tickets to secure all future communication. The privacy of the user can be endangered, if LTCA and PCA work together to trace and link long-term certificate and authorization tickets.

3.4   Service Concepts

In a service-oriented architecture, as proposed in this chapter, it is essential to think about the usage of services. Services should be easily accessible for service users, service providers should be able to introduce new services without great effort, and it should be possible to search for a service (Fig. 7). An architecture should be able to support all of the above and support the users and the providers of the services. To propagate the services, two concepts are introduced: the service directory and the service announcement.

3.4.1   Service Usage

A Service is always a form of information exchange, mostly in the context of some kind of applications. From an architecture's perspective, this means communication has to take place between two parties. These parties are known as service provider (SP), the party offering the service, and service user (SU), the party using the service. To interact with each other and at first to find each other, the concept of a service directory (SD) is introduced. In short, service providers publish their services in the service directory so that service user can look up services in the service directory.

Before service providers can publish their services in the service directory, they must get their services tested. This is done by the service test and authorization institution (STA-I) [17]. The service must be compliant with the rules set by the system. This includes service interaction possibilities, service quality, service availability, security, and so on. When a service passes the evaluation, the service provider is eligible to request a certificate for the service, so that it can be used in the system. If the service is not validated by the STA-I, the service provider must adapt its service, so that it is consistent with the rules. With the certificate in hand, a service provider can place its service in the service directory and make it possible for the service user to find and use the service. In addition, a direct service provider to service user relation via standard communication channels (like e-mail) can be established to inform potential service customers about new services.

To use a service, the service user needs a valid certificate as a service user. After obtaining the certification, the authorization process can take place. (Hint: The financial and legal aspects of the transactions between service providers and service users are out of scope of this communication architecture, but these have to be addressed in a real-world implementation.) The usage itself is done via a direct communication between the SP and the SU. Since all communication links in the systems network are encrypted, this communication is also secured against eavesdropping. More about the secure communication is provided in Sect. 3.3. To ensure privacy especially of the service user, a concept for pseudonym service usage is also available.

3.4.2   Pseudonym Service Usage

With the implementation of pseudonym service usage, a service user can access and use a service without compromising its pseudonymity. This concept allows a uniform, simple process for the service provider. It even supports decentralized billing. It is scalable and most importantly ensures that the service users cannot be tracked by third parties regarding the services they use or the usage pattern for a specific service. This concept can be used regardless whether a mobile or a stationary service is used.

This concept utilizes the PKI infrastructure as described in Sect. 3.3 and reuses the security concepts of pseudonyms and application identifiers (AID) and the

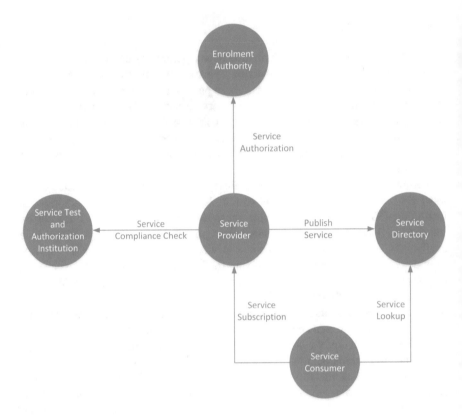

Fig. 7 Service concepts

infrastructure created for vehicle-to-x ETSI ITS G5 communication. Every service gets a dedicated identifier for the service: the AID [1]. For the service usage, a dedicated pool of pseudonyms is created especially for the AID of the service. Those pseudonyms can additionally be extended with service-specific permissions (SSP). These permissions are used to define exactly what kind of messages belonging to the service are allowed to be signed with a pseudonym [6]. The service users use the pseudonym and the associated private key to authenticate against the service provider.

Figure 8 illustrates the concept. On the top are four certificate authorities (CA). The root CA (RCA) is the source of the certificate trust chain and the security anchor. At least one RCA has to exist, but more than one is possible. The RCAs have to be known upfront or at least have to be trusted explicitly. The RCA certifies the underlying LTCA, PCA, and SPCA. The long-term CA (LTCA) is responsible for the identity of the service user. It issues the identity of the user and ties this cryptographic identity to the user. The pseudonym CA (PCA) issues a set of pseudonyms for the user. These pseudonyms are used by the user to secure the communication and at the same time protect its privacy. The service provider CA

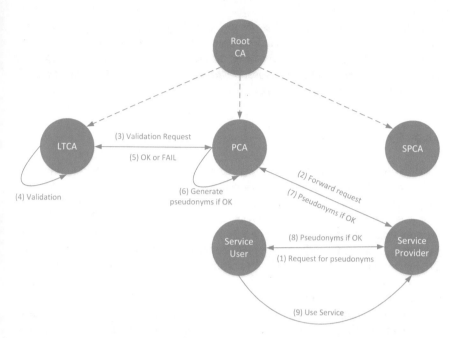

Fig. 8 Pseudonym service usage

(SPCA) issues certificates to the service provider. Finally, the service user and the service provider are on the bottom of the image.

From a privacy point of view, the mutual authentication for communication is a little drawback because, in contrast to the Internet where mostly only server-side authentication is used, the client also has to authenticate itself. On the other hand, in the C2X context, mutual authentication is used because of the ad hoc nature of the communication.

When a service user wants to use a specific service, it sends a request for pseudonyms (1) to the service provider. This request contains, among other things, the keys that should be certified. The service provider forwards the request, its services AID, and the requested SSP to the PCA (2). The PCA sends the request to the LTCA for verification, if the requesting service user is allowed to get new pseudonyms to use the service (3). The LTCA validates the request that is signed by the service user and encrypted for the LTCA (4). The encryption of the request using LTCA keys prevents the PCA from linking the real (long-term) ID, which is contained in the encrypted request, to the newly requested pseudonym IDs. The result of the validation is sent back to the PCA (5), which can either be "OK" if the service user is allowed to demand pseudonyms or "FAIL" otherwise. If the request was granted, the PCA creates the pseudonyms (6) and sends the encrypted pseudonyms back to the service provider (7). The service provider then forwards the pseudonyms to the service user (8). With these pseudonyms, the service user is capable to use the service identified by the AID and SSP in the pseudonyms (9).

Today, the Internet does not provide such a concept. Users are traceable and have to identify themselves to use bookable services. However, many concepts have been developed to protect the user's privacy on different levels such as the Tor network or do-not-track features and many more. Nevertheless, no concept is widely accepted and used by a majority of the users.

3.4.3 Service Directory

Service providers are interested in ensuring that many service users find and use their service. For this purpose, a service directory (SD) should be used. The idea of the SD is based on the DNS concept. Service providers can publish their services in so-called service entries (SE). Service users can then search (look up) the SD for services that match their criteria. The SE contains a set of information that describes the service. The set consists of human-readable information for a manual search and machine-readable information for automated processes. This includes but is not limited to name of the service, service provider identification, textual description, related categories and keywords, interface specification and identifier, data format, encoding requirements, registration and billing information, and a service-level agreement (SLA). The latter provides information about the rights and duties of the service user that utilizes the service.

All system blocks should be designed to provide independence, flexibility, and reliability. To accomplish this, multiple service directories from different service providers and on different levels should be implemented. To provide all suitable results to a service user's queries, the SD must be able to exchange and synchronize information and to provide an interface for other SD to query them. A distributed SD system, which synchronizes all, entries provides a robust structure and can handle downtimes of individual SDs. It can be placed where needed and implements a fast service search algorithm [7].

Distribution is only one aspect for the service directory. To strengthen the flexibility, to shorten response times, and also to enhance usability, a hierarchical approach is proposed. A service provided by a SP can be classified as either global service (GS) or local service (LS). This accounts for the fact that services in an ITS environment often have only a (geographical) local significance. For example, parking area information, petrol prices, or roadworks information are only relevant for a specific geographical region, whereas traffic information or weather information can be relevant for a larger area up to an entire country. This implies that not all services are available in all SD. Additionally, service could be local in terms of a limited user base, for example, a vehicle manufacturer has his own service directory for the vehicles of its brands and provide special services (e.g., for maintenance) only to its customers. He can also make global services available throughout its SD via synchronization to higher-level SDs. But the manufacturers' SD would not synchronize its local services to the SDs above. In another example, a roadside unit could provide services via IEEE 802.11p [8]-based VANET (vehicular ad hoc network) communication. This roadside unit could announce its services via a service announcement [13] mechanism.

In a service directory architecture with a distributed structure, not all nodes hold all services but only a relevant subset. For the synchronization process, intelligent algorithms have to be used, so that databases of the service directories are all up to date. In addition, intelligent query algorithms have to be implemented, so that local service directories (including part of global and all local services) can query global service directories (including all global services and no local services).

To utilize the global and local services and to minimize the traffic between service directories, a hierarchical SD structure is proposed. Figure 9 shows a possible setup of this structure. On the top are service providers (SPs). Those SPs publish services to the global service directories on the top level, so-called top-level service directory (TSD). Those service directories are all completely synchronized with each other. Consequently, every TSD contains the full set of global services (GS).

The next level is the communication network connecting service providers and service users. In this level, SDs with a subset of the global service directory and local services exist. They are consequently called network service directories (NSDs). In a network, several NSDs can exist either in parallel or in hierarchical structure. A NSD can be the contact point for a number of SDs located near the user for specific services. NSD can provide entries for services of a greater region. This could be an entire city or province but also service providers located in the network area, for example, services provided by a mobile network operator like billing.

The lowest level of SD is represented by the local service directories (LSDs). Those are connected to a NSD or a TSD, respectively. In the LSD, local services like parking information or geometry of a construction site for automated driving can be located. It also provides a look-up service, so that service users can search for services. It could be located on a roadside unit or a mobile edge cloud of a cellular network.

The services must be registered with the corresponding SD. In the case of a global service (GS), it has to be publicized in a TSD. Depending on their characteristics, LS can be registered with NSD or LSD. As a rule, services will only be propagated to SD in the same layer or below its own layer. So a service registered in a NSD cannot reach the TSD. But a service publicized in the TSD can be propagated to a LSD.

On the bottom, the service user (SU) has two ways to get information about available services. The first one is that the lowest service directory (LSD) environment provided a look-up service. With this service, it is possible to search for other services. The search is not limited to the services available in the LSD but can include services on network or global level, if the LSD provider supports the availability of those services; for example, an original equipment manufacturer (OEM) or a vehicle manufacturer might provide a value-added service not only for its customers but also for all vehicles. In this case, another OEM could decide that it will make this particular (global) service available for its customers, instead of providing its own service. Anyway, the service directory answering the service query should provide a result set of all services available for the service user. The

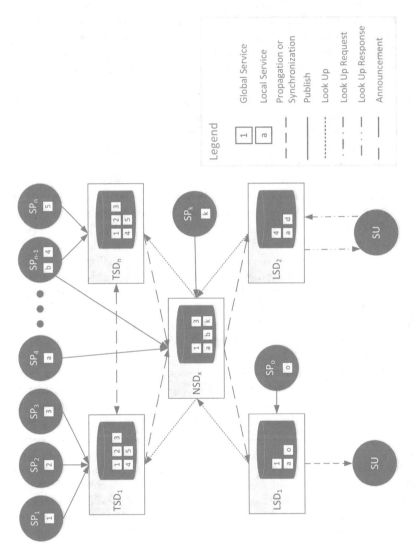

Fig. 9 Service directory architecture

user itself can then choose which service suits most and use that service. This look-up service should be available on all service directories with the same interface. In terms of the Internet, it would be like a DNS request. The ability to search for services should not only be limited to the classical service users (e.g., human beings, vehicles, and smartphones) but could also help other service providers to use particular services from other service providers. For example, a service provider for traffic conditions could summarize information gathered from different other service providers, like traffic control centers, mobile network operators, and navigation and map providers. In this case, the service providers become, from the service directory point of view, service users.

There are different ways to propagate information between the different levels of service directories. On the global level, the service directories have to be fully synchronized; on the levels below, this could only be true for SDs of the same organization, for example, a mobile network operator. To propagate services to the levels below, for consistency reasons, only a propagation below and not above is recommended. Therefore, two mechanisms are proposed:

- **Persistent preemptive propagation (PPP).** Service information on the higher layer are automatically and without querying pushed to the layers below. Hence, service information from the TSD are pushed to the NSD and from the NSD to the LSD. The information in the layer where the information is pushed should be stored with a long caching time or even persistently. In the same way, the service directories are informed if a service is no longer available. This has the advantage wherein local service directories never have to query service directories above for services but can assume that they have all the services that are currently available. On the other hand, a lot of propagation traffic is produced between the SD levels, and also a lot of services are stored and have to be analyzed if a service user searches for a service [7].
- **Volatile reactive propagation (VRP).** To adapt the Internet mechanism, this propagation mechanism is very similar to the one used within DNS. If a user searches for service information that is not available on the local service directory, this SD will query the NSD for this information. And if the NSD has not the necessary information available, it will query the global service directories. The answer is then propagated from the GSD to the NSD and from the NSD to the LSD. On the NSD and LSD, the information is cached for a certain amount of time but not persistently. This should ensure that the information in the lower service directories is correct and no false information is sent to the service users. When the caching time is expired for an entry, it is deleted from the SD. If a service user is querying again for this kind of service, the whole process has to start again. The advantage is that only the real necessary information is transmitted throughout the network. Additionally, the local SD database could be much smaller. On the contrary, the answer for a service inquiry of a user could take more time if the whole chain has to be passed through [7].

3.4.4 Service Announcement

To advertise available services to potential service customers, the concept of service announcement in the ITS context is in development [5, 9, 11]. Service announcement can be seen as a management functionality of SD environments. It pushes information about services toward the potential service users. The push is conventionally done with the service announcement message (SAM). The mechanisms currently developed focus on ad hoc V2X communication. In this case, a service provider (e.g., a roadside unit) can push information to passing vehicles about services it offers. With the abovementioned concept of service directories, the services in the local service directory can be advertised. In addition, a service should be available, so that the vehicles can look up and search for other services currently not announced by the roadside station. But this concept should not be limited to the ad hoc communication and should not only include services available via ad hoc communication. In the first step, the service announcement feature should be extended to carry information on what channel and on what communication technology a service can be reached. For example, a roadside station could provide the information that an event booking service for events in this region is available via cellular communication and additionally provide the name of the server providing the service.

To go one step further, the SAM should not only be sent via ad hoc communication but can also be used in on other communication links such as cellular communication, consumer Wi-Fi, or digital audio broadcast (DAB). Also these channels can provide information about services on the channel itself but also about services on other communication channels. It is important to keep the number of advertised services on an acceptable level and provide only the information needed to use the service; otherwise, the announcements could congest the channels. Additionally, the service user has to process all information in the message. What information on what services will be announced is up to the actor that sends the actual SAM. But some guidelines about the structure of SAM and the amount of services announced in a SAM have to be adopted by the standards developing organizations (SDO). As a conclusion, a SAM should include information about the service (name, topic, usage conditions) and on which communication link and on which channel the service can be used.

3.5 Role Models

Usually, when creating software architectures, creating roles to specify logical elements springs easily to mind. These roles group a certain set of functionality and describe what an actor who is fulfilling this role needs to provide to the architecture. Roles also describe the responsibilities of each actor, without being tied to a specific entity. By abstracting the behavior from a concrete entity, it is easier to identify which set of functionality is necessary for the architecture. Also, it is simple to replace the actor or introduce a second actor to also fulfill this role if needed.

The behavior/responsibility described by roles is called "action." Characteristics of actions are as follows:

- Actions are among themselves complementary and/or neutral. If an action is assigned to a role, further assigned actions may not contradict previously defined actions. Conflicting actions cannot be part of a role, but these must be placed into their own role.
- The contribution of actions for goal fulfillment can be measured.
- The summarized actions can be implemented by an actor. A role can be oversized, so that an actor is not able to implement all actions. This role must then be redefined and further roles have to be introduced.

As ubiquitous and straightforward as this approach is in the technical world, it is relatively rare and seldom used in economic fields. However, it has various advantages, if a role model, then called "institutional role model" [12], is used to describe the responsibilities of economic institutions. Economic institutions are firms, public authorities, federations, courts, and fixed institutions. They are equipped with full-action rights, property rights, and obligations to act as a social subsystem, which regulates behavior, communication, and action of individuals in a certain way.

In traditional economics and architectures, technical roles get fulfilled by an institution. If, for example, a firm alone cannot take on the responsibility of fulfilling the role, two or more institutions might collaborate and create a new organization. This approach is well proven, but cumbersome. The foundation of a new organization, often acting as an own legal entity, is time- and resource-consuming. Furthermore, this approach is not very flexible and dramatically fails, if one of the partners of such a joint venture decides to leave. It is easily understandable, that institutions from different fields and with different interests will be differently active in the various market phases. Traditionally, research institutions like universities are more heavily invested during research and development phases but reduce their investment during market growth.

To have the flexibility to cope with this changing interest of acting institutions, it is wise to use the role model approach in a twofold manner: first, to identify and specify technical roles and, second, to specify roles for economic activities. The possible economic activities, as defined by [14], can be grouped in so-called meta roles [15]:

- **Business management** takes place via a corporate management.
- **Sales** which contains the actions of sales and service offer.
- **Procurement** which contains the actions of procurement and data acquisition.
- **Production** which summarizes research and development, manufacturing, store-keeping, and administration.
- **Human resources** which contains the actions of human resource management.
- **Financial Management** covers external and internal financing and interior investment.

- **Controlling** does not contain the monitoring system, but all activities to make accounting and controlling possible.

With those meta roles in mind, one can specify how those roles should be played by the different actors in the various market phases.

4 Outlook

The introduction of C-ITS will bring new applications that need a different kind of architecture. The communication architecture presented in this chapter provides possibilities to overcome the challenges that will arise in this domain. The concepts for service management, hybrid communication, geographical message distribution, and especially the solutions for security and privacy are key techniques to develop a reliable, secure, adaptable, and sellable environment. Besides the technological aspects, the organizational and economic ideas are equally important. The concept of role models allows the definition of responsibilities and technological boundaries. A sophisticated technology solution will only be deployed if it is possible to find an economic feasible concept. The proposed economic solution reduces the possibility of deficits or failure and provides the foundation for a sustainable business.

The technological concepts can help to define a real-world technical solution. Further steps have to be taken to elaborate the technical details. Interfaces and data formats have to be defined, working relationships need to be established, and a legal framework for collaboration has to be institutionalized. The stakeholders in the C-ITS sector (industry and government) have to decide whether they want to go the next step and set up a system for traffic-related services. This system could provide the framework for the deployment of service that delivers crucial safety information to all road users.

References

1. ETSI (2011) Intelligent Transport Systems (ITS), Classification and management of ITS application objects. Standard, European Telecommunications Standards Institute, Sophia Antipolis Cedex. https://www.etsi.org/deliver/etsi_ts/102800_102899/102860/01.01.01_60/ts_102860v010101p.pdf
2. ETSI (2011) Intelligent Transport Systems (ITS); Vehicular communications; GeoNetworking; Part 4: Geographical addressing and forwarding for point-to-point and point-to-multipoint communications; Sub-part 1: Media- Independent Functionality. Standard, European Telecommunications Standards Institute, Sophia Antipolis Cedex
3. ETSI (2012) Intelligent Transport Systems (ITS), Framework for Public Mobile Networks in Cooperative ITS (C-ITS). Standard, European Telecommunications Standards Institute, Sophia Antipolis Cedex
4. ETSI (2013) Intelligent Transport Systems; Facilities layer function; Part 2: Services Announcement specification. Standard, European Telecommunications Standards Institute, Sophia Antipolis Cedex

5. ETSI (2013) Intelligent Transport Systems (ITS), Security, Security header and certificate formats. Standard, European Telecommunications Standards Institute, Sophia Antipolis Cedex
6. ETSI (2014) Intelligent Transport Systems (ITS), Vehicular Communications, Basic Set of Applications, Part 2: Specification of Cooperative Awareness Basic Service. Standard, European Telecommunications Standards Institute, Sophia Antipolis Cedex
7. Fünfrocken M et al (2014) CONVERGE - Future IRS-infrastructures As Open Service Networks. In: 21st World Congress on Intelligent Transportation Systems, Detroit
8. IEEE (2010) IEEE Standard for Information technology – Local and metropolitan area networks – Specific requirements – Part 11: Wireless LAN Medium Access Control (MAC) and Physical Layer (PHY) Specifications Amendment 6: Wireless Access in Vehicular Environments. Standard, Institute of Electrical and Electronics Engineers, New York
9. IEEE (2010) IEEE Standard for Wireless Access in Vehicular Environments (WAVE) — Networking Services. Standard, Institute of Electrical and Electronics Engineers, New York
10. IETF (2008) The transport layer security (TLS) protocol, version 1.2. Standard. Internet Engineering Task Force, Fremont. https://tools.ietf.org/html/rfc5246
11. ISO (2013) Intelligent transport systems – communications access for land mobiles (CALM) – ITS station management – part 5: Fast service advertisement protocol (FSAP). Standard, International Organization for Standardization, Vernier, Geneva
12. Mann M et al (2015) Deliverable D1.2: Final Operational Requirements and Role Models. Project deliverable, CONVERGE Project, Saarbrücken
13. Pillado Quintas M et al (2014) A vision for an V2X service announcement concept. In: 2014 International conference on connected vehicles and expo (ICCVE), pp 343–344. http://dx.doi.org/10.1109/ICCVE.2014.7297568
14. Schneider D (1995) Betriebswirtschaftslehre: Band 1: Grundlagen, 2nd edn. De Gruyter Oldenbourg, München
15. Schneider D (1997) Betriebswirtschaftslehre: Band 3: Theorie der Unternehmung. De Gruyter Oldenbourg, München
16. Space Foundation (2014) The space report 2014: The authoritative guide to global space activity. Space Foundation, Colorado Springs, CO
17. Spinner A et al (2015) Deliverable D4.3: Architecture of the Car2X Systems Network. Project deliverable, CONVERGE Project, Saarbrücken

Part VI
Future Trends

Future Trends in Electric Vehicles Enabled by Internet Connectivity, Solar, and Battery Technology

Ben Rutten and Roy Cobbenhagen

Abstract The personal car has been one of the most defining inventions of the past century. Ranging from satisfying the human need for mobility to the design of cities, cars are ubiquitous and dominant in our daily lives. It is therefore very much of interest to analyze the trends of the automotive industry to predict how personal mobility might change in the future. However, we look not only at trends that occur within the automotive industry but also at other global technology trends that are related to the domain of automotive technology, such as generation and distribution of renewable energy and the rise of the "Internet of Things" (IoT). The focus of this chapter will be on the relationships between these various trends and how they might interact.

We will elaborate that the main changes in the future automotive ecosystem are enabled by strong digitization resulting in three dominant trends that are mutually benefitting each other, possibly resulting in disruptive change in mobility: firstly, the electrification of the vehicle drivetrain, strongly influenced by the take-up of sustainable energy production by solar and wind farms; secondly, the uptake of sharing economy stimulating the change from car ownership to car usage by all kinds of mobility services; thirdly, the general known trend (and not discussed in this chapter) of the automation of vehicle driving itself.

This disruptive change of the whole road mobility system toward a mobility service-oriented system will be fueled by further penetration of digitization at all aspects of mobility systems and components.

B. Rutten (✉)
Strategic Research Area Smart Mobility, Eindhoven University of Technology (TU/e),
Eindhoven, The Netherlands
e-mail: b.j.c.m.rutten@tue.nl

R. Cobbenhagen
Department of Mechanical Engineering, Eindhoven University of Technology, Eindhoven,
The Netherlands
e-mail: A.T.J.R.Cobbenhagen@tue.nl

© Springer Nature Switzerland AG 2019
Y. Dajsuren, M. van den Brand (eds.), *Automotive Systems and Software Engineering*, https://doi.org/10.1007/978-3-030-12157-0_15

1   Introduction

The personal car has been one of the most defining inventions of the past century. Ranging from satisfying the human need for mobility to the design of cities, cars are ubiquitous and dominant in our daily lives. It is therefore very much of interest to analyze the trends of the automotive industry to predict how personal mobility might change in the future. However, we look not only at trends that occur within the automotive industry but also at other global technology trends that are related to the domain of automotive technology, such as generation and distribution of renewable energy and the rise of the "Internet of Things" (IoT). The focus of this chapter will be on the relationships between these various trends and how they might interact.

We first look in Sect. 2 how the automotive sector itself looks at the most important trends in its sector, resulting in increased use of mobility services replacing partly private car ownership. Sections 3 and 4 present developments in the generation and distribution of renewable energy, respectively. The focus is the trend of increasing solar energy performance and how that affects the charging of battery electric vehicles (BEVs). Section 5 presents reasoning, based on first principles (namely the well-to-wheel efficiency), as to why the BEV is the most obvious choice for future sustainable personal transport in the near future. Section 6 presents developments that are reducing the energy consumption of a vehicle that are not directly related to the powertrain. In Sect. 7, it is argued that (B)EVs make the ideal ride and car-sharing vehicles. Section 8 demonstrates how the trends in the energy industry and automotive sector could be combined into a solar-powered electric vehicle (EV). Section 9 argues that hybrid cars are a temporary solution that cannot compete with the long-term promises of the fully electric powertrain. The chapter will be concluded with some illustrations on EV technology implementations of the Automotive Student Teams at Eindhoven University of Technology (TU/e), Netherlands.

2   The Evolution of the Automotive Ecosystem in the Coming Decade

The *automotive ecosystem* will probably undergo big changes due to the ICT capabilities entering the transport sector [1–3]. The KPMG yearly survey analyzes the current state and future prospects of the worldwide automotive industry. It is executed across nearly 1000 senior executives from the world's leading automotive companies by interviewing them, including automakers, suppliers, dealers, financial service providers, rental companies, mobility service providers, and companies from the information and communication technology (ICT) sector, referred to as the *automotive ecosystem*. Additionally, KPMG surveyed over 2400 consumers from around the world to give their perspective and compared their opinions against the opinions of the world's leading auto executives.

Automotive executives [1, 2] rank ICT as key trend disrupting the auto industry in a survey requesting for most important trends affecting the industry until 2025. Ranked #10 in the KPMG 2015 survey, connectivity has moved up to #1 in the KPMG 2016 survey and #2 in the KPMG 2017 survey. Battery EV (BEV) is now ranked #1.

The Internet has entered the mobility arena with key technology enablers like (wireless) broadband, positioning, mobile (smartphone) platforms, and cloud computing. IoT will make happen that like persons also machines and sensors will be connected. The new EU data protection act will enable horizontal data applications and stimulate the sharing economy enabled by the key technology enablers. And the vehicle itself will become one big computation platform, with more and more advanced driving support system (ADAS) functions, changing the vehicle to highly automated systems, connected to the cloud. "The car is a gigantic data generating engine" [1, 2], and the electric drivetrain will enter massively the automotive sector, driven by traditional big innovators like BMW and Toyota but also heavily influenced by strong upcoming new market entrants like Tesla [1, 2]. Things will change faster than before, and similar disruptions as in other sectors that were confronted with digitization are to be expected in the mobility sector, as we have seen in the hotel booking sector (from booking in travel shops to booking on your mobile phone), flight ticketing sector (from booking in shops to booking on your mobile phone), printed photo business (the Kodak case), music sector (from vinyl via CD, downloading, to streaming services).

Accessing information about a car from a distance by connecting the car to the Internet shows to be an enabling technology. This technology is well on its way and does not see any major problems that would stop it. The only issue that needs be addressed and proves to be difficult is the security of the connection. This is a very important factor as it involves both safety and privacy. Even though this is a stressing subject, it seems not to pose any problems with the rollout of the connected car and the services needed for comfortable EV driving [4].

This disruptive digitization trend will probably change the automotive ecosystem into five new archetypes in which service providers will take control of the value chain and push device manufacturers into the background [3]. Although less strongly reported, this phenomenon is also identified by [1]. The five archetypes are service mobility providers, device manufacturers, infrastructure players, device component manufacturers, and infrastructure component manufacturers. The first two archetypes will be described more in detail as these are affecting most mobility and (electric) vehicle developments.

1. *Mobility Service Providers.* "Service Providers are at the fat end of the value chain—meaning they are the ones who are in touch with mobility customers of all kinds, the ones to reap the lion's share of revenue and profits" [3]. As we see in other Internet-business-dominated areas, the competitive advantage of mobility service providers is based on algorithms and platforms providing access to customized mobility services. Like in other service-dominant businesses, growth is easier to achieve in an asset-light setup, combining existing assets

of different stakeholders in the service platform, which can be relatively easily changed due to changing market conditions [5]. Mobility as a Service (MaaS) is an example of a new upcoming concept loaning the principles as we see in the telecommunications and consumer electronics industry: service providers take control of the value chain and push device manufacturers into the background. Even Apple is more or less moving from being a device manufacturer to being an (integrated) service and device provider; Apple is earning more and more revenue from services, 20% higher than last year, second revenue stream (12%) after the iPhone revenue stream (65%) [6]. BMW and Daimler are mitigating the risk of service providers owning the user by the introduction of their mobility service concepts like DriveNow and Car2Go [7, 8].

2. *Device Manufacturers.* "These are the OEM's as we know today which may well become mere Device Manufacturers. These companies' business model would be limited to developing, manufacturing and selling vehicles. Most of them would sell, or more likely lease, to Mobility Service Providers as they are the ones with the direct access to the customer. White Label and Contract Manufacturers deliver commoditized vehicles for different needs and customer groups. The Mobility Service Provider defines specifications. The focus of innovation would have to shift from product to process and manufacturing technologies" [3]. But of course, there is also a place for branded original equipment manufacturers (OEMs) [3], especially for the ones that move more to the mobility service provider side; for instance, BMW is already experimenting on its DriveNow mobility concept [9]. The new entrant Tesla is also moving toward an integrated service and device manufacturer, as explained later in this chapter. Customer relationship can be retained by the OEMs, but the center of gravity will shift toward tech (ICT) companies and mobility service providers [1, 2]. However, to meet these challenges, it is clear that the branded OEMs' informational engineering needs to become a core competence [1, 2] so that they will not move to being a white-labeled or even contract manufacturer. It is too early to state in which direction existing OEMs in majority will move; service innovator examples like BMW and Daimler show that they will stay strong in the market by their vertically integrated service concepts, but at least newcomers do have a fair chance to take over parts of the existing OEM market, like what Tesla is showing. Electric vehicles will be an upcoming important market segment, but also ICT capabilities are essential, as we will elaborate in the next chapters.

3 Solar Energy Will Disrupt the Energy Market and Vehicle Energy Source

For automotive executives, the top concern for the last 5 years was growth in emerging markets, which recently has dropped to #4 and #5 for the last 2 years; see Fig. 1 [2]. Alternative powertrain technologies are high on the list of trends in

**Regulatory pressure pushes awareness for electrification:
Battery electric vehicles are this year's #1 key trend.**

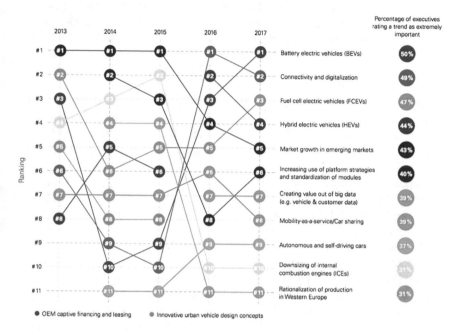

Source: KPMG's Global Automotive Executive Survey 2017

Fig. 1 Year-on-year key automotive industry trends until 2025, expressed by over 1000 senior executives of the automotive industry, lastly surveyed by KPMG in 2017 [2]

2017, with hybrid electric vehicles at #2 and #4, battery electric vehicles up from #9 to #3 to #1, and fuel cell electric mobility staying fairly constant and last year light ascending to #3 [2]. The formerly high-ranked trend of downsizing internal combustion engines has dropped from #2 in 2015 to #10 in 2016 and 2017, which can be seen as a reaction to the "dieselgate" scandal of unreliable emission test results [1].

Solar energy will play a dominant role in the upcoming 15 years with wind energy as strong supportive carrier. Already 15 years ago it was investigated that renewables could completely supply the energy demand in the USA [4]. Only taking photovoltaic (PV) solar panels with 10% efficiency would take an area of 10,000 square miles, what is quite an area, but still less than one quarter of surface used by all roads and streets in the USA. Combining it with other renewables, like wind, geothermal, and hydroelectric energy, this area for PV panels would be quite smaller [10]. However, even at that time it was also already noted that one of the major drawbacks to many forms of renewable energy is their intermittency. Since Turner's publication, nearly 20 years ahead we know that PV efficiency is 15% or higher;

solar panel costs went down with big steps [11] and continued going down strongly until now. Furthermore, many technologies are being developed or are already on the market for solving this intermittency challenge [12]. The increase of renewable energy production, of which solar and wind are main components, will rise strongly without any doubt. In a global expert investigation commissioned by the German government, half of them is stating that around 2050 about 60% or larger of the energy system will rely on renewables and 90% is stating that the renewable energy production will be doubled from 20 to 40% [12].

The intermittence challenge, which needs to be solved, is also addressed [12], and it reported that there is an overwhelming agreement by the consulted experts on how to solve this. "Storage technologies are key, and will be the central pillar of energy supply with a high share of renewables; Several different storage technologies are required—there is no 'one-size-fits all' solution; Lithium batteries in e-mobility will predominate over the next decade; The question of which storage technologies will ultimately dominate the market is still open; Research and development in storage technologies are vital; Improving storage technologies will turn the current energy market entirely on its head; Development of new business models and supporting policies for storage will be at the top of the priority list for the coming years" [12].

These trends can be illustrated by the system approach thinking energy-mobility service provisioning company Tesla in its second 10-year master plan "part deux," consisting of four major elements [13]: "Create solar roofs with seamlessly integrated battery storage; Expand the electric vehicle product line to address all major segment; Develop a vehicle self-driving capability that is 10X safer than manual via massive fleet learning; Enable the vehicle to make money for you when you aren't using it by subscribing the vehicle to a (Tesla) car sharing service."

This master plan shows that Tesla is moving toward being a branded OEM and mobility and even energy service provider, or at least empowers the consumers as local energy producers, using Tesla products. To execute this, Tesla acquired already Solarcity (which is producing solar panels) and has a strong cooperation with SolarEdge [14], one of the global leading inverter companies. Another aspect of becoming an energy service provider is the fact that Tesla is building its own production capacity for batteries. With its Gigafactory opened in 2017, it doubles the worldwide production capacity for lithium-ion batteries. From that year onward, Tesla has been planning to build up to half a million EVs per year [15]. And with increasing EV numbers, Tesla could expand the battery production capacity by building new Gigafactories in Nevada or other locations abroad [15] (Fig. 2).

This master plan of Tesla fits in another business trend, that solar energy will be a driving force in the automotive industry. "Already, in many regions, the lifetime cost of wind and solar is less than the cost of building new fossil fuel plants, and that trend will continue. But by 2027, something remarkable happens. At that point, building *new* wind farms and solar fields will often be cheaper than running the existing coal and gas generators. This is a tipping point that results in rapid and widespread renewables development" [16]. In high solar-irradiation regions, it is already the case that solar is cheaper than coal or oil. In Dubai the price is 0.03 $/kWh (reported in May 2016), and in Chile it is (reported in August 2016)

Fig. 2 Gigafactory of Tesla in Nevada [15]

already down to 0.029 $/kWh, half the price compared to coal in Chile [17]. French electric power company Engie expects that production price in 2025 will go down to 0.01 $/kWh, stating that "The promise of quasi-infinite and free energy is here" [18].

The sudden rise of electric vehicles is on the verge of disrupting oil markets as well [19], and that has big implications for electricity markets as more cars plug in. "The adoption of electric cars will vary by country and continent, but overall they'll add 8% to global total electricity use by 2040" [16].

Solar and wind energy and electric vehicles will strengthen each other in growth. Scaling these new energy technologies drives toward lower prices both for solar panels and batteries [16]. The main reason for this sharp decline in costs for solar panels and batteries is that it is about technology and not about fuel: "Solar is a technology, not a fuel, and as such it gets cheaper and more efficient over time. This is the formula that's driving the energy revolution" [16]. The same yields for batteries: 1 kg of batteries can be recharged over 1000 times with renewable energy. One kilogram of fuel can be burned only once (Fig. 3).

In another report by UBS, the outlook for EVs, solar power, and stationary batteries was analyzed in a holistic fashion [20]. UBS concludes that these three technologies combined in one household energy-mobility system could leverage each other, with the potential to bring disruptive changes to the electricity sector. "According to their calculations, this combination will reach a tipping point somewhere around 2020 (Fig. 4). By that time, solar arrays, EVs and stationary battery technology will be sophisticated and affordable enough that an investment in such a combination will be worthwhile. In a detailed example of a German household in 2020, the purchase would take roughly 8 years to pay back, and

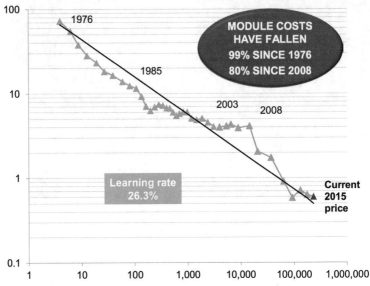

Fig. 3 When solar power doubles, solar panel costs decline by 26% [16]

afterwards the consumer will enjoy over 12 years of free electricity for their EV's (Fig. 4)" [20].

4   Grid Connection Stays Important

On top of that, vehicle batteries will be used for the storage of peaks in solar energy and can "fuel" your house overnight, solely, or together with a residential storage system. In general, the fluctuating behavior of solar (and wind) energy on a daily base can be solved using batteries. Seasonal buffering may be difficult in countries further away from the equator as (1) there is less solar energy, (2) the solar energy is unevenly distributed between summer and winter, and (3) the energy demand is higher in the winter due to heating [4]. Flexible capacity technology, primarily large batteries for the home and grid, that will smooth out the peaks inherent in wind and solar power will be as important by 2028 as solar systems [16].

However, it will not be possible to have a 100% residential loading capacity for all EVs due to the fact that only a part of the households have a rooftop suited for it or live in such high dense populated areas, that there is simply no room enough for roof-topped solar panel systems on the neighborhood and street level.

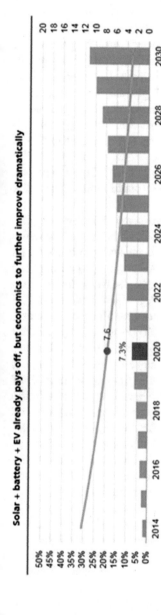

Solar + battery + EV already pays off, but economics to further improve dramatically

Source: UBS estimates
Note: Chart shows economics in Germany.

ROI (pre-interest) —— Payback time (# of years) • 2020

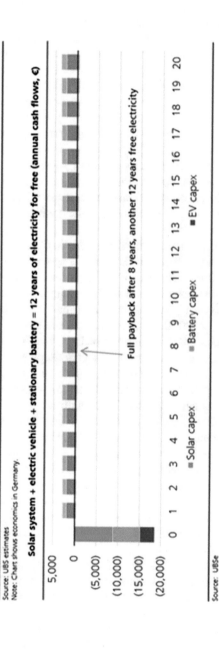

Solar system + electric vehicle + stationary battery = 12 years of electricity for free (annual cash flows, €)

Full payback after 8 years, another 12 years free electricity

Solar capex ■ Battery capex ■ EV capex

Source: UBSe

Fig. 4 EV + solar + battery combination economics (pay-off years and ROI) and detailed pay-off/cash flow projection by 2020 for a German household [20]

This could be a major drawback for BEVs as grid usage is still needed for EV charging points, what will pose significant strain on the grid when EV usage is growing. These networks are not designed to cope with a large-scale deployment of BEVs. This can be solved by smart grid solutions, both technically and by operational incentives and new business models. Examples are incentives to charge during the day at the workplace as the solar energy will be much cheaper than the energy price at home at night. This will still strain the grid, but the grid around offices and factories is more easily adapted than every single street [4]. There will probably be a significant price difference between the daytime solar energy and the buffered energy. Buffering of rooftop solar energy will probably not be sufficient to fully charge the EV, so charging from the grid (at home or the office) still needs to be done, but it will be done less. Even households that own property to place a charger and will have a solar energy system with a battery as buffer need to charge every now and then at other locations and still rely on the grid when ad hoc charging at public charging stations. And people who do not own property to install their own charger will be dependent on (semi-)public chargers; this could limit seriously the rollout of BEVs.

A rough calculation for the Netherlands has shown that public fast charging points are needed when EV numbers grow. Only 25% of Dutch households have the possibility to connect a charging point at home, which is mostly not fast charging, but that is also not needed when charging overnight. But for charging during your trip, you need shorter charging times. As an example, in the Netherlands, there are now 4200 fuel stations and a same kind of network needs to build up in line with increasing the number of EVs. Starting with a network along main routes, Fastned is now setting up 50 locations, planned to be 200 (Fastned 2016), and the company is to build fast charging stations in cities [21]. At short notice, a number of 500 public fast charging locations (on average equipped with 5–10 charging points) is needed to be aligned with the policy goals of increasing the number of EVs [22]. In mid-2016, about 500 fast charging points have been counted in the Netherlands, public and private combined, clustered in less charging locations, on a total of 24,000 charging points (9000 public + 15,000 private) [23]. Especially for a vehicle with higher battery storage capacity, so that a range of over 300 km is feasible, fast charging is a must [24]. This type of vehicles will come massively on the market in 2017 and 2018 as many OEMs have announced their plans, among others Daimler, Volkswagen, Tesla, Opel, Audi, Renault. Mercedes has announced to plan for several new EVs, one of them the new SUV EQ launched in September 2016 at Paris motor show, which will have a range of 500 km and will be for sale in the course of 2019. Mercedes will create a new subbrand for the EVs [25]. Volkswagen (VW) announced about 30 new EV models until 2025 as a reaction to the "dieselgate" emission scandal. Recently, Volkswagen announced to launch a new model with a 600 km range [26]. Also, Volkswagen has announced the first full battery EV, midsize VW Golf-like type, to be launched early 2019, with a range of 480 km and just a 15-minute charging time, about three times faster than state-of-the-art Tesla models at the moment. This VW vehicle possibly cannot benefit on a network level using this impressive charging technology due to necessary built-up for this type of charging infrastructure [27]. BMW launched the i-brand a few years ago. The full electric BMW i-3 model has a range of 200 km, and higher driving range is expected in the years after [28]. Opel revealed at the September 2016 Paris motor show the full electric Ampera-e with a range of 380 km [29, 30].

Audi has announced the A9 E-Tron, probably to be commercially on the market in 2020. The vehicle will be equipped with a 95 kWh battery resulting in 500 km driving range [31]. And last but not least, Tesla has revealed its midrange Model-3 vehicle, with a range of 350 km. Production started in 2017 with 100,000 vehicles, scaling up in 2018 to a production of 400,000 EVs [32]. Tesla models S and X have extended range over 500 km from an extended (from 90 to 100 kWh) battery pack [33]. Also Tesla is working on better battery cell chemistry, which should result in higher storage capacity over 100 kWh without getting bigger battery packs [33]. Renault enhanced its Renault Zoe with extended driving range up to about 300 km [34].

5   Battery Electric EV Powertrain Best Efficiency

With solar energy going to be so dominant, the question is what this means for the vehicle. Several technologies have been compared that can provide propulsion for a car on sustainability and performance [4]. A sustainability index was created, called the "sun-to-wheel efficiency," which measures how well a certain type of propulsion could transform sunlight into motion. From the six selected technologies, the electrically propelled battery electric vehicle (BEV) and the solar car came out on top, with hydrogen fuel cell and solar fuels in the middle and algae fuel and fossil fuels at the very end. The BEV came out on top due to its highly efficient powertrain and the fact that electricity could be generated with high efficiency from solar panels [4].

In terms of driving comfort, BEVs have been found favorable over internal combustion engine (ICE) cars due to less noise, higher acceleration, and energy sustainability. The main problem with the BEV and the solar car is that the battery they use currently has a very poor energy density when compared to the fuels (roughly a factor 20 lower), which results in a limited range. Further analysis shows that the battery prices are going down exponentially and the performance increases linearly. This results in the fact that by 2020, it is a no-brainer to buy an EV: that is, when the performance and cost of the BEV are on par with the combustion car [4]. A global market analysis predicts that in 2022 the BEV will cost the same as the vehicle with combustion engine [19], resulting in strong increase in EV sales; see Fig. 5.

"The future of EVs lies with the ability of science and industry to discover innovations that will significantly reduce the cost of batteries. Fortunately, a wide consensus of investment banks and consultancy firms expect a significant decline in battery prices over the coming decade. If their predictions hold true and oil prices recover to more sustainable levels, we can expect EVs to become cost-competitive

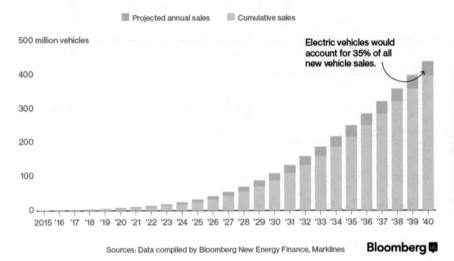

Fig. 5 Annual sales of BEVs [19]

versus ICE vehicles around the year 2020" [20]. Bosch has announced that it has technology developed that will halve the price of batteries in 2020 [35]. Tesla is hinting that the batteries produced in its Gigafactory will have a price of 125 $/kWh, which is quite near the 100 $/kWh price that is widely assumed at the tipping point where EVs become cheaper compared to ICE vehicles [36].

Many countries have policy measures on EVs, like purchase tax incentives and programs on the rollout of charging points. In Netherlands, the vehicle tax scheme on purchase is nearly down to zero and companies have additional possibilities on investment tax benefits. The Dutch policy goal is to have 2,00,000 EVs on the road in 2020 and 1 million in 2025, compared to 8 million total number of vehicles in 2016. Discussion is going on in Dutch parliament to have only EV new purchases as of 2025 [22]. This discussion is inspired by Norway's parliament, where legislation is being prepared that in 2025 only EVs can be sold [37]. Prof. Steinbuch from TU/e predicts that in 2025 annual sales in the Netherlands could be 400,000 EVs, the number of new vehicle that is now sold every year [24], when assuming as of 2018 a yearly increase of sales of only 3%.

6 Lightweight Urban Vehicle and Aerodynamic Highway Vehicle

Batteries are the main issue today in terms of the driving range of EVs, and battery development alone will possibly solve this problem in the coming decade. However, there are other physical vehicle properties, which will not only increase the range but also increase the performance: lightweight construction and better aerodynamics. The mass of the vehicle can be reduced by 30–50% by using carbon-fiber-reinforced plastics (CFRP) [4]. Lighter vehicles are especially suited in city traffic as acceleration and rolling resistance are dominant there and they both are linearly dependent on the mass of the car. Weight reduction also has the benefit of a positive spiral: the lighter one component gets, the lighter other components can be. If the chassis were to be reduced in weight, then the brakes, wheels, and suspension can be lighter (as the strength requirements become lower), which results in the fact that the battery can have a smaller capacity and thus be lighter, etc. CFRP is ready to be used in mass-market vehicles in 5–10 years. The price of CFRP in production will continue to go down (if not faster) due to the fact that this is a major new market for the CFRP and because lighter vehicles require smaller (expensive) batteries [4].

At higher speeds, aerodynamics is of more importance as the energy consumption due to aerodynamics is proportional to the speed cubed. There is a lot of unused potential in aerodynamics. We see that due to the fact that BEVs currently on the market have a lower drag coefficient than the combustion cars. Even though electric powertrains are different from combustion cars, the exterior packaging of the cars is still the same. The absence of a big ICE compartment under the hood gives room for designers to design new form factors. The advent of the electric car unlocks a new world for designers to make cars, which are aesthetically pleasing and aerodynamic [4]. A very good example is the TU/e solar vehicle; read further below.

7 Battery EV Is Ideal for Ride and Car Sharing

The battery EV (BEV) is the ideal vehicle for ride or car sharing. This is due to the very low usage costs of the BEV, which is key for a ride- or car-sharing concept where usage is quite extensive and requires more hours per day compared to (privately) owned fossil-fueled cars. The durability, life span, and thus maintainability of an EV are potentially quite higher due to less moving parts and components compared to the ICE vehicle. So for ride and car sharing, the main costs are usage costs. It was computed that for BEVs, this could be 0.03 [€/km] compared to 0.10–0.13 [€/km] for fossil fuel cars [4]. This difference could be even higher in the future as solar energy makes electricity cheap and fuels will remain expensive. The investment costs of the BEV itself would be 1.3 times higher compared to a privately owned fossil-fueled car but would become lower in the future due to cheaper batteries.

That the BEV is the ideal vehicle for car and ride sharing can be illustrated by the Tesla master plan for the coming 10 years [13]. Although it is coupled with vehicle self-driving capabilities, Tesla sees it as the enabler for car sharing, even by its own fleet. "You will be able to add your car to the Tesla shared fleet just by tapping a button on the Tesla phone app and have it generate income for you while you're at work or on vacation, significantly offsetting and at times potentially exceeding the monthly loan or lease cost. This dramatically lowers the true cost of ownership to the point where almost anyone could own a Tesla. Since most cars are only in use by their owner for 5–10% of the day, the fundamental economic utility of a true self-driving car is likely to be several times that of a car, which is not. In cities where demand exceeds the supply of customer-owned cars, Tesla will operate its own fleet, ensuring you can always hail a ride from us no matter where you are" [13].

BMW is already experimenting with new urban mobility concepts, with its new i-3 EVs in the heart of its urban mobility concept and with already products on the market like car sharing, locating and using public charging stations, and finding parking spaces [9].

8 Solar Cars Are Most Energy Efficient and Can Have a Driving Range Up to 1500 Km

The advantage of the solar vehicle equipped with its own solar panels is that the production, distribution, and buffering of energy is done at the same place where the energy is spent. This increases the energy efficiency of the BEV as the grid, public charging, and external buffering is skipped. It also means that the square meters of solar panel on the household's rooftop that were otherwise used for the BEV can now be used for other energy demand or, more importantly, to charge the solar vehicle when the solar vehicle's panel did not provide enough [4]. Or even the other way around, the vehicle can charge the household when the solar vehicle is plugged in at home at night. In this way, at night the vehicle (partly) provides for the household's energy demand. During driving, the solar vehicle requires hardly any recharging and thus increases the ease of use for the driver. During the World Solar Challenge 2015 in Australia, it was proven that a four-person family solar vehicle could drive 1500 km on a single charge [38] with two-passenger payload. This Stella Lux solar-powered EV concept car proves that designing from scratch with the goal of minimum energy usage while providing user comfort and keeping the legally allowed maximum speed level of 150 km/h pays out. The TU/e solar team initiated the Stella concept car in the new cruiser class of the World Solar Challenge in 2013. Before the race, many competitors stated that the Stella vehicle form factor was really bad, which turned into a big applause during the race and was awarded with the first price as the winning concept at end of the race. The new concept was a real breakthrough, as proven by the many look-alikes in the second cruiser class challenge 2 years later in 2015, where TU/e's Stella Lux was again the world champion. It is still too early to predict when the first solar vehicles will come to market, but it was proven that a brand-new design from scratch pays off, focusing on minimum vehicle weight, minimum aerodynamic drag, and an integrated E-architecture for solar energy power, battery power, and power control. Chinese company Hanergy has launched the first four commercial prototypes [39]. The vehicle concept of one of them is strongly inspired by TU/e's Stella vehicle. Hanergy claims 350 km on a single charge and with an empty battery a range of 80 km per day. Sono motors announced to bring its solar car Sion to the market in 2018 [29] (Fig. 6).

9 Hybrid Vehicles

Will hybrid vehicles stay part of the product portfolio of OEMs, or will they fade away in the coming decade because of superior BEVs? Difficult to predict; it depends on whether the trends as described above will really break through, like the solar-battery-vehicle concept, improving efficiency and lowering battery costs, as well as an up-to-date charging location infrastructure. However, the main advantage of the BEV is the elegance of its architectural simplicity. Compared to the normal car, the BEV contains an absolute minimum of moving parts. It can also be made very small, and the energy storage can be conveniently designed in the floor of the vehicle, enabling the overall vehicle design to be optimized for people, luggage, handling, and safety.

This simplicity results also in lowest service costs over time, although after the vehicle's first dozen years of service, it might be needed to change the battery pack, which could be a very expensive proposition. Nevertheless, upcoming solutions are a second life for these batteries for stationary applications, where weight and space are less demanding than for vehicle application.

Because of the simplicity of BEVs caused by having removed the internal combustion engine (ICE) driveline, it is proven that OEMs can create a high-performance vehicle optimized for a low center of gravity and maximum acceleration and handling.

The BEV might have four important negatives [40], compared to hybrid vehicles: operating costs due to high vehicle investment cost, range and refueling time, and environmental temperature sensitivity. However, this was reported 2 years ago, and the market has dramatically changed since then. Many OEMs have announced new EVs, and the price for a midrange EV will be around 30 k euro in 2019, which means that operating costs are near to break even with ICE and hybrid vehicles due to lower investment levels and very low electricity cost compared to fuel cost [4, 16, 19, 20].

The second argument on low range and refueling time is expected to be less negative than reported in 2014 in [40]. OEMs are announcing vehicles with driving ranges up to 300–500 km. That is for daily use for the great majority of the users by far enough. The question is how to deal with 300–500 km range in more exceptional

Fig. 6 Stella Lux (source: Bart van Overbeeke) and Hanergy solar concept vehicle [39]

circumstances like a yearly holiday trip. Here small behavioral change could help out: by hiring a dedicated car for this purpose or by adapting a bit of your travel user behavior and experience by changing your travel pattern and combining every 2.5 hours of your rest area time at a fast charging location [41].

The third argument is being solved for high temperatures by battery management and cooling systems. For low temperatures, there is still a challenging design opportunity. Battery capacity and thus driving range drops quite a bit at lower temperatures. In a Norwegian winter, a Tesla Model S was tested extensively and a 40% drop in range was reported [42]. For moderate Dutch winters, this is not a big deal, but in regions with severe low temperatures, you need to be prepared in planning your trip by charging more frequently.

Summarizing, hybrid vehicles will stay for a while for sure, but BEVs are on their way to be the optimum solution, assuming battery performance will increase further, especially under low temperature conditions.

10   TU/e Automotive Teams

TU/e's strategic research area Smart Mobility focuses on solving societal challenges through research and with the help of students who represent the ideas of tomorrow.

Scientific education and research blossom within the Automotive student teams, which develop and build different vehicles to compete internationally with other university teams, vehicles in which the students put theoretical solutions to societal challenges into practice, with a good understanding and plenty of enthusiasm and together with equal-minded sponsors and professors.

TU/e has seven Automotive Student Teams and one honors student team, of which four are working on EVs: Solar Team Eindhoven (STE), University Racing Eindhoven (URE), TU/ecomotive, Storm; Team Fast is working on an electric bus, focusing on the solar fuel formic acid powering the fuel cell; two teams are working on hybrid and cooperative vehicles: InMotion and A-Team; the honors Autonomous Eindhoven Team is working on an autonomous self-driving EV. We will focus here on the four teams that have the electric drivetrain as core competence.

The main common characteristics of these automotive teams are that they are completely self-propelled, bottom-up initiatives by a group of students. They are by nature multidisciplinary, demanded by vehicle design, which covers many aspects, both hard-core technology (mechanical and electric engineering) and essential disciplines like computer science, industrial design, and industrial engineering. The budgets that the teams need are collected by attracting sponsors. These sponsorships are both in kind (industrial knowledge, engineering, and components) and in cash. TU/e sponsorship includes dedicated housing for all teams in one building, stimulating knowledge exchange between the teams, as well as knowledge by professors, advice on communication strategy, and in-cash sponsorship. In this way, young students are stimulated to apply theoretical knowledge into practice, learning teamwork and entrepreneurial skills.

10.1   University Racing Eindhoven

University Racing Eindhoven (URE) stands for technological innovation and
engineering. By competing in the Formula Student competition, the world's largest
engineering design competition, URE is pushing itself and its car to new limits!
Using a combination of the engineering skills of all the members and help from their
sponsors, URE is able to build a high-tech revolutionary electric Formula-style race
car every year. During the Formula Student events, all over the world, the result of
the hard work will be put to the test. The drive for being the very best comes from
a passion for engineering. The events are not only about winning; they are also a
perfect place to share information with other teams and technological companies to
develop new ideas (Fig. 7).

The concept behind the Formula Student competition is that a fictional manufacturing company has contracted a student design team to develop a small
Formula-style race car. The prototype race car is to be evaluated for its potential
as a production item. Each student team designs, builds, and tests a prototype based
on a series of rules, whose purpose is both ensuring on-track safety (the cars are
driven by the students themselves) and promoting clever problem solving.

The competition was introduced in 1978 in the USA. Since then, it expanded to
other countries and continents. Nowadays, nearly 600 teams participate in Formula
Student events all over the world!

In addition to these events, various sponsors of the competition provide awards
for superior design accomplishments. For example, innovative use of electronics,
recyclability, crash worthiness, analytical approach to design, and overall dynamic

Fig. 7 URE's racing car 2016 (source: Bart van Overbeeke)

performance are some of the awards available. At the beginning of the competition, the vehicle is checked for rule compliance during the Technical Inspection. Its braking ability, rollover stability, electrical safety, and noise levels are checked before the vehicle is allowed to compete in the dynamic events (Acceleration, Skidpad, Autocross, and Endurance).

Formula Student encompasses all aspects of a business, including R&D, manufacturing, testing, marketing, management, and fundraising. Formula Student takes students out of the classroom and puts them in the real world.

10.2 TU/ecomotive

TU/ecomotive is focusing on the modular urban EV and biomaterial-enforced and -fabricated car body and chassis. Nowadays, every car is adjusted to a certain need. Some vehicles are developed specifically for long journeys on a highway, while others are optimized for usage in an urban area. The modular vehicle can be an all-purpose-built vehicle in one. It can be changed, at every given moment, in such a way that it is applicable for every drive. This significantly improves the vehicle's efficiency. People do not have to buy a vehicle every couple of years; they can just update their vehicle regularly, in such a way that they are comfortable and happy with it. This significantly decreases waste production. The vehicle TU/ecomotive is envisioning contributing to sustainability as well by being highly energy efficient. Lightweight design and a highly efficient electric drivetrain should make the vehicle consume significantly less energy than conventional vehicles.

To prove efficiency, the vehicle will be submitted to the Urban Concept class challenge of the Shell EcoMarathon. This yearly competition for hyper-efficient vehicles, powered by a range of different fuels, has also a battery electric class.

In order to prove that the vehicle is not just a prototype, the vehicle has been successfully submitted to the vehicle conformity test by the Dutch vehicle admission and compliancy authority RDW, to get a license plate. In the examination by the RDW, safety plays a leading role. TU/ecomotive's concept vehicle proves that even very lightweight vehicles can be made to the same degree of safety. Having a license plate allows TU/ecomotive to show that its vehicle has range and practicality in the environment it is designed for.

TU/ecomotive likes to show to the public the validity of its vision. It hopes to change the mindset of people, the way people view mobility. It will continue to do so with its next vehicle, of which the design will be presented in the fall of 2016 (Fig. 8).

Fig. 8 TU/ecomotive's concept vehicle Nova (source: author)

Fig. 9 STE's solar concept vehicle Stella Lux (source: Bart van Overbeeke)

10.3 Solar Team Eindhoven

Solar Team Eindhoven (STE) has developed Stella: a new solar-powered vehicle. Stella is a fully working prototype and is a step forward to solar mobility. STE is aiming for a clean future where driving is safe and simple. It likes to see Stellas on the road everywhere (Fig. 9).

Stella is an energy positive family car. This means that it is so efficient that it generates more energy than it consumes during the entire year, even in Dutch weather conditions! In Australian weather conditions, it was proven during the World Solar Challenge 2015 that the driving range is 1500 km on a single charge with two-passenger payload. The aerodynamic design has an important role in this: consider, for example, the tunnel, which runs through the center of the car. Furthermore, Stella Lux has an extended roof on both sides of the car. Because of this, it was possible to place an extra row of solar panels on the car. Stella Lux was designed to be extremely light by using materials like carbon fiber and aluminum.

STE has redesigned the vehicle concept from the ground up, leveraging to a new type of architecture. Unveiling Stella, world's first solar-powered family car, changed the standard of driving. A mobility platform produces energy instead of using it. This will transform society tremendously. STE will continue developing solar mobility. It searches for new possibilities, new applications, and new technologies to improve the way we drive. The World Solar Challenge is the stage to show its proof of concepts.

STE is sharing its enthusiasm about technology and a sustainable future with the public. STE is actively involving large groups of people in its developments through events and activities in the region and by educating students. By this approach, it shows the world that the value of sustainable energy is tremendous and that the possibilities are immense. It wants to inspire other people to create surprising concepts to improve the future.

10.4 STORM

STORM Eindhoven has developed the world's first electric touring motorcycle traveling around the world in 80 days. STORM wants to prove its innovation and show the true potential of electric transport to the world in the STORM World Tour. STORM has completed the tour on its STORM Wave motorbike. With a stunning range of 380 km, it can truly replace the existing fuel-driven equivalent. With the innovative modular battery pack, STORM can reenergize the motorcycles in a matter of minutes by swapping the batteries. This also allows you to make the motorcycle lighter and even easier to handle by leaving part of the battery pack at home (Fig. 10).

The STORM World Tour, which started on August 14, 2016, has made a big impact all over the world. The team went around the world in exactly 80 days, powered solely by the existing electricity grid. To show the possibilities for electric vehicles today, STORM recharged its batteries using only the existing electricity grid. Recharging was done at companies, households, or educational institutes that took part in its network: the STORM GRID. This way people around the world can experience electric mobility.

Having to cover 23,000 km in 80 days has challenged STORM Eindhoven to create two motorcycles that hold great amounts of energy but also enable the driver

Fig. 10 The STORM Pulse e-motorcycle (source: Bart van Overbeeke)

to easily replenish this energy. After developing the prototype, STORM Pulse, STORM has taken the next step and developed the STORM Wave.

STORM Wave is not the only electric motorcycle around. There are plenty of electric motorcycles, but none of them offer the freedom of driving a whole day. This is where the STORM Wave stands out. A 28.5 kWh battery pack gives the motorcycles a range of 380 km without charging. When empty, the energy can be renewed within 7 minutes.

The STORM World Tour has made an impact on the world, showing the capabilities of e-mobility and our innovation. During the tour events that have been hosted, shows have been given and a documentary has been made to maximize reach and impact.

11   Conclusions

The main changes in the future automotive ecosystem are enabled by strong digitization resulting in three dominant trends that are mutually benefitting each other, resulting in a disruptive change in mobility.

In this chapter 2, dominant trends have been elaborated: firstly, the electrification of the vehicle drivetrain, strongly influenced by the take-up of sustainable energy production by solar and wind farms and, secondly, the uptake of sharing economy stimulating the change from car ownership to car usage by all kind of mobility services. Combining this with the third general-known trend (and not discussed in this chapter) of the automation of vehicle driving itself, this will result in disruptive change of the whole road mobility system toward a mobility service-oriented system, fueled by further penetration of digitization at all aspects of mobility systems and electronic/software components.

Acknowledgements This chapter is for some parts based upon a study executed at TU Eindhoven (TU/e) by two MSc students Lex Hoefsloot and Roy Cobbenhagen [4], who were amongst others the founders of the first automotive solar team at TU/e, as team manager and technical manager respectively. They lead the team in 2012–2013 toward the world championship of the World Solar Challenge (WSC) 2013 in Australia in the newly initiated cruiser class challenge. This title was prolonged by their successors in the WSC 2015 and WSC 2017.

References

1. Becker D (2016) Global Automotive Executive Survey 2016. KPMG. January 2016
2. Becker D (2017) Global Automotive Executive Survey 2017. KPMG
3. Berger R. A CEO agenda for the (r)evolution of the automotive ecosystem, Munich. Roland Berger. March 2016
4. Hoefsloot L, Cobbenhagen R (2015) The car of the future. Internship report. TU Eindhoven. June 2015
5. Grefen P (2014) Achieving business agility through service engineering in extended business networks. TU Eindhoven. July 2014
6. Gassée JL (2016) http://qz.com/674049/the-second-largest-part-of-apples-revenue-now-comes-from-something-called-services/. May 2016
7. DriveNow. https://www.drive-now.com/en
8. Car2go. https://www.car2go.com/US/en/
9. BMW. https://www.bmwgroup.com/en/innovation/technologies-and-mobility/mobility-services.html. BMW. 2016
10. (1999) A realizable renewable energy future. Science 285:687–689
11. A vision for crystalline silicon photovoltaics (2006) Prog Photovolt Res Appl 14:443–453
12. REN21 (2017) Renewables Global Futures Report: Great debates towards 100% renewable energy
13. Musk E (2016) https://www.tesla.com/nl_NL/blog/master-plan-part-deux. Tesla. July 2016
14. SolarEdge (2016) http://www.solaredge.com/us/products/storedge#/
15. Stewart J (2018) Wired. https://www.wired.com/story/tesla-model-3-production-elon-musk/
16. Randall T (2016) Bloomberg. http://www.bloomberg.com/news/articles/2016-06-13/we-ve-almost-reached-peak-fossil-fuels-for-electricity. June 2016
17. Davis L (2017) Energy Institute at Haas. https://energyathaas.wordpress.com/2017/03/27/four-reasons-why-chile-is-the-biggest-solar-market-in-latin-america/
18. Fitzgerald Weaver J (2016) Electrek. https://electrek.co/2016/12/28/solar-power-at-1¢kwh-by-2025-the-promise-of-quasi-infinite-and-free-energy-is-here/
19. Randall T (2016) Bloomberg. http://www.bloomberg.com/features/2016-ev-oil-crisis/. February 2016
20. Blokland HW (2015) Foresight Investor. http://foresightinvestor.com/articles/554942-startfragment-the-rise-of-ev-amp-hybrid-cars-endfragment-nbsp-nbsp-br. February 2015
21. Brandes E (2016) Duurzaambedrijfsleven. http://www.duurzaambedrijfsleven.nl/mobiliteit/16319/fastned-opent-eerste-snellaadstation-in-de-stad
22. NRC. http://www.nrc.nl/nieuws/2016/08/15/maak-opladen-net-zo-makkelijk-als-tanken-3697868-a1516226

23. Steinbuch M (2016) Steinbuch. https://steinbuch.wordpress.com/2016/08/18/cijfers-elektrisch-rijden-nl-tm-juli-2016/
24. Verlaan J. http://www.nrc.nl/nieuws/2016/04/14/iedereen-elektrisch-rijden-eerst-snellere-laadpa-1611573-a903206
25. Muoio D (2016) Businessinsider. https://www.businessinsider.nl/mercedes-electric-suv-should-worry-tesla-2016-9/?international=true&r=US
26. APA/Reuters (2016) Diepresse. http://diepresse.com/home/5067990/VW-kundigt-Elektroauto-mit-bis-zu-600-Kilometer-Reichweite-an
27. Etherington D (2016) TechCrunch. https://techcrunch.com/2016/08/26/volkswagens-2019-electric-car-said-to-get-300-miles-on-a-15-minute-charge/. August 2016
28. Wikipedia (2016) https://en.wikipedia.org/wiki/BMW_i
29. Middelweerd H (2016) Duurzaambedrijfsleven. http://www.duurzaambedrijfsleven.nl/mobiliteit/17543/elektrische-auto-laadt-zichzelf-op-door-ingebouwde-zonnepanelen?utm_source=nieuwsbrief&utm_medium=e-mail&utm_campaign=Weekly%20Updates%2014%20September
30. Slump R (2016) Telegraaf. http://m.telegraaf.nl/autovisie/article/26699271/elektrische-opel-ampera-e-komt-380-kilometer-ver
31. King D (2016) Autoblog. http://www.autoblog.com/2016/08/31/audi-planning-a9-electric-vehicle-to-challenge-tesla/. August 2016
32. Wikipedia (2016) https://en.wikipedia.org/wiki/Tesla_Model_3
33. Tesla (2016) http://www.teslaupdates.co/2016/08/100kwh-battery-pack-coming-to-model-s-and-x.html?m=1. August 2016
34. Adrian W, Markus H (2016) InsideEVs. http://insideevs.com/new-2017-renault-zoe-ze-40-400-km-range-41-kwh-battery/
35. Murgia M (2016) Telegraph. http://www.telegraph.co.uk/technology/news/11885801/Bosch-invents-new-electric-car-battery-to-double-mileage.html
36. Lambert F (2017) Electrek. https://electrek.co/2017/02/18/tesla-battery-cost-gigafactory-model-3/
37. ANP (2016) http://www.nu.nl/buitenland/4307658/verbod-benzineautos-in-noorwegen-lijkt-stap-dichterbij-.html
38. STE (2015) Solar Team Eindhoven. http://www.solarteameindhoven.nl/solar-team-eindhoven-reveals-stella-lux/
39. Hanergy (2016) Allcarindex. https://www.allcarindex.com/auto-car-model/China-Hanergy-Solar-L/. July 2016
40. Wahlman M (2014) TheStreet. https://www.thestreet.com/story/12682853/4/the-great-debate%2D%2Dall-electric-cars-vs-plug-in-hybrids.html. April 2014
41. Steinbuch M (2016) Steinbuch. https://steinbuch.wordpress.com/2016/06/27/onze-eerste-100-000-km-elektrisch-rijden/. August 2016
42. Teslarati (2015). http://www.teslarati.com/tesla-battery-range-sub-zero-snowy-conditions/. February 2015

Autonomous Vehicles: State of the Art, Future Trends, and Challenges

Piergiuseppe Mallozzi, Patrizio Pelliccione, Alessia Knauss, Christian Berger, and Nassar Mohammadiha

Abstract Autonomous vehicles are considered to be the *next big thing*. Several companies are racing to put self-driving vehicles on the road by 2020. Regulations and standards are not ready for such a change. New technologies, such as the intensive use of machine learning, are bringing new solutions but also opening new challenges. This chapter reports the state of the art, future trends, and challenges of autonomous vehicles, with a special focus on software. One of the major challenges we further elaborate on is using machine learning techniques in order to deal with uncertainties that characterize the environments in which autonomous vehicles will need to operate while guaranteeing safety properties.

1 Introduction

During recent years, the automotive sector has been experiencing major transformations. Today's vehicles have a lot of software with millions of lines of code and more processing power than early NASA's spacecraft. Over the past 20 years, the size

P. Mallozzi (✉) · A. Knauss · C. Berger
Chalmers University of Technology, Gothenburg, Sweden

University of Gothenburg, Gothenburg, Sweden
e-mail: mallozzi@chalmers.se; alessia.knauss@chalmers.se; christian.berger@gu.se

P. Pelliccione
Chalmers University of Technology, Gothenburg, Sweden

University of Gothenburg, Gothenburg, Sweden

Università degli Studi dell'Aquila, L'Aquila, Italy
e-mail: patrizio.pelliccione@cse.gu.se

N. Mohammadiha
Chalmers University of Technology, Gothenburg, Sweden

University of Gothenburg, Gothenburg, Sweden

Zenuity, Gothenburg, Sweden
e-mail: nasser.mohammadiha@zenuity.com

© Springer Nature Switzerland AG 2019
Y. Dajsuren, M. van den Brand (eds.), *Automotive Systems and Software Engineering*, https://doi.org/10.1007/978-3-030-12157-0_16

347

of software has grown by a factor of 10 every 5–7 years.[1] Gigabytes of software run inside the ECUs, which are small computers embedded in the vehicle. In the same period of time, the amount of Electronic Control Units (ECUs) has grown from around 20 to more than 100 pushing companies such as Volvo to renovate the electrical architecture of their vehicles in order to cope with such complexity [45]. With this amount of software embedded in the vehicles, car manufacturers have to become more of software companies in order to deal with today's challenges.[2]

Connected vehicles will benefit from intelligent transport systems (ITSs), smart cities, and the Internet of Things (IoT). They will combine data from inside the vehicle with external data coming from the environment (other vehicles, the road, signs, and the cloud). In such a scenario, different applications will be possible: smart traffic control, better platooning coordination, and enhanced safety in general.

Intelligent vehicles will reach full automation, freeing the driver from performing any task. This is a path that will only be reached gradually. There are six levels of vehicle automation (Sect. 1.1 details these levels). To date, almost all of the vehicles in circulation settle on the levels comprised between levels 0 and 2 (level 2 is defined as "partial automation"), in which the systems are limited to assist the driver without replacing him.

Both companies and academia are putting an enormous effort into developing technical solutions in order to increase the levels of autonomy in vehicles. The business landscape is in a profound transformation as we describe in Sect. 1.2. Companies such as Google and Apple, with no tradition at all in car manufacturing, are racing to put their autonomous vehicles on the road in the near future. Software volume in vehicles has been increasing for years; it is expected to increase by 50% by 2020 [22]. 80–90% of the innovation within the automotive industry is based on electronics, and a big part of electronics is software [30, 45].

1.1 Levels of Vehicle Automation

SAE International[3] has defined six levels of vehicle automation as shown in Fig. 1. In today's vehicles, the driver plays a fundamental role. Besides controlling the vehicle, she/he is also expected to recover the vehicle from failures. With full autonomy (SAE level 5), the driver is no longer in the loop. The vehicle, besides controlling the basic driving maneuvers, should handle all possible situations during driving. As the driver is not monitoring the vehicle and is not a fallback option anymore as in the lower levels of automation, it is expected that nothing goes wrong.

---

[1] "Surviving in an increasingly computerized and software driven automotive industry"—Martin Hiller, keynote at ICSA 2017 conference: http://icsa-conferences.org/2017/attending/keynotes/.

[2] This Car Runs on Code http://spectrum.ieee.org/transportation/systems/this-car-runs-on-code.

[3] SAE International is a professional association and standards organization for engineering professionals in transportation industries. http://www.sae.org.

	SAE level	Name	Description
The human driver monitors the environment	0	No Automation	Human driver is responsible for steering, throttle and breaking.
	1	Driver Assistance	The vehicle can perform some control function but not everywhere.
	2	Partial Automation	The vehicle can handle steering, throttle and breaking but the driver is expected to monitor the system and take over in case of faults.
The driving system monitors the environment	3	Conditional Automation	The vehicle monitors the surroundings and notifies the driver if manual control is needed.
	4	High Automation	The vehicle is fully autonomous but only in defined use cases
	5	Full Automation	The driver has only to set the destination. The vehicle will handles any surrounding and make any kind of decision on the way.

Fig. 1 SAE levels of autonomy as defined in standard J3016. Source: SAE

A fatal error in the vehicle is a potential threat to the safety of the vehicle and the passengers inside the vehicle. Hence, fully autonomous vehicles have to support fault tolerance. For this reason, there is a need to investigate requirements to allow the system to continue operating safely during faults. Vehicles have to transition from being *fail safe* to becoming *fail operational*.

1.2 Autonomous Vehicles Ecosystem

The ecosystem landscape is changing when looking at the lower levels of automation to the higher levels of automation [31]. The players in the ecosystem of levels 1–2, where the driver is supported through different vehicle functions (e.g., active safety features), contain the companies that develop the vehicle (i.e., OEMs), suppliers, the government, and organizations that are responsible to define test methods/test catalogs, as well as legislative organizations. Furthermore, certification bodies are taking the defined test standards as an input and are running different tests based on the standards. Test facilities provide the ground to run such tests in a safe environment. Another big player in the ecosystem are customers who buy the final product. Other players are, for example, researchers, insurance companies, and journalists [31].

For the higher levels of automation (i.e., levels 3–5), additional players enter the ecosystem: companies that focus on communication aspects will play a big role in this ecosystem, as communication will be an important basis for highly automated driving [31]. Communication allows the vehicles to ensure that they have redundant information sources (i.e., several sources of information) to ensure safety, reliability, and quality of autonomous vehicles. As can be seen in this example, data plays an essential role in autonomous vehicles. Hence, companies that focus on data will be additional players in the ecosystem [31]. When it comes to OEMs for highly automated driving, we notice that companies need to shift to being software companies instead of developing software just as a side product, as already mentioned above.

The production of a vehicle requires the integration of thousands of parts. Until now, suppliers have been extremely important in providing different components to the vehicle manufacturers that have become more like assemblers. OEMs are interested in investing in autonomous vehicles because it is such a new and high-impact area. New suppliers are joining the autonomous vehicles business, providing hardware and software solutions, such as Nvidia, Mobileye, and Intel. OEMs are establishing partnerships with suppliers focusing on autonomous vehicles. For example, Zenuity is a joint venture on self-driving cars between Volvo Cars and the supplier Autoliv. BMW announced the alliance with Intel and Mobileye with the plan to bring self-driving cars on the road by 2021. Mercedes joins the forces with Bosch, one of the biggest automotive supplier, to develop level 4 and level 5 vehicles.

Google and Tesla are among the pioneers to push the autonomous driving-related software into the market, but at least 33 companies are working on autonomous vehicles.[4] Most of them had promised to put the autonomous car on the road by 2021, assuming that the necessary regulations will be put in place by then.

In addition to autonomous cars, there is a lot of interest in autonomous trucks. The companies working on developing autonomous trucks include Volvo AB, Daimler, Scania, Iveco, Komatsu, and Caterpillar. Autonomous trucks might have to deal with technology and social challenges even more than self-driving cars. From the technology perspective, autonomous trucks have to coordinate their movements with other trucks so that they can drive closely in a *platoon formation*.

Truck platooning not only reduces road congestion but also dramatically increases the vehicle efficiency by reducing fuel consumption. It is believed that it will also save on labor costs, due to relaxed governmental regulation on obliged driver resting time. This concept has been studied for years, and in April 2016, a convoy of connected trucks completes for the first time cross-border trip in Europe.[5] The trucks platoon involves a leader truck that decides the route and speed and several follower trucks driving in automated mode.

2 Autonomous Driving: State of the Art

In recent years, automation has been playing a big role in reducing the need for human intervention for a variety of tasks. In the automotive domain, new technologies have been developed to increase the autonomy of the vehicles resulting

---

[4]https://www.cbinsights.com/blog/autonomous-driverless-vehicles-corporations-list/.

[5]https://www.theguardian.com/technology/2016/apr/07/convoy-self-driving-trucks-completes-first-european-cross-border-trip.

in advanced driver-assistance systems (ADAS). Applications such as the adaptive cruise control (ACC), advanced emergency braking system (AEBS), and lane-keeping assist (LKA) are already available in modern cars controlling part of the vehicles and assisting humans. Vehicles have acquired functionalists to help or replace humans in the most common driving tasks, such as identifying road lanes, detecting pedestrians, and avoiding collisions, as further explained in Sect. 2.1.

Another aspect of autonomous vehicles is the transition from the hardware being the main focus of vehicles to software becoming the main enabler in fully autonomous vehicles. Software updates will become as easy and frequent as those on our smartphones. The software will be upgraded over the air, and it will bring more functionality to the vehicle or enhance important aspects such as safety. Fleets of vehicles will collect data in the environment that is then analyzed and redeployed as a new software update. In order to address such increasing complexity in the software, new challenges need to be addressed in the electronic architecture of the vehicles as further explained in Sect. 2.2.

2.1   Vehicle Functionality

One of the main catalysts of building such autonomous vehicles is to enhance the overall safety. In recent years the attention of vehicle manufacturers has been extended from the already consolidated passive safety systems (airbags, seat belts, impact resistance, etc.) to active safety advanced systems, mainly designed to prevent dangerous situations and accidents rather than protect humans in case of a crash. More recently, with developing of connected cars, safety will take another major step, preventing crashes by taking into account information coming from outside the vehicle [46], for example, from a bicycle coming from a side of a building.

In traditional vehicles, the car uses data locally stored in the vehicle and the communication based on signals between the different ECUs (Electronic Control Units) of the car. In autonomous vehicles, the control can rely on aggregated data coming from multiple sensors but also from the infrastructure outside the vehicle such as the cloud. The communication rather than being signal-based within the vehicle is service/IP based with the outside world. Sharing and controlling sensitive information could be of crucial importance in dangerous situations. For example, recently Volvo Cars have developed a mechanism to inform road users and road maintenance of slippery roads [46]. When a car detects slippery conditions on a road, it sends the information to the Volvo Cloud. This information can then be used to predict the road condition for later times [43]. When another car is approaching the slippery part of the road, Volvo Cloud notifies the approaching vehicle that will automatically reduce its speed.

These technology advancements keep bringing new functionality to vehicles that become more and more autonomous. The path toward fully autonomous vehicles is marked by incremental advances in different categories of assistance. Reaching level

5 of autonomy is more an evolutionary path rather than a revolution from one day to another. Assisting the driver is the first step toward automation. ADAS (advanced driver-assistance systems) aims to assist the driver or to automate some functionality in order to enhance safety and driving comfort, from assisting the driver during parking maneuvers to enhancing the vision during the night. Some of the systems already available in modern vehicles are:

- *ACC (adaptive cruise control)* allows maintaining the safety distance between the car and the vehicle in front.
- *FCW (forward collision warning)* uses camera and radar sensors to identify obstacles on the path of the vehicle.
- *LKAS (lane-keeping assist system)* in addition to warning the driver when the vehicles are going out of its lane also automatically steers the vehicle back to the center.
- *ISA (intelligent speed assistance)* helps the driver to maintain the speed, thanks to a software that recognizes the road signs indicating the limits or more simply by setting an own choice.

ADAS functions are mostly limited to assist the driver without replacing it, but the direction is to create more functions and integrate them in order to reach higher levels of autonomy. Several pilot assistance systems are already available in the market bringing the vehicles forward in the automation level 3 (partial automation, the driver has to be ready to take over at all times). Tesla keeps updating his software for the autopilot adding new functionalities. Besides keeping the lane, it can also overtake other vehicles on the highway and park the car autonomously. Volvo Cars has developed the Pilot Assist system that is able to drive automatically when certain conditions are met. However, the system demands that the driver does always hold her/his hands on the steering wheel.

In order to reach higher levels of autonomy, we need more sophisticated applications that make the most out of the sensors and communication channels available in connected vehicles. Significant developments in the information and communication technologies (ICT) have enabled the advancement of new safety applications. Bila et al. [11] categorize different safety-critical applications that have been developed for intelligent vehicles. *Vehicle detection* application aims to detect and track surrounding vehicles with respect to the host vehicle. Such applications are useful for collision avoidance but also for decision-making based on the movements of other vehicles. *Road detection* and *lane detection* are fundamental tasks to adequately steer the vehicle in "drivable" regions. *Pedestrian detection* is of critical importance for reducing the number of traffic accidents and road fatalities. *Drowsiness detection* is a feature that is already offered by some vehicle companies with the aim of reducing accidents caused by drowsy drivers. Once drowsiness is detected, the vehicle should take over and bring itself into a safe state by, for example, pulling over to the side of the road. More generally, *collision avoidance* is the main system that can include all of the above applications. It can either warn the driver or, in more autonomous systems, take control of the vehicle by steering and breaking.

2.2 Vehicle Architectures

Self-driving vehicular systems in robotic cars, automated trucks, and the like share similarities in their fundamental software architectures. The predominant design pattern in robotic vehicles participating in the 2004 and 2005 DARPA Grand Challenge [59] as well as the 2007 DARPA Urban Challenge [3, 4, 41, 48] was following the well-known *Observe-Orient-Decide-Act* (OODA) pattern, typically in a *pipes-and-filters*-based realization. In practice, these robotic vehicles have software modules dealing with:

- Gathering data from the sensors like cameras, radars, laser scanners, ultrasound sensors, GPS positioning sensors, and the like (*Observe*)
- Data fusion and analysis using sensor fusion, for example (*Orient*)
- Interpretation and decision-making (*Decide*)
- Acting by selecting appropriate driving trajectories to derive the set points for the control algorithm (*Act*)

These software modules get data streams as input, apply filters for data processing, and produce output data that in turn is serving as input data for subsequent software modules. This pattern is still dominant for other robotic applications [6, 7] and a well-accepted engineering approach [8].

Along with the pipes-and-filters pattern and a system architecture dating back to the DARPA Grand and Urban Challenges, software systems distributed across several physically separated computing nodes interconnected over networks were dominantly in use. Therefore, the typical software design that can be found in such systems is based on the *publish/subscribe* communication principle. While this pattern quickly enables a distributed software system whose distribution can even be dynamically adjusted at runtime following the recently emerging *microservices* design pattern [9, 25, 39]. However, the fundamental network design enabling such distributed systems is putting the fundamental software architecture design at risk considering the increasing demand for handling more and more data streams within self-driving vehicles originating from various sensors to redundantly cover 360° of a vehicle's surroundings, especially today, when CPUs are becoming more and more powerful with recent chips providing 32 cores that can be operated next to powerful specialized processing units like GPUs, FPGAs, or TPUs. Under such circumstances, a publish/subscribe network-oriented communication is unnecessarily introducing communication overhead for serializing and deserializing the messages to be exchanged on local systems. Consequently, a more centralized information exchange pattern is needed. The *blackboard* pattern might be the optimal pattern since it is able to cope with increasing data loads. This is an important aspect when several software agents running in parallel can independently and instantaneously access and process large amounts of data while meeting real-time requirements.

Behere and Törngren [5] have defined a functional reference architecture for autonomous driving. They divide the architectural components into three main categories: (1) perception, (2) decision and control, and (3) vehicle platform manipulation. All the components are also distributed in those belonging to the *driving intelligence* and to the *vehicle platform*. They recommend a clear separation between these two: the first should take high-level decision, while the latter is composed of low-level control loops and actuators that apply the decisions.

Two Swedish projects, called NGEA and NGEA2, are investigating the challenges and providing recommendations for the electrical and software architecture of the next generation of vehicles. The projects are coordinated by Volvo Cars and involve some universities, including Chalmers, and research centers in Sweden and many suppliers of the OEM, including Autoliv, Arccore, Combitech, Cybercom, Knowit, Prevas, ÅF-Technology, Semcom, and Qamcom. The projects released an initial version of an architecture framework for Volvo Cars [45]. According to the ISO/IEC/IEEE 42010:2011 standard [27], an architecture framework prefabricated knowledge structure, identified by architecture viewpoints, used to organize an architecture description into architecture views. The framework is based on the conceptual foundations provided by the standard [27] and currently focuses on three new viewpoints that need to be taken into account for future architectural decisions: continuous integration and deployment, ecosystem and transparency, and the car as a constituent of a system of systems. Another natural viewpoint that will be part of the framework is the viewpoint about autonomous driving.

A self-driving car should be able to deliver safe driving in all planned conditions, that is, under conditions and use cases where the autonomous functionality is available. In this regard, it is of vital importance that a self-driving car correctly reacts to all usual and unusual behavior of other road users and traffic conditions. Moreover, the system should be reliable and fault-tolerant considering the hardware. A well-known fault-tolerant architecture that is widely used in the avionics systems is the redundant architecture [5, 26]. In a simple setup, all the components, including perception, decision and control, and actuators, are duplicated, and, therefore, the system becomes more tolerant to the software and hardware failures. Moreover, one way to handle high-ASIL (Automotive Safety Integrity level) components can be realized using ASIL decomposition via a combination of a monitor/actuator architecture and redundancy [32].

3 Autonomous Driving: Trends and Future Direction

Autonomous vehicles have been considered science fiction only a few years ago. Technological advancements together with increases of computational power and a huge amount of available data are making autonomous driving less of a vision and more of a reality in the near future.

Artificial intelligence will play a big role in recognizing situations and, ultimately, in making optimal decisions as explained in Sect. 3.1. Furthermore, autonomous vehicles must adapt their behavior at runtime (Sect. 3.2) and continuously evolve (Sect. 3.3). In the remaining of this section, we discuss the trends and future directions of autonomous driving.

3.1 Artificial Intelligence

An autonomous vehicle can be seen as a rational agent: it perceives information from the environment, computes a decision, and, finally, applies actions to its environment [49]. Artificial intelligence is the ultimate concept that will allow machines to perform any sort of task autonomously and in any situation. Under the umbrella of artificial intelligence, applications such as machine learning (ML) are being increasingly used to implement solutions in different fields, one of which is self-driving vehicles. Techniques such as deep learning have recently become popular thanks to their outstanding results in a variety of different tasks, such as image recognition and decision-making.

Machine learning (ML) techniques can be used with fuzzy information; it can deal with uncertainty using a huge amount of data coming from the vehicle's sensors. They build models based on the data and later use the same models to make predictions. For example, the information captured by the camera and lidar sensors is collected. In the next step, a semantic representation of the raw data is produced with techniques such as sensor fusion. Later, machine learning techniques are used to classify such data into surrounding objects, supporting vehicles with their detection and collision avoidance.

Fully autonomous vehicles require different machine learning techniques to be applied to several components of the vehicle. Recently, deep neural networks (DNNs) have been adopted as one of the main techniques to solve several important tasks related to the perception part of autonomous driving. This includes, for example, object detection and classification, road detection, and semantic segmentation of images. DNNs are powerful methods and can learn complex representations from raw data and hence eliminate the need for hand-crafted features, which is difficult for other machine learning methods. In order to build an autonomous agent (i.e., an entity that is able to make decisions based on what it observes), there are three main tasks to be taken into consideration: perception, prediction, and planning. Machine learning can be applied in all of them [19].

1. *Perception*: Perception is the main area where deep learning is used at the moment. It consists in extracting semantic knowledge from raw data coming from vehicle's sensors. In autonomous vehicles, it can be found, for example, in object detection and tracking, pedestrian detection, road signs recognition, and location and mapping, supporting functions for intelligent vehicles, such as prediction of friction, destination, and traffic flow [15, 18, 36, 40].

   Identifying objects in the environment can be achieved with advanced deep learning algorithms. Convolutional neural networks (CNNs) [34] are considered one of the best deep learning models to classify images. Since 2012, ImageNet has launched a competition where different groups try to set the new state of the art for classification and detection of millions of images. In 2016 Google has proposed GoogLeNet v4 [57], an algorithm that achieves 3.08% error rate in image classification (ImageNet has reported the human error rate to be 5.1%).

2. *Prediction*: In order to build models that can predict the future states of the environment, recurrent neural networks [47] can be used. Furthermore,

supervised learning can be used to predict the near future based on the current state [53]. Convolutional neural networks can be used to directly predict the steering angle of the vehicle from raw input signals such as the pixel of the front-facing camera [12].

3. *Planning*: Perception and prediction are combined to plan a sequence of actions. Frameworks such as reinforcement learning serve such a purpose. It can be used with both deep neural networks and recurrent neural networks. Reinforcement learning [56] involves an agent that acts on an environment by choosing predefined actions with the goal of maximizing a numerical reward. The environment issues these rewards to the agent for every action it receives. The agent keeps track of the rewards and uses them to improve its decision-making policy. Another paradigm to teach a system how to perform tasks is *learning from demonstration* [1], where policies are learned from a teacher. Imitation learning, or behavioral cloning, directly copies the teacher and can be achieved by supervised learning alone. Apprenticeship learning instead tries to infer the goal of the teacher that once formalized as a reward function can be used by reinforcement learning.

3.2 Self-adaptive Systems

An interesting direction that is emerging is self-adaptive systems, that is, systems that are able to adapt their behavior at runtime without human intervention [16, 17, 50] in response to changes in the environment or in their internal state. An autonomous vehicle can be framed as a self-adaptive system: it gathers knowledge from the environment, contextualizes it considering also its internal state, and adapts itself to achieve its goals. This process is performed continuously at runtime.

All self-adaptive systems implement some sort of feedback loop that drives their adaptations [13]. This basic mechanism for adaptation has been applied for years in control engineering; it consists of four main activities: collect, analyze, decide, and act. A well-known reference model for describing the adaptation processes is the MAPE-K loop (consisting of the parts model, analyze, plan, execute, and the knowledge base) [29]. Furthermore, the system is collecting new data from the environment, learns from it, and continuously improves. An example of such a system is the never-ending language learning [14].

Because of the complexity of such systems, maintaining them is a problem similar to large-scale distributed systems. To manage their complexity, they should maintain themselves through a variety of different scenarios [10]. They should self-optimize, continuously seeking for ways to improve its actions on the environment. Self-protect from malicious attack or an error in one system should not propagate to other systems. Self-heal, after detecting that a fault or error has occurred, should recover autonomously. Ultimately, in order to achieve all these self-properties, autonomous vehicles should reach self-awareness, but it is hard to reach it without a breakthrough in artificial intelligence.

Self-adaptive systems and online learning are promising concepts but are still far from practical application in the automotive industry. Collecting data at runtime, self-adaption, and continuous evolution are a fascinating concept. When dealing with safety-critical systems, the adaptation of the system must be certified as *safe* before it can be applied. A more realistic scenario is that automotive companies collect data from their customer, and based on this data, they adapt their method and software which will be pushed back to the vehicles once it has been tested and proved safe.

However, with the intensive use of machine learning techniques, it is hard to test the software and be sure that it will act always as intended. Methods such as *runtime monitoring* can help to prevent the system from violating its requirements. On the one side, we can let the autonomous vehicle reconfigure itself at runtime following high-level policies that specify *what* to achieve and let the system determine *how*. On the other side, a monitor can help to check that its decisions do not violate any predefined invariant of the system [37].

3.3   Continuous Software Engineering

A vehicle is a mechatronic system composed of hardware, software, and mechanical parts. The development of such complex systems has traditionally followed the V-model. This process follows a V-shape starting from high-level requirements of each component going down to the implementation and, then, up again to the verification of the integrated components. The size of the software in the vehicle has grown tremendously in recent years, and for this reason, the development cycle has moved from a rigid V-model to more flexible agile methodologies where the iterations are much more frequent, thus enabling a fast feedback. Moreover, as anticipated above, OEMs are becoming software companies, and they would like to support continuous evolution of software, even after the vehicles are already on the road, through continuous integration and continuous deployment (CI&D) practices. CI&D practices promise to shorten integration, delivery, and feedback cycles [54]. These techniques have been applied by pure software companies, such as Facebook, and now automotive companies have a strong motivation to embrace organization-wide continuous integration and delivery practices to yield improvements in flexibility and cycle time despite the challenges [30]. Challenges include tooling, which works well in a V-model context but that might cause friction in a continuous engineering environment, especially because it facilitates silos thinking and does not support cross-functional aspects [30]. The same paper recommends also to establish suitable transparency between the actors in the automotive value chain, to avoid bottlenecks, as well as information overload, and to address the need for scalability and cross-functional collaboration.

Moreover, construction of shared plant models in a particular value chain as a continuous delivery target can be valuable in future research. Model-driven engineering can bring great benefits in the design and development of automotive systems [20]. In continuous integration and delivery, models need to be integrated on the system level, thus bringing together knowledge about application software, ECU hardware and basic software, and mechanics. This might allow domain experts to be directly involved in the development of the software and have running prototypes from early stages of development. Furthermore, simulation environments enable to perform tests on the system before the mechanical and software components from the suppliers are available, for example, using executable Simulink models of the actual ECU of the vehicle and testing them in a Model-in-the-Loop (MIL) environment where the other models (so-called plant) are used for simulating the surroundings of the ECU.

Continuous evolution of a vehicle should additionally exploit the knowledge collected by other vehicles. In this sense, assuming that mistakes are inevitable also for autonomous vehicles, not every vehicle has to make the same mistake to learn from it. Once the software has been fixed, all other vehicles will be updated based on this knowledge and no other vehicles will do the same mistake again. As Tesla says: *as more real-world miles accumulate and the software logic accounts for increasingly rare events, the probability of injury will keep decreasing.*[6]

3.4 User Aspects

For the lower levels of automation, the driver is responsible for the driving task and takes all risk. For fully automated vehicles, the vehicle itself is responsible for that task and does not require the driver to interact with the system. For the levels in between, the driver is expected to interact with the vehicle. For level 1 and level 2, the vehicle is supporting the driver. If it fails in any of its tasks, the driver is still responsible for the monitoring of the driving task and its own safety. Hence, the vehicle can be only seen as a support and not the final point of decisions. For levels 3–5, the vehicle itself executes the driving task as well as monitors the driving environment. For level 3, the driver is still expected to be the fallback option, in case the vehicle fails in any situation. Then the driver is expected to take over the driving task.

This is one of the reasons why user-related aspects are gaining increased importance in autonomous driving. In 2013 the Asiana Flight 214 crashed at San Francisco International Airport.[7] The CBS News aviation and safety expert Capt. Chesley "Sully" Sullenberger hypothesized that "the pilots may have not fully understood the workings of the automation and that they assumed that the auto-thrust was controlling the speed when in fact it was not." Even though this example comes from avionics, it clearly shows that the interplay between humans and

---

[6]https://www.tesla.com/blog/tragic-loss.

[7]http://www.cbsnews.com/news/asiana-crash-hearing-role-of-pilots-automation-to-be-questioned/.

machines is not easy, and it is not enough to just assume that the driver will need to have the hands on the steering wheel and will take control if something will go wrong. Humans have to be aware of the current situation, monitor the vehicle, and be aware of system limitations [38]. Humans are not good at remaining vigilant in the long run while monitoring the system, and this is partly due to under-stimulation. Moreover, unlike airplanes, we cannot assume to have professional drivers in cars, and most probably in the future, as a side effect of automation, we might even have drivers with degraded performance.

While for the lower levels of automation functional aspects are very important, with increased automation, nonfunctional aspects play an important role as well. Recent work has been focusing on aspects like driver taking over the steering wheel after automated driving and driver handing back the steering wheel [60], the different factors that play a role for the time that it takes to take over the driving task [61], and the factors besides the take-over time that plays a role in the take-over quality [62]. Furthermore, six types of transitions between the driver and vehicle were defined: (1) optional driver-initiated driver-in-control, (2) mandatory driver-initiated driver-in-control, (3) optional driver-initiated automation-in-control, (4) mandatory driver-initiated automation-in-control, (5) automation-initiated driver-in-control, and (6) automation-initiated automation-in- control [35].

However, many questions remain open and more research is needed in order to provide an answer. Examples of questions are:

- Which human behavior model to consider while developing systems? We cannot develop systems according to how humans behave under ideal circumstances since humans are not always taking decisions rationally. Often we make decisions based on context, feedback, culture, stress, workload, perception, expectations, training, feelings, etc.
- Will users behave the same over time?
- Can a driver do something wrong? Should the system try to correct human behaviors, or should we assume that humans are always correct?
- Do the drivers have the needed information in order to actively make the right decisions? In various modes of automation, user involvement decreases, and this can put the driver out-of-the-loop.
- How to establish a sympathetic cooperation between the driver and the vehicle, by making sure that it covers all situations and all different (possibly unpredictable) behaviors?

4 Verification of Autonomous Driving: Challenges for Guaranteeing Safety

The main challenges for deploying autonomous vehicles are raised by the implicit nature of the vehicle: its connectivity to the external world and its self-adaptability situations in the environment. This openness allows vehicles to exchange information with other vehicles (vehicle-to-vehicle (V2V) communication) or to the environment (vehicle-to-infrastructure (V2I) communication). Vehicles should be able to adapt to a different scenario in the environment without human intervention. Such scenarios are usually defined at design time, but one of the challenges is to deal with the uncertainty of the environment as explained in Sect. 4.2. Vehicles become then part of an open system, whose boundaries might be subject to dynamic changes.

On one side, the vehicle autonomy depends on the information it receives from its sensors. Information, in this case, rely on controllable devices, and it is possible to compute their reliability and other quality attributes. On the other side, the self-adaptation capabilities can be triggered also by information coming from other vehicles or infrastructure. Information, in this case, is coming from devices that could be completely uncontrollable, and it is often difficult to compute their reliability. This is completely different from the classical control of vehicles.

Autonomous vehicles bring enormous benefits to the society in terms of reducing traffic, saving fuel consumption, and especially increasing the overall safety. However, at the same time, they open new challenges that need to be addressed before we can see them driving around our cities. In the following, we describe some of the major challenges for guaranteeing safety in autonomous vehicles as pointed out also by Koopman [33] and Schmittner [51].

4.1 Safety Standards Are Not Ready for Autonomous Vehicles

Safety standards in the automotive domain, such as ISO26262, have been used to address the safety requirements of the system throughout its entire life cycle. The standard sets out detailed requirements for processes, methods, and tools used during the design and development of the electrical and electronic system of the vehicle. The requirements related to functional safety represent a challenge for the designers, who must integrate them from the earliest stages of the development process.

The life cycle begins with the description of an item, identifying its functions and interfaces. It follows the hazard analysis and risk assessment, which determine the formulation of a safety goal to be achieved by the item. An Automotive Safety Integrity Level (ASIL) is associated with the item estimating severity, probability, and controllability of hazards, based on item functional behavior. The evaluation of the ASIL is divided into four levels, from ASIL A (the lowest degree of hazard) to ASIL D (the highest degree hazard). For the safety functions with a relatively low degree of criticality, the most suitable level is probably ASIL A. On the other hand, for safety functions with a high degree of criticality, the ASIL C level may be necessary or ultimately ASIL D.

Verification and validation of the system requirements are done according to a V-model that ties each requirement to a different kind of testing. Furthermore, in an environment where systems continuously evolve, such standards may fail to identify all safety-related issues of the system. While there are clear regulations for the safety aspects regarding isolated car system, there is a lack of safety requirements for systems composed of connected cars. Furthermore, in order to be safe, the vehicle must be secure from attacks.

Traditionally, the software in the car was mainly involved in improving the vehicle's internal functions. The vehicle has been considered a closed system with no communication with the external environment. In a connected world bounding the vehicle communication with the environment is no longer possible. The attention is now moving toward connecting the car with the outside world.

Vehicles are no longer single monolithic systems but they form a connected system of systems [46]. A malfunction in one of the system can propagate to other systems. However, the standard assumes that every other system is functioning correctly, viewing hazards and malfunctions only of a single vehicle level. This obviously cannot be always the case in a system of system scenario.

From a security point of view, vehicles are opened at their boundaries, and this increases their attack surface. Security is a new factor to consider when engineering the system. The standard does not include security as a risk factor, hence not considering security breaches as a possible cause of hazards. Furthermore, although it does not assume misuse of the system, it does not consider people with malicious intentions (e.g., hackers). The deployment of vehicular networks is rapidly approaching, and their success and safety will depend on good security solutions; this will also raise privacy concerns that need to be addressed [44].

4.2   Uncertainty Is Everywhere

Self-driving vehicles will have to operate in all sorts of situations. They will be requested to operate in unpredictable and uncontrollable environments that are dominated by uncertainty. Sources of uncertainty are the environment around the vehicle, the availability of the resources that the vehicle can access at a given time, or the difficulty of predicting the other vehicle's behavior [2, 21, 24]. Uncertainty can require the system to dynamically evolve its goals and adapt itself while it is running.

As autonomous vehicles have to handle all faults or exceptions, the requirements for each scenario must be taken into consideration. The problem here is that the requirements are not fully known. The requirements should model all possible situations and unforeseeable events (e.g., extreme weather conditions, wrong traffic signs, or different kinds of animals). Enumerating all such requirements is impractical if not impossible [55].

4.3   The Use of Machine Learning

Different types of machine learning (ML) techniques are being used in various parts of autonomous vehicles, for example, to classify the detection of objects from sensors data. Such components require a set of training data which has to be independent of the validation data to avoid overfitting. One main problem with ML methods is that they are optimized for average cost function and they do not guarantee for corner cases. Challenges in this area are comprised of the fact that when using methods such as neural networks, it is difficult for humans to understand the rules that have been learned by simply looking at its weights. This is one of the hot research areas at the moment, and researchers are investigating different ways to visualize and understand the logic behind the learned neural networks in solving various tasks. Brute force testing is a widely used technique to validate the network resulting in an expensive and not always the best validation method. Furthermore, because the neural network learns the rules from a training set, if certain data is missing or wrongly correlated in the training data, this can cause the network to fail to cause safety hazards. In other words, if there is a special case that the system has not experienced, it cannot correctly predict such case; this is known as the black swan problem [42]. Hence, it is hard to detect and isolate bugs where the behavior is not expressed through traditional lines of codes but entrusted to a neural network. The network would need to be retrained with the risk to *unlearn* correct behaviors. This also motivates the safe AD architectures, where the safety of the complete system can be guaranteed.

Integrating machine learning components with traditional software is a challenge that risks leading to technical debt [52]. Technical debt indicates the long-term cost that arises when implementing a quick solution that works in the short run. Besides having all the maintainability problems of traditional software, machine learning components have special issues. Design principles, such as separation of concerns and strict abstraction boundaries, might fail to be applied in machine learning software as it uses signals from a variety of components and it has dependencies on external data. Some of the problems that the use of machine learning can cause are:

- The use of machine learning erodes boundaries and diminishes encapsulation and modular design, leading to *CACE principle*: changing anything changes everything. This causes great difficulty to maintain the code, isolate changes, and make improvements.
- Unstable data dependencies are caused when input data is changing behavior over time. This can be *implicit*, that is, a machine learning system that is being updated over time; the output of such a system is used as input to another machine learning system. For example, let us consider the case in which an input signal was previously miscalibrated. The model consuming it likely fits these miscalibrations, and a silent update that corrects the signal will have sudden ramifications for the model. In order to avoid problematic situations, a suggestion is to freeze the earlier models before using their output in a new ML system.
- The amount of needed supporting code to use generic ML packages results in glue code system. In such conditions, trying other methods or making improvements become very expensive.
- The absence of good abstraction to support ML systems (similar to what exists for database technology) makes it easy to blur the lines between components.

4.4   Validation Process Is Not Clear

A common way to assess safety in autonomous vehicles is through extensive testing, by test-driving the vehicles and evaluating the vehicle's performance. The more the vehicles drive in autonomous mode, the more experience they gather, and this will result in continuously improving the system. Companies like Waymo[8] advertises that their fleet of autonomous vehicles has driven 2.5 million miles accumulating the equivalent of 400 years human driving experience. Although such numbers seem impressive, in order to demonstrate their reliability, an autonomous vehicle might need to drive for hundreds of millions or in some cases billions of miles [28]. New methods to establish the safety of autonomous vehicles are needed. Big data analysis becomes very important in this regard; statistical signal processing and ML methods have to be developed to analyze the large amounts of data that is collected from the test vehicles or customer vehicles. Examples of such analysis includes (1) the detection of the use cases for which the sensors or the complete system provide poor performance, (2) anomaly detection in the sensor data [58], and (3) the creation of realistic simulation frameworks through the use of sensor models where the logged sensor data are used to improve the quality of the simulations performed in a cluster of computers. To be able to do so, sensor comparison frameworks [23] are required to be able to compare the data obtained from two different sensors, where one of the sensors could have, for example, significantly higher accuracy and can be used as a reference sensor.

Furthermore, the use of probabilistic models (as in the object detection) and stochastic algorithms (as in planning) poses new challenges in the validation process. Having a probabilistic system passing the test once does not guarantee that the same test will succeed every time. Testing becomes difficult for two reasons. The first reason is that due to the non-repeatability of such algorithms, it can be difficult to exercise a particular corner case. The second reason is that it is difficult to evaluate whether a result is correct or not in the case where there are multiple correct behaviors for each test case.

4.5   Nontechnical Challenges

The issues of liability in the event of an accident, the insurance coverage, and the moral dilemmas are all aspects which constitute nontechnical challenges. As the levels of automation increase, the liability in case of accident shifts from being totally a responsibility of the driver (levels 1 and 2) to the OEMs of the intelligent vehicle (levels 4 and 5). Laws are emerging today in order to approve the tests of self-driving cars on the road,[9] but laws and regulations that can allow autonomous vehicles to drive on the road are still missing.

---

[8]https://waymo.com.

[9]https://www.engadget.com/2017/05/12/germany-self-driving-car-test-laws/.

Another challenge of autonomous driving is played on trust: convincing people to trust the technology remains the most difficult obstacle to tackle. There are also ethical issues to be taken into consideration. When an accident happens, the outcome might have been already decided months in advance by someone else. It is still not clear who will take the decisions in the society among the programmers, the policymakers, the government, or other stakeholders. These are surely not easy decisions to make, and the decision should involve different stakeholders with different expertise going beyond the technical ones.

Even choosing an ethical framework that *minimize harm* could lead to morally arguable issues. For example, consider a scenario where a vehicle has to take the decision of either turning left or right in order to save its passengers. By turning left, he would hit a motorcyclist with the helmet on and by turning right a motorcyclist without any helmet. By following the principle of *minimizing harm*, the vehicle should choose to hit the motorcyclist with the helmet penalizing him for being responsible.

5 Conclusions

Autonomous vehicles are increasingly attracting attention, and they are investigated from several different points of views. In this chapter, we have presented an overview of the topic from several perspectives. First, we have categorized autonomous vehicles according to their levels of automation, and have described the ecosystem composed of companies and organizations working on them. Subsequently, we have analyzed the state of the art of the vehicle functionalities and how these functionalities are integrated with the overall vehicle architecture. Afterward, we have discussed the current trends including machine learning, self-adaptive systems, and continuous software engineering and how these trends fit in the automotive industry while taking into consideration the humans that will interact with the vehicle. Finally, we have pointed out some of the major challenges we believe need to be addressed in order to guarantee the safety of autonomous vehicles due to the fact that they will operate in unknown and uncontrollable environments.

References

1. Argall BD, Chernova S, Veloso M, Browning B (2009) A survey of robot learning from demonstration. Robot Auton Syst 57(5):469–483
2. Autili M, Cortellessa V, Di Ruscio D, Inverardi P, Pelliccione P, Tivoli M (2011) Eagle: engineering software in the ubiquitous globe by leveraging uncertainty. In: Proceedings of the 19th ACM SIGSOFT symposium and the 13th European conference on foundations of software engineering, ESEC/FSE '11. ACM, New York, NY, pp 488–491. https://doi.org/10.1145/2025113.2025199

3. Bacha A, Bauman C, Faruque R, Fleming M, Terwelp C, Reinholtz C, Hong D, Wicks A, Alberi T, Anderson D, Cacciola S, Currier P, Dalton A, Farmer J, Hurdus J, Kimmel S, King P, Taylor A, Covern DV, Webster M (2008) Odin: team VictorTango's entry in the DARPA urban challenge. J Field Robot 25(9):467–492. https://doi.org/10.1002/rob.v25:8
4. Baker CR, Dolan JM (2008) Traffic interaction in the urban challenge: putting boss on its best behavior. In: IROS, pp 1752–1758
5. Behere S, Törngren M (2016) A functional reference architecture for autonomous driving. Inf Softw Technol 73:136–150
6. Berger C, Dukaczewski M (2014) Comparison of architectural design decisions for resource-constrained self-driving cars – a multiple case-study. In: Plödereder E, Grunske L, Schneider E, Ull D (eds) Proceedings of the INFORMATIK 2014, Gesellschaft für Informatik e.V. (GI), Stuttgart, pp 2157–2168. http://subs.emis.de/LNI/Proceedings/Proceedings232/2157.pdf
7. Berger C, Rumpe B (2012) Autonomous driving – 5 years after the urban challenge: the anticipatory vehicle as a cyber-physical system. In: Goltz U, Magnor M, Appelrath HJ, Matthies HK, Balke WT, Wolf L (eds) Proceedings of the INFORMATIK 2012, Braunschweig, pp 789–798. https://arxiv.org/pdf/1409.0413v1.pdf
8. Berger C, Rumpe B (2012) Engineering autonomous driving software. In: Rouff C, Hinchey M (eds) Experience from the DARPA urban challenge. Springer, London, pp 243–271. https://doi.org/10.1007/978-0-85729-772-3_10
9. Berger C, Nguyen B, Benderius O (2017) Containerized development and microservices for self-driving vehicles: experiences & best practices. In: Proceedings of the third international workshop on automotive software architectures (WASA), p 6
10. Berns A, Ghosh S (2009) Dissecting self-* properties. In: Third IEEE international conference on self-adaptive and self-organizing systems, 2009. SASO'09. IEEE, Piscataway, pp 10–19
11. Bila C, Sivrikaya F, Khan MA, Albayrak S (2017) Vehicles of the future: a survey of research on safety issues. IEEE Trans Intell Transp Syst 18(5):1046–1065
12. Bojarski M, Del Testa D, Dworakowski D, Firner B, Flepp B, Goyal P, Jackel LD, Monfort M, Muller U, Zhang J, et al (2016) End to end learning for self-driving cars. arXiv preprint arXiv:160407316
13. Brun Y, Serugendo GDM, Gacek C, Giese H, Kienle HM, Litoiu M, Müller HA, Pezzè M, Shaw M (2009) Engineering self-adaptive systems through feedback loops. In: Software engineering for self-adaptive systems, vol 5525. Springer, Berlin, pp 48–70
14. Carlson A, Betteridge J, Kisiel B, Settles B, Hruschka ER Jr, Mitchell TM (2010) Toward an architecture for never-ending language learning. In: AAAI, vol 5, p 3
15. Chavez-Garcia RO, Aycard O (2016) Multiple sensor fusion and classification for moving object detection and tracking. IEEE Trans Intell Transp Syst 17(2):525–534
16. Cheng BH, Giese H, Inverardi P, Magee J, de Lemos R, Andersson J, Becker B, Bencomo N, Brun Y, Cukic B, et al (2008) Software engineering for self-adaptive systems: a research road map. In: Dagstuhl seminar proceedings, Schloss Dagstuhl-Leibniz-Zentrum für Informatik
17. de Lemos R et al (2013) Software engineering for self-adaptive systems: a second research roadmap. In: Software engineering for self-adaptive systems II. Springer, Berlin, pp 1–32
18. Dollar P, Wojek C, Schiele B, Perona P (2012) Pedestrian detection: an evaluation of the state of the art. IEEE Trans Pattern Anal Mach Intell 34(4):743–761
19. El Sallab A, Abdou M, Perot E (2017) Deep reinforcement learning framework for autonomous driving. In: Electronic imaging. Autonomous vehicles and machines
20. Eliasson U, Heldal R, Lantz J, Berger C (2014) Agile model-driven engineering in mechatronic systems-an industrial case study. In: International conference on model driven engineering languages and systems. Springer, Cham, pp 433–449
21. Esfahani N, Malek S (2013) Uncertainty in self-adaptive software systems. Springer, Berlin, pp 214–238. https://doi.org/10.1007/978-3-642-35813-5_9
22. Fleming B (2014) An overview of advances in automotive electronics [automotive electronics]. IEEE Veh Technol Mag 9(1):4–9. https://doi.org/10.1109/MVT.2013.2295285

23. Florbäck J, Tornberg L, Mohammadiha N (2016) Offline object matching and evaluation process for verification of autonomous driving. In: International conference on intelligent transportation systems (ITSC), pp 107–112. https://doi.org/10.1109/ITSC.2016.7795539
24. Garlan D (2010) Software engineering in an uncertain world. In: Proceedings of the FSE/SDP workshop on future of software engineering research, FoSER '10. ACM, New York, NY, pp 125–128. https://doi.org/10.1145/1882362.1882389
25. Giaimo F, Berger C (2017) Design criteria to architect continuous experimentation for self-driving vehicles. In: Proceedings of the international conference on software architecture (ICSA), pp 203–210. http://arxiv.org/abs/1705.05170
26. Hammett R (2001) Design by extrapolation: an evaluation of fault-tolerant avionics. In: 20th DASC. 20th Digital avionics systems conference (Cat. No.01CH37219), vol 1, pp 1C5/1– 1C5/12. https://doi.org/10.1109/DASC.2001.963314
27. ISO/IEC (2011) ISO/IEC/IEEE 42010:2011 Systems and software engineering – architecture description. https://www.iso.org/standard/50508.html
28. Kalra N, Paddock SM (2016) Driving to safety: how many miles of driving would it take to demonstrate autonomous vehicle reliability? Transp Res A Policy Pract 94:182–193
29. Kephart JO, Chess DM (2003) The vision of autonomic computing. Computer 36(1):41–50
30. Knauss E, Pelliccione P, Heldal R, Ågren M, Hellman S, Maniette D (2016) Continuous integration beyond the team: a tooling perspective on challenges in the automotive industry. In: Proceedings of ESEM '16. ACM, New York. https://doi.org/10.1145/2961111.2962639
31. Knauss A, Schroeder J, Berger C, Eriksson H (2017) Paving the roadway for safety of automated vehicles: an empirical study on testing challenges. In: Proceedings of intelligent vehicle symposium (IV)
32. Koopman P, Wagner M (2016) Challenges in autonomous vehicle testing and validation. Technical report, Carnegie Mellon University; Edge Case Research LLC
33. Koopman P, Wagner M (2016) Challenges in autonomous vehicle testing and validation. SAE Int J Transp Saf 4(1):15–24
34. Krizhevsky A, Sutskever I, Hinton GE (2012) Imagenet classification with deep convolutional neural networks. In: Advances in neural information processing systems, pp 1097–1105
35. Lu Z, Happee R, Cabrall C, Kyriakidis M, de Winter J (2016) Human factors of transitions in automated driving: a general framework and literature survey. Transp Res F 43:183–198
36. Maldonado-Bascon S, Lafuente-Arroyo S, Gil-Jimenez P, Gomez-Moreno H, López-Ferreras F (2007) Road-sign detection and recognition based on support vector machines. IEEE Trans Intell Transp Syst 8(2):264–278
37. Mallozzi P (2017) Combining machine-learning with invariants assurance techniques for autonomous systems. In: Proceedings of the 39th international conference on software engineering companion. IEEE, Piscataway, pp 485–486
38. Martens M, van den Beukel A (2013) The road to automated driving: dual mode and human factors considerations. In: Proceedings of conference on intelligent transportation systems, pp 2262–2267
39. Masek P, Thulin M, Andrade H, Berger C, Benderius O (2016) Systematic evaluation of sandboxed software deployment for real-time software on the example of a self-driving heavy vehicle. In: Proceedings of the 19th IEEE intelligent transportation systems conference (ITSC), pp 2398–2403. https://doi.org/10.1109/ITSC.2016.7795942, http://ieeexplore.ieee.org/abstract/document/7795942/
40. Montemerlo M, Thrun S, Koller D, Wegbreit B, et al (2002) Fastslam: a factored solution to the simultaneous localization and mapping problem. In: Aaai/iaai, pp 593–598
41. Montemerlo M, Becker J, Bhat S, Dahlkamp H, Dolgov D, Ettinger S, Haehnel D, Hilden T, Hoffmann G, Huhnke B, Johnston D, Klumpp S, Langer D, Levandowski A, Levinson J, Marcil J, Orenstein D, Paefgen J, Penny I, Petrovskaya A, Pflueger M, Stanek G, Stavens D, Vogt A, Thrun S (2008) Junior: the Stanford entry in the urban challenge. J Field Robot 25(9):569–597. https://doi.org/10.1002/rob.v25:9, http://ieeexplore.ieee.org/xpl/articleDetails.jsp?arnumber=6681152
42. Nassim NT (2007) The black swan: the impact of the highly improbable. Random House, New York

43. Panahandeh G, Ek E, Mohammadiha N (2017) Road friction estimation for connected vehicles using supervised machine learning. In: IEEE intelligent vehicles symposium (IV)
44. Parno B, Perrig A (2005) Challenges in securing vehicular networks. In: Workshop on hot topics in networks (HotNets-IV)
45. Pelliccione P, Knauss E, Heldal R, Ågren SM, Mallozzi P, Alminger A, Borgentun D (2017) Automotive architecture framework: the experience of Volvo cars. J Syst Archit 77:83–100. https://doi.org/10.1016/j.sysarc.2017.02.005, http://www.sciencedirect.com/science/article/pii/S1383762117300954
46. Pelliccione P, Kobetski A, Larsson T, Aramrattan M, Aderum T, Ågren M, Jonsson G, Heldal R, Bergenhem C, Thorsén A (2017) Architecting cars as constituents of a system of systems. In: Software-intensive systems-of-systems. ACM, New York
47. Pinheiro P, Collobert R (2014) Recurrent convolutional neural networks for scene labeling. In: International conference on machine learning, pp 82–90
48. Rauskolb FW, Berger K, Lipski C, Magnor M, Cornelsen K, Effertz J, Form T, Graefe F, Ohl S, Schumacher W, Wille JM, Hecker P, Nothdurft T, Doering M, Homeier K, Morgenroth J, Wolf L, Basarke C, Berger C, Gülke T, Klose F, Rumpe B (2008) Caroline: an autonomously driving vehicle for urban environments. J Field Robot 25(9):674–724. https://doi.org/10.1002/rob.20254
49. Russell S, Norvig P, Intelligence A (1995) A modern approach. Artificial Intelligence Prentice-Hall, Englewood Cliffs, pp 25–27
50. Salehie M, Tahvildari L (2009) Self-adaptive software: landscape and research challenges. ACM Trans Auton Adapt Syst 4(2):14
51. Schmittner C, Ma Z, Gruber T (2014) Standardization challenges for safety and security of connected, automated and intelligent vehicles. In: 2014 International conference on connected vehicles and expo (ICCVE), pp 941–942
52. Sculley D, Holt G, Golovin D, Davydov E, Phillips T, Ebner D, Chaudhary V, Young M, Crespo JF, Dennison D (2015) Hidden technical debt in machine learning systems. In: Advances in neural information processing systems, pp 2503–2511
53. Shalev-Shwartz S, Ben-Zrihem N, Cohen A, Shashua A (2016) Long-term planning by short-term prediction. arXiv preprint arXiv:160201580
54. Ståhl D, Bosch J (2014) Modeling continuous integration practice differences in industry software development. J Syst Softw 87:48–59. https://doi.org/10.1016/j.jss.2013.08.032
55. Sutcliffe A, Sawyer P (2013) Requirements elicitation: towards the unknown unknowns. In: 2013 21st IEEE international requirements engineering conference (RE). IEEE, Piscataway, pp 92–104
56. Sutton RS, Barto AG (1998) Reinforcement learning: an introduction, vol 1. MIT Press, Cambridge
57. Szegedy C, Ioffe S, Vanhoucke V, Alemi A (2016) Inception-v4, inception-resnet and the impact of residual connections on learning. arXiv preprint arXiv:160207261
58. Tashvir A, Sjöberg J, Mohammadiha N (2017) Sensor error prediction and anomaly detection using neural networks. In: The first Swedish symposium on deep learning (SSDL)
59. Thrun S, Montemerlo M, Dahlkamp H, Stavens D, Aron A, Diebel J, Fong P, Gale J, Halpenny M, Hoffmann G, Lau K, Oakley C, Palatucci M, Pratt V, Stang P, Strohband S, Dupont C, Jendrossek LE, Koelen C, Markey C, Rummel C, van Niekerk J, Jensen E, Alessandrini P, Bradski G, Davies B, Ettinger S, Kaehler A, Nefian A, Mahoney P (2006) Stanley: the robot that won the DARPA grand challenge. J Field Robot 23(9):661–692. https://doi.org/10.1002/rob.20147
60. Wintersberger P, Green P, Riener A (2017) Am I driving or are you or are we both? A taxonomy for handover and handback in automated driving. In: Proceedings of driving assessment conference
61. Zeeb K, Buchner A, Schrauf M (2015) What determines the take-over time? An integrated model approach of driver take-over after automated driving. Accid Anal Prev 78:212–221
62. Zeeb K, Buchner A, Schrauf M (2016) Is take-over time all that matters? The impact of visual-cognitive load on driver take-over quality after conditionally automated driving. Accid Anal Prev 92:230–239

Printed in the United States
By Bookmasters